Practical Problems in VLSI Physical Design Automation

Sung Kyu Lim

Practical Problems in VLSI Physical Design Automation

Springer

Sung Kyu Lim

Georgia Institute of Technology
School of Electrical and Computer Engineering
777 Atlantic Drive NW
Atlanta GA 30332-0250
USA
limsk@ece.gatech.edu

ISBN 978-1-4020-6626-9 e-ISBN 978-1-4020-6627-6

Library of Congress Control Number: 2008930560

Printed on acid-free paper.

9 8 7 6 5 4 3 2 1

springer.com

To Mina, Yuna, and Jeanie

Contents

List of Figures

List of Tables

Foreword

The modern integrated circuit is among the most complex engineering products ever built by mankind. Since its invention some four decades ago (the 1960s), the number of transistors per integrated circuit has doubled every two years, following the now-famous Moore's Law. I began working on integrated circuit design in late 1980s, and have witnessed an over-1000x increase in complexity—for example, from Intel's 80486 processor with 1.2 million transistors introduced in 1989 to Intel's dual-core Itanium-2 processor with 1.7 billion transistors introduced in 2006. The exponential growth in such "electronic brains" has transformed practically all areas of modern society, making possible all the recent revolutions in information technology: personal computing, telecommunications, bioinformatics, digital imaging, electronic commerce, and more.

The design of very large-scale integrated (VLSI) circuits, however, has become very challenging, involving hundreds of designers and extensive use of computer-aided design (CAD) tools. One of the oldest, yet most important CAD problems for VLSI circuits is physical design automation, where one needs to compute the best physical layout of millions to billions of circuit components on a tiny silicon surface (no more than $5cm^2$), a process similar to solving a highly complex jigsaw puzzle with nano-scale pieces. The early objective was on area minimization, that is, to come up with the most compact layout. More recent focus includes optimization of circuit performance, power, and manufacturability, as the physical arrangement of these circuit components defines the amount of interconnects or wires needed to connect them, and the interconnects play an increasingly important role (relative to the transistors) in determining the overall circuit performance, power, area, and cost in today's nanometer process technologies.

I have been teaching a graduate course on VLSI circuit physical design automation at UCLA for eighteen years (in fact, the author of this book was a student in this class thirteen years ago). It is a rather demanding course, as it integrates the knowledge of VLSI circuits, algorithm design, combinatorial optimization, mathematical programming, and software implementation in order to come up with an effective solution to a problem. Yet this has been a fascinating subject to study and to teach, and it has been rewarding to see students learn how to automate the design of one of the most complex man-made systems on earth.

I have used and consulted several textbooks in developing my course materials, such as *Physical Design Automation of VLSI Systems* by Preas and Lorenzetti in the late 1980s and multiple editions of *Algorithms for VLSI Physical Design Automation* by Sherwani in the 1990s. These books are very helpful, as they capture a large body of research results distributed in various publications and organize them in a systematic and unified way for algorithm description and complexity analysis. However, given the complexity of most of the algorithms, I have always recognized a need to have well-constructed, meaningful examples to illustrate the execution of these algorithms so that students can gain a deeper understanding and appreciation.

This book by Professor Sung Kyu Lim meets this need well. It selects several classical algorithms in each key topic of physical design automation, including clustering, partitioning, floorplanning, placement, and routing, and explains each of these algorithms through a carefully constructed example showing a step-by-step execution of the algorithm. Professor Lim intentionally leaves out the detailed, pseudo-code type of algorithm description and formal complexity analysis, but focuses on allowing students to observe and practice the algorithm on practical examples. This style clearly differentiates this book from others on this subject, and makes it a valuable additional resource to help students and practitioners grasp key concepts and techniques in VLSI physical design automation.

After forty years, VLSI physical design automation still remains a vibrant field for both research and commercial development. Over a hundred research papers are presented each year, with new ideas addressing the ever-increasing design complexity and constraints. The modern physical design tools command hundreds of thousands of dollars per license, as offered by major VLSI CAD tool vendors, and multiple start-up companies emerge every year. This book will be a valuable addition to help you advance into this exciting and rapidly moving field.

Jason Cong, Ph.D.
Professor and Chairman
Department of Computer Science
University of California at Los Angeles
Los Angeles, California
April 2008

Foreword

As VLSI technology advances into the nanometer era, physical effects have to be considered in all stages of design. In the past, the audiences of physical design books were primarily limited to students and researchers who were interested in the development of CAD tools for physical layout. This has changed. Today, there is a strong demand of fundamental knowledge in physical design by CAD tools developers across the board. Basic techniques for physical design can be modified to generate estimations of physical effects to be used in higher level CAD tools for power minimization and timing closure. These techniques can also be adapted to mitigate process variations cause by downstream manufacturing steps (e.g., CMP, lithography). Thus an up-to-date textbook on physical design would certainly be a welcome contribution to the entire CAD community.

Prof. Lim's book addresses the algorithmic aspects of the physical design of VLSI circuits and systems. Although there have been several textbooks written on this subject, this book distinguishes well from the others. First, the last major book on this subject was published over 10 years ago, so a new book obviously has the advantage of its ability to include most up-to-date developments. Second, Prof. Lim includes some classical materials that are not available in previous textbooks. Finally, the most unique aspect of this book is its presentation style. I cannot agree more with Prof. Lim that one learns the best with examples. This book's central theme indeed is to use small examples to illustrate all steps of an algorithm in details. Moreover, examples are clearly accompanied by many figures, taking full advantage of the power of visualization. After all, a picture is worth a thousand words! In my opinion, Prof. Lim has succeeded. CAD researchers and students will find this book a valuable resource, as either a text or a reference.

Martin D. F. Wong, Ph.D.
Professor
Department of Electrical and Computer Engineering
University of Illinois at Urbana-Champaign
Urbana, Illinois
April 2008

Preface

If you are like me, I learn the best with examples. This is especially true when it comes to understanding CAD papers with complex algorithms and deep mathematical concepts. I remember how much I appreciated the examples in the calculus book when I was in college. The best part of "Introduction to Algorithms" by Cormen, Leiserson, Rivest, and Stein to me was (and still is) the examples. A good example with coherent figures goes a long way for me: it not only helps me understand (or recall) how the algorithm itself works, but I am able to follow the complexity analysis, discussions of limitations and extensions, etc. This is why I use a small example of, say Fiduccia and Mattheyses algorithm, and solve it "by hand" in my graduate course on physical design.

Most of the textbooks on physical design provide examples on some papers, but I always wanted more. In addition, most of the popular papers that are frequently used in physical design classes do not provide examples. This was why I started designing them myself since I started offering a graduate course on physical design at Georgia Institute of Technology in 2001. The students in my class, mostly from the School of Electrical and Computer Engineering, do not usually have background on algorithms, graph theory, and combinatorial optimization. This is where the usefulness of good examples is revealed again. It is usually not until they see an example that they begin to understand and appreciate why the paper is so influential and the algorithm behind it is beautiful and useful.

Overview of this Book

This book consists of 6 chapters on various topics of physical design including clustering, partitioning, floorplanning, placement, Steiner routing, and multinet routing. Each chapter starts with the definition of the problem to be solved, followed by sample problems and the solutions of some of the well-known works. Each section contains a brief introduction and an overview of each work in addition to its sample problems. Some additional problems are provided at the end of each chapter, mostly slight variations of the original sample problems. Each chapter is then concluded with a section that presents some of the well-known follow-up works published in the literature. This book contains a huge amount of figures and tables to help understand the algorithms in an intuitive and organized fashion. A major effort was made to provide a sample problem that (a) is not overwhelmingly difficult or trivial, (b) reveals the crux of algorithm, and (c) contains enough explanation and illustration that are easy to follow.

To the Teachers

This book is designed for a one semester-long introductory course on physical design. This book can be used as a supplement to other textbooks. I also found that this book together with the original research papers were equally effective. The additional practice problems at the end of each chapter can be assigned as homework. It would be interesting and educational to provide programming projects that involve implementing any of the 25 algorithms presented in this book.

A website "`http://users.ece.gatech.edu/limsk/book`" is provided to offer the following resources:

1. Source files of all figures used in the book (in EPS, PNG, and JPG formats)
2. Source files of all the LP (linear programming) and QP (quadratic programming) formulations used in the book
3. Presentation slides for all sample problems in the book (in PPT format)
4. Answers to the additional practice problems (in PDF format)
5. Electronic copies of the original research papers (in PDF format)
6. Bug report and errata

Some of these items are freely available on the web, but please contact me at "`limsk@ece.gatech.edu`" for items #3 and #4. My future plan (or hope I should say) is to provide source codes and/or binaries of all of the algorithms presented in this book together with circuits to experiment with. This is inspired by the widely-known GSRC Bookshelf that already provides similar resources for state-of-the-art academic physical design tools.

To the Students

Physical design is an exciting and highly rewarding area, and students with motivation and skills are always in huge demand. My goal is to see you realize that the learning curve is not so steep after all. Thus, a huge effort was poured in to make this book easy and intuitive to follow. I tried to reduce the amount mathematical notations as much as possible and use plain explanation instead. But, in some cases, reading the original paper first before you read the book (or the other way around) would still be helpful. The best way to learn algorithms to me was to work out a small example by hand. Many students have shared the same experience with me. You do not learn calculus just by reading the book, but by solving the problems in the book by yourself. The same rule applies here for physical design algorithms.

Another effective way to master the concept is in fact to implement the algorithms using C or C++. Many people that I know in the physical design community, including myself, started with an implementation of an existing algorithm, which was then later developed into research projects.

Errors and Omissions

Despite our best attempt, there are many well-known works that are not included in this first edition. In addition, other topics in physical design such as clock and power routing, physical design for FPGA and MCM, layout and interconnect optimization such as buffering and sizing, etc, are missing as well. Please allow me to say that I have saved these items for the next edition!

Again, despite our effort, this book may still contain errors. We will be truly grateful if you could help us correct those mistakes. Please send any reports of bugs, misprints, and other errata to me at `"limsk@ece.gatech.edu"`. Your suggestion on the paper and topic selection as well as any other aspect of this book will be greatly appreciated. In the meantime, please visit our website for errata: `"http://users.ece.gatech.edu/limsk/book"`

Atlanta, May 2008
Sung Kyu Lim

Acknowledgments

My first thanks go to the foreword writers: Prof. Jason Cong at the University of California at Los Angeles, and Prof. Martin D. F. Wong at the University of Illinois at Urbana-Champaign. I also thank Prof. Andrew B. Kahng at the University of California at San Diego for his helpful comments on various parts of this book. They are undoubtedly world leaders of research, education, and commercialization in the field of physical design automation. Several sections of this book are dedicated to their influential works.

I am thankful for the current members of Georgia Tech Computer-Aided Design (GTCAD) laboratory, who spent days and nights thoroughly verifying the answers in this book, thanks to their careless advisor: Faik Baskaya, Michael Healy, Dae Hyun Kim, Young Joon Lee, Mohit Pathak, Ye Tao, and Xin Zhao. Without their help, the errata would have been much lengthier than it is now.

Thanks are due to the 21 students who took my graduate course "ECE6133: Physical Design Automation of VLSI Systems" in spring 2008. They were forced to participate in the alpha/beta testing of this book and had to suffer from many mistakes in the earlier version. They also had to solve most of the additional practice problems at the end of each chapter as their homework. I am sure many future students will benefit from their sacrifice.

My personal thanks go to Mark de Jongh at Springer. He was the first one to see the potential of this book and went an extra mile to convince his colleagues at Springer. He remained supportive in the midst of my numerous requests for deadline extensions. I thank Cindy Zitter at Springer for her patience with my endless questions. Everyone at the production team deserves my thanks for their hard work.

My sincere thanks go to my two daughters Mina and Yuna, and my wife Jeanie. My daughters were too young to understand (4 and 2 years old) why their daddy never had time on weekends and weekdays. I am not sure if they will choose their career (or take a course) in physical design (I will try), but if they do, I am curious to hear what they have to say about this book. I am very fortunate to have a wife who designed such a professional cover for this book. I sincerely believe that she deserves much more than what I can express with my words. Last, but not least, this book would not even exist without our parents.

Atlanta, May 2008
Sung Kyu Lim

Chapter 1

CLUSTERING

Given a gate-level circuit, the goal of circuit clustering is to group gates into clusters and obtain the network of the clusters. The size of the cluster-level network in terms of the number of clusters and the number of connections is smaller than that of the original circuit. A typical objective is to minimize the number of inter-cluster connections (= maximize the number of intra-cluster connections), the maximum number of inter-cluster connections on any path, etc. Typical constraints include the maximum cluster size, the maximum number of external connections for a cluster, etc. The number of clusters to be obtained is not specified, and the area balance among the clusters is usually not required. Circuit clustering is usually performed as a pre-process of circuit partitioning and placement to reduce the complexity of the problem. This chapter presents sample problems related to the following works:

- Rajaraman and Wong algorithm [Rajaraman and Wong, 1995]
- FlowMap algorithm [Cong and Ding, 1992]
- Multi-level Coarsening algorithm [Karypis et al., 1997]

The first two are delay-oriented algorithms, where the longest path delay in the clustered-level circuit is minimized under a certain node/edge delay model. The last is a cutsize-oriented algorithm, where the number of inter-cluster connections is minimized.

S.K. Lim, *Practical Problems in VLSI Physical Design Automation*

1. Rajaraman and Wong Algorithm

Given a directed acyclic graph (DAG) that represents a gate-level circuit, Rajaraman and Wong presented the first algorithm [Rajaraman and Wong, 1995] that produces a delay-optimal clustering solution under the "general delay model". In this model each node has a unique delay, the inter-cluster edge has a constant delay, and the intra-cluster edge does not incur any delay. The size of each cluster is bounded by another constant. The maximum delay from any PI (primary input) to PO (primary output) in the clustered network is minimized. Some nodes in the original DAG may be duplicated in the clustering solution.

Quick Overview

The algorithm consists of two phases, namely, labeling and clustering phase. During the labeling phase, we compute the labels for each node in topological order. This label denotes the longest path delay from any PI to each node, where the path delay includes both the node and the inter-cluster edge delay. During the labeling process, the clustering information is also collected, where we compute which subset of nodes is clustered together for each node. In case a node is included in multiple clusters, we duplicate this node. During the clustering phase, the actual grouping and duplication occur while visiting the nodes in the opposite topological order.

We first compute a $n \times n$ matrix Δ that contains all-pair maximum delay values. Each entry at row x and column v, denoted by $\Delta(x, v)$, is the longest path delay from the output of x to the output of v in the DAG using node delay values only (= ignoring all interconnect delays). Next, we initialize the label of all PI nodes to their delay values and all other nodes to zero. We then visit non-PIs in a topological order to compute their labels. Given a node v, we do the following to compute $l(v)$, its label:

1. We compute the sub-graph rooted at v, denoted G_v, that includes all the predecessors of v.

2. We compute $l_v(x)$ for each node $x \in G_v \backslash \{v\}$, where $l_v(x) = l(x) + \Delta(x, v)$. $l(x)$ denotes the current label for x, and $\Delta(x, v)$ is an entry of the Δ matrix mentioned above.

3. We sort all nodes in $G_v \backslash \{v\}$ in decreasing order of their l_v-values and put them into a set S.

4. We remove a node from S one-by-one in the sorted order and add it to the cluster for v, denoted $cluster(v)$, until the size constraint is violate.

5. We compute two values l_1 and l_2. If $cluster(v)$ contains any PI nodes, the maximum l_v value among these PI nodes becomes l_1. If S is not empty after

filling up $cluster(v)$, the maximum $l_v + D$ among the nodes remaining in S becomes l_2, where D is the inter-cluster delay.

6. The new label for v is the maximum between l_1 and l_2.

During the clustering phase, we first put all PO nodes in a set L. We then remove a node from L and form its cluster. Given a node v, we form a cluster by grouping all nodes in $cluster(v)$, which was computed during the labeling phase. We then compute $input(v)$, the set of input nodes of $cluster(v)$. Next, we remove a node x from $input(v)$ one-by-one and add it to L if we have not formed the cluster for x yet. We repeat the entire process until L becomes empty.

Practice Problem

Consider the directed acyclic graph in Figure 1.1. Assume that the delay of the nodes is 1, and the inter-cluster delay is 3. The clustering size limit is set to 4.

1. Compute $\Delta(x, y)$, the maximum delay matrix.

   Recall that each delay value is computed from the output of the source node to the output of the destination node. See Figure 1.2.

2. Obtain a topological order T of the non-PI nodes.

$$T = [d, e, f, g, h, i, j, k, l]$$

3. Compute the label and clustering information for node d.

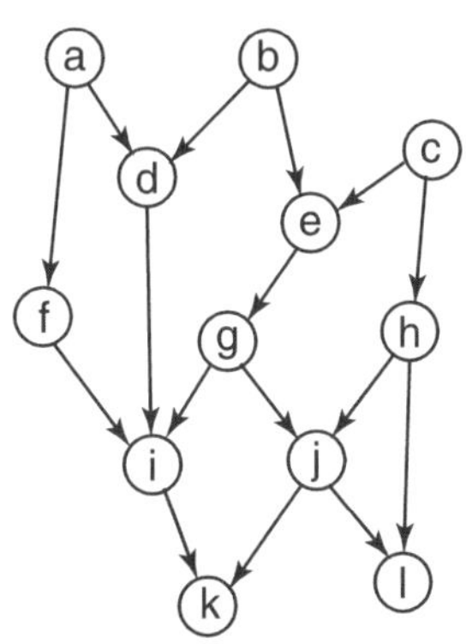

Figure 1.1. A directed acyclic graph, where $PI = \{a, b, c\}$, and $PO = \{k, l\}$.

	a	b	c	d	e	f	g	h	i	j	k	l
a	0	0	0	1	0	1	0	0	2	0	3	0
b	0	0	0	1	1	0	2	0	3	3	4	4
c	0	0	0	0	1	0	2	1	3	3	4	4
d	0	0	0	0	0	0	0	0	1	0	2	0
e	0	0	0	0	0	0	1	0	2	2	3	3
f	0	0	0	0	0	0	0	0	1	0	2	0
g	0	0	0	0	0	0	0	0	1	1	2	2
h	0	0	0	0	0	0	0	0	0	1	2	2
i	0	0	0	0	0	0	0	0	0	0	1	0
j	0	0	0	0	0	0	0	0	0	0	1	1
k	0	0	0	0	0	0	0	0	0	0	0	0
l	0	0	0	0	0	0	0	0	0	0	0	0

Figure 1.2. Maximum delay matrix of the DAG in Figure 1.1.

First, $G_d = \{a, b, d\}$. By definition $l(a) = l(b) = 1$. Thus,

$$l_d(a) = l(a) + \Delta(a, d) = 1 + 1 = 2$$
$$l_d(b) = l(b) + \Delta(b, d) = 1 + 1 = 2$$

Then we have $S = \{a, b\}$ (recall that S contains $G_d \backslash \{d\}$ with their l_d values sorted in a decreasing order). Since both a and b can be clustered together with d while not violating the size constraint of 4, we form

$$cluster(d) = \{a, b, d\}$$

Since both a and b are PI nodes, we see that

$$l_1 = \max\{l_d(a), l_d(b)\} = 2$$

Since S is empty after clustering, l_2 remains zero. Thus,

$$l(d) = \max\{l_1, l_2\} = 2$$

4. Compute the label and clustering information for the remaining nodes. We visit the nodes in topological order T:

   (a) Node e: $G_e = \{b, c, e\}$. $l(b) = l(c) = 1$. Thus,

   $$l_e(b) = l(b) + \Delta(b, e) = 1 + 1 = 2$$
   $$l_e(c) = l(c) + \Delta(c, e) = 1 + 1 = 2$$

   We see that $S = \{b, c\}$, and we form $cluster(e) = \{e, b, c\}$. Since b and c are PI nodes, $l_1 = \max\{l_e(b), l_e(c)\} = 2$. Since S is empty after clustering, l_2 remains zero. Thus, $l(e) = 2$.

(b) Node f: $G_f = \{a, f\}$, and $l(a) = 1$. Thus,

$$l_f(a) = l(a) + \Delta(a, f) = 1 + 1 = 2$$

$S = \{a\}$, and $cluster(f) = \{a, f\}$. Since a is a PI node and S is empty, $l_1 = \max\{l_f(a)\} = 2 = l(f)$.

(c) Node g: $G_g = \{b, c, e, g\}$. Thus,

$$l_g(b) = l(b) + \Delta(b, g) = 1 + 2 = 3$$
$$l_g(c) = l(c) + \Delta(c, g) = 1 + 2 = 3$$
$$l_g(e) = l(e) + \Delta(e, g) = 2 + 1 = 3$$

$S = \{b, c, e\}$ and $cluster(g) = \{b, c, e, g\}$. Since b and c are PI nodes, $l_1 = l_g(b) = l_g(c) = 3$. Since S is empty after clustering, $l(g) = l_1 = 3$.

(d) Node h: $G_h = \{c, h\}$, and $l(c) = 1$. Thus,

$$l_h(c) = l(c) + \Delta(c, h) = 1 + 1 = 2$$

$S = \{c\}$, and $cluster(h) = \{c, h\}$. Since c is a PI node and S is empty, $l_1 = \max\{l_h(c)\} = 2 = l(h)$.

(e) Node i: $G_i = \{a, b, c, d, e, f, g, i\}$ (see Figure 1.3). Thus,

$$l_i(a) = l(a) + \Delta(a, i) = 1 + 2 = 3$$
$$l_i(b) = l(b) + \Delta(b, i) = 1 + 3 = 4$$
$$l_i(c) = l(c) + \Delta(c, i) = 1 + 3 = 4$$
$$l_i(d) = l(d) + \Delta(d, i) = 2 + 1 = 3$$
$$l_i(e) = l(e) + \Delta(e, i) = 2 + 2 = 4$$
$$l_i(f) = l(f) + \Delta(f, i) = 2 + 1 = 3$$
$$l_i(g) = l(g) + \Delta(g, i) = 3 + 1 = 4$$

$S = \{g, e, c, b, a, d, f\}$, and we form $cluster(i) = \{i, g, e, c\}$.[1] Note that c is PI, so $l_1 = l_i(c) = 4$. Since $S = \{b, a, d, f\} \neq \emptyset$ after clustering, we have $l_2 = l_i(m(S)) + D = l_i(b) + D = 4 + 3 = 7$ (recall that $m(S)$ is the node in S with the maximum value of l_i value). Thus, $l(i) = \max\{l_1, l_2\} = 7$.

[1] $cluster(i) = \{i, g, e, b\}$ is another possible solution.

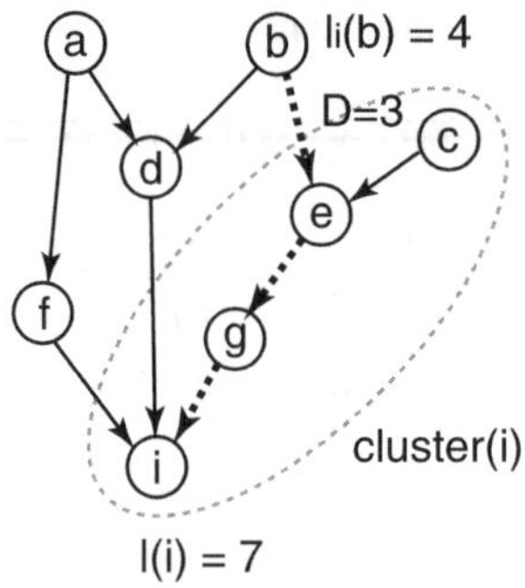

Figure 1.3. The sub-tree rooted at i $(= G_i)$ and $cluster(i)$. The timing critical path, shown in dotted path, contains four nodes and a inter-cluster interconnect $(D = 3)$. Thus, the delay is 7, which agrees with $l(i) = 7$.

(f) Node j: $G_j = \{b, c, e, h, g, j\}$. Thus,

$$\begin{aligned} l_j(b) &= l(b) + \Delta(b, j) = 1 + 3 = 4 \\ l_j(c) &= l(c) + \Delta(c, j) = 1 + 3 = 4 \\ l_j(e) &= l(e) + \Delta(e, j) = 2 + 2 = 4 \\ l_j(h) &= l(h) + \Delta(h, j) = 2 + 1 = 3 \\ l_j(g) &= l(g) + \Delta(g, j) = 3 + 1 = 4 \end{aligned}$$

$S = \{g, b, c, e, h\}$, and we form $cluster(j) = \{b, e, g, j\}$.[2] b is PI, so $l_1 = l_j(b) = 4$. Since $S = \{c, h\} \neq \emptyset$ after clustering, we have $l_2 = l_j(m(S)) + D = l_j(c) + D = 4 + 3 = 7$. Thus, $l(j) = 7$.

(g) Node k: $G_k = \{a, b, c, d, e, f, g, h, i, j, k\}$. Thus,

$$\begin{aligned} l_k(a) &= l(a) + \Delta(a, k) = 1 + 3 = 4 \\ l_k(b) &= l(b) + \Delta(b, k) = 1 + 4 = 5 \\ l_k(c) &= l(c) + \Delta(c, k) = 1 + 4 = 5 \\ l_k(d) &= l(d) + \Delta(d, k) = 2 + 2 = 4 \\ l_k(e) &= l(e) + \Delta(e, k) = 2 + 3 = 5 \\ l_k(f) &= l(f) + \Delta(f, k) = 2 + 2 = 4 \\ l_k(g) &= l(g) + \Delta(g, k) = 3 + 2 = 5 \\ l_k(h) &= l(h) + \Delta(h, k) = 2 + 2 = 4 \\ l_k(i) &= l(i) + \Delta(i, k) = 7 + 1 = 8 \\ l_k(j) &= l(j) + \Delta(j, k) = 7 + 1 = 8 \end{aligned}$$

[2] $cluster(j) = \{c, e, g, j\}$ is another possible solution.

$S = \{i, j, g, b, c, e, a, d, f, h\}$, and we form $cluster(k) = \{g, i, j, k\}$. There is no PI in $cluster(k)$, so $l_1 = 0$. Since $S = \{b, c, e, a, d, f, h\} \neq \emptyset$ after clustering, $l_2 = l_k(m(S)) + D = l_k(b) + D = 5 + 3 = 8$. Thus, $l(k) = 8$.

(h) Node l: $G_l = \{b, c, e, h, g, j, l\}$. Thus,

$$\begin{aligned} l_l(b) &= l(b) + \Delta(b, l) = 1 + 4 = 5 \\ l_l(c) &= l(c) + \Delta(c, l) = 1 + 4 = 5 \\ l_l(e) &= l(e) + \Delta(e, l) = 2 + 3 = 5 \\ l_l(h) &= l(h) + \Delta(h, l) = 2 + 2 = 4 \\ l_l(g) &= l(g) + \Delta(g, l) = 3 + 2 = 5 \\ l_l(j) &= l(j) + \Delta(j, l) = 7 + 1 = 8 \end{aligned}$$

$S = \{j, g, b, c, e, h\}$, and we form $cluster(l) = \{e, g, j, l\}$. There is no PI in $cluster(l)$, so $l_1 = 0$. Since $S = \{b, c, h\} \neq \emptyset$ after clustering, we have $l_2 = l_l(m(S)) + D = l_l(b) + D = 5 + 3 = 8$. Thus, $l(l) = 8$.

Table 1.1 summarizes the labeling and clustering results.

5. Generate the clusters and draw the clustered-level graph.

   We initially set $L = \{k, l\}$ and $S = \emptyset$.[3] We use a short-hand notation $cl(v)$ to denote $cluster(v)$.

Table 1.1. Summary of the labeling and clustering process of the Rajaraman and Wong algorithm.

Node	Label	Clustering
a	1	$\{a\}$
b	1	$\{b\}$
c	1	$\{c\}$
d	2	$\{a, b, d\}$
e	2	$\{b, c, e\}$
f	2	$\{a, f\}$
g	3	$\{b, c, e, g\}$
h	2	$\{c, h\}$
i	7	$\{c, e, g, i\}$
j	7	$\{b, e, g, j\}$
k	8	$\{g, i, j, k\}$
l	8	$\{e, g, j, l\}$

[3]Note that this S is different from the S used during the labeling phase. Despite the confusion it may cause, we try use the same notation as the one that is used in the original paper [Rajaraman and Wong, 1995].

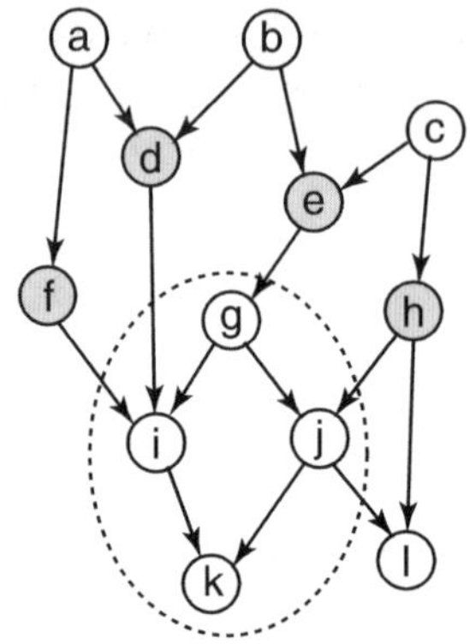

Figure 1.4. Illustration of $cluster(k)$ and its input nodes $\{f, d, e, h\}$ shown in gray.

(a) Remove k from L, and add $cl(k)$ to $S = \{cl(k)\}$. According to Table 1.1, we see that $cl(k) = \{g, i, j, k\}$. Then, $I[cl(k)] = \{f, d, e, h\}$ as illustrated in Figure 1.4. Since S does not contain clusters rooted at f, d, e, and h, we have $L = \{l\} \cup \{f, d, e, h\} = \{l, f, d, e, h\}$.

(b) Remove l from L: $S = \{cl(k), cl(l)\}$. Then, $I[cl(l)] = \{b, c, h\}$. We have $L = \{f, d, e, h\} \cup \{b, c, h\} = \{f, d, e, h, b, c\}$.

(c) Remove f from L: $S = \{cl(k), cl(l), cl(f)\}$. Since $I[cl(f)] = \emptyset$, $L = \{d, e, h, b, c\}$.

(d) Remove d from L: $S = \{cl(k), cl(l), cl(f), cl(d)\}$. Since $I[cl(d)] = \emptyset$, $L = \{e, h, b, c\}$.

(e) Remove e from L: $S = \{cl(k), cl(l), cl(f), cl(d), cl(e)\}$. Since $I[cl(e)] = \emptyset$, $L = \{h, b, c\}$.

(f) Remove h from L: $S = \{cl(k), cl(l), cl(f), cl(d), cl(e), cl(h)\}$. Since $I[cl(h)] = \emptyset$, $L = \{b, c\}$.

(g) Remove b from L: $S = \{cl(k), cl(l), cl(f), cl(d), cl(e), cl(h), cl(b)\}$. Since $I[cl(b)] = \emptyset$, $L = \{c\}$.

(h) Remove c from L: $S = \{cl(k), cl(l), cl(f), cl(d), cl(e), cl(h), cl(b), cl(c)\}$. Since $I[cl(c)] = \emptyset$, $L = \emptyset$.

From the final S we see that 8 clusters are generated. Table 1.2 shows the list of these clusters. Figure 1.5 shows the clustered-level graph. Note that eight nodes are duplicated, where b and c are duplicated twice.

6. What is the maximum delay in the clustered graph? Give an example of the longest path.

Table 1.2. List of clusters generated by the Rajaraman and Wong algorithm.

Root	Elements
k	$\{g, i, j, k\}$
l	$\{e, g, j, l\}$
f	$\{a, f\}$
d	$\{a, b, d\}$
e	$\{b, c, e\}$
h	$\{c, h\}$
b	$\{b\}$
c	$\{c\}$

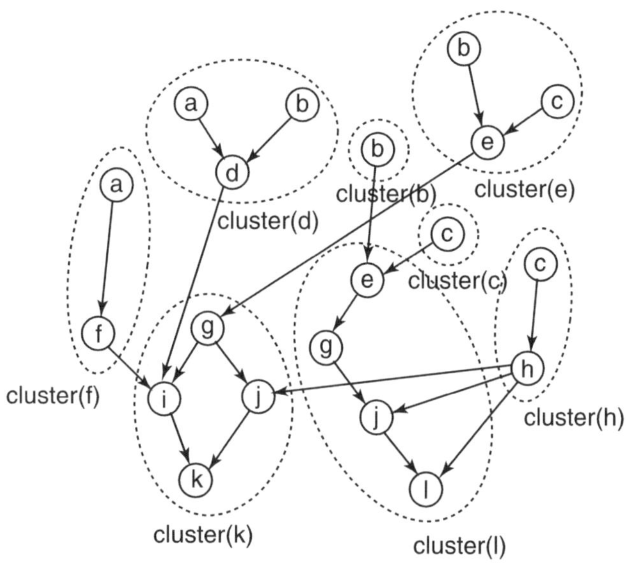

Figure 1.5. Clustered-level graph.

According to Table 1.1, the maximum label is 8, which corresponds to the maximum delay. The path $c - e - g - i - k$ shown in Figure 1.5 has the delay of 8. The path $c - e - g - j - l$ also has the delay of 8. Note that there exist several other paths with the delay of 8.

2. FlowMap Algorithm

Cong and Ding presented the first algorithm named FlowMap [Cong and Ding, 1992] that produces a delay-optimal clustering solution under the "unit delay model". In this model the inter-cluster edge has a unit delay while the nodes and intra-cluster edges do not incur any delay. The difference between the clustering problem that Cong and Ding solved compared with Rajaraman and Wong [Rajaraman and Wong, 1995] is that the number of external connections, not the area, for each cluster is bounded by a constant. This so called "pin constraint" is applicable to the K-input look-up table (LUT) mapping problems for field-programmable gate array (FPGA) designs. The maximum delay from any primary input (PI) to primary output (PO) in the clustered network is minimized in the solution. Some nodes in the original directed acyclic graph (DAG) may be duplicated in the clustering solution.

Quick Overview

FlowMap algorithm consists of two phases, namely, labeling and mapping phase. During the labeling phase, two values are computed for each node v in topological order: clustering $\overline{X}_v$ and label $l(v)$. $\overline{X}_v$ denotes the set of nodes to be clustered together with v, and $l(v)$ denotes the longest path delay measured from the PI nodes to v, where only the inter-cluster edges incur a unit delay. During the mapping phase, the actual grouping and duplication occur while visiting the nodes in the reverse topological order.

During the labeling phase, we first initialize the label for all PI nodes to zero. We then visit non-PIs in topological order. Given a node t, we do the following to compute its new label, $l(t)$: (it is assumed that the readers are familiar with the network flow concept):

1. We compute the sub-graph rooted at t, denoted N_t, that includes all of the predecessors of t. We then add a source node s to N_t and connect it to all PIs in N_t.

2. We compute p, which is the maximum label among all fan-in nodes of t.

3. We obtain N_t', where all nodes with their labels equal to p are collapsed into t.

4. We obtain a flow-network N_t'', where each node x in N_t' except s and t is split into two nodes (x, x') and connected via a "bridging edge" $e(x, x')$. We assign the capacity of 1 to all bridging edges and ∞ to all non-bridging edges in N_t''.

5. We compute a cut $C(X'', \overline{X}'')$ that separates s and t in N_t'' with the cutsize not larger than K, the pin constraint. This is done by finding augmenting paths from s to t (= if N_t'' contains more than K augmenting paths, we

conclude that the s-t mincut has a cutsize larger than K.) If multiple feasible cuts are found, we choose the one with the "minimum height", where the height of a cut $C(X, \overline{X})$ is defined to be the maximum label among the nodes in X, the source-side partition.

6. If C (= min-height K-feasible cut) is found, we first include all nodes in $\overline{X}''$ (= sink-side partition) into $\overline{X}_t$ (= cluster for t). If a node x is split into (x, x') during the N_t to N_t' transformation in step 4, and $e(x, x')$ is cut in C, we remove x' from $\overline{X}_t$ (= in this case x becomes an input node of $\overline{X}_t$). Lastly, $l(t) = p$.

7. If C is not found, $\overline{X}_t$ contains t only, and $l(t) = p + 1$.

During the mapping phase, we first put all PO nodes in a set L. We then remove a node from L and form its cluster as follows: given a node v, we form a cluster, named v', by grouping all non-PI nodes in $\overline{X}_v$, which was computed during the labeling phase. We then compute $input(v')$, the set of input nodes of v', and include them in L. A node x is an input node of v' if $e(x, y)$ exists in the original DAG and $y \in v'$. We repeat the entire process until L becomes empty.

Practice Problem

Consider the 2-bounded Boolean network (= all gates have up to 2 inputs) and its directed acyclic graph representation in Figure 1.6. Assume that the pin constraint K is set to 3.

1. Compute the label (= $l(v)$) and clustering (= $\overline{X}_v$) for all nodes in the graph.

   First, all PIs are assigned zero for their label. We then visit the remaining nodes in topological order $T = [a, b, c, d, e, f, g, h, i, j, k]$.

   (a) Node a: We first build N_a as shown in Figure 1.7(a). We see that $p = 0$. This helps us build N_a' and N_a'' as shown in Figure 1.7(b) and Figure 1.7(c). Note that it is not possible to find a cut in N_a'' with a cutsize smaller or equal to $K = 3$. Thus, $\overline{X}_a = \{a\}$ and $l(a) = p + 1 = 1$.

   (b) Node b and c: This is similar to the case for node a, so $\overline{X}_b = \{b\}$, $l(b) = 1$, $\overline{X}_c = \{c\}$, and $l(c) = 1$.

   (c) Node d: Figure 1.8 shows N_d, N_d', and N_d'' under $p = 1$. There is a possible cut in N_d'' as shown on Figure 1.8(c), where the maximum flow value and the cutsize is 3. The height of this cut is zero because the label for all nodes in the source-side partition is zero. Nodes a and d are partitioned to the sink-side. Thus, $\overline{X}_d = \{a, d\}$, and $l(d) = p = 1$.

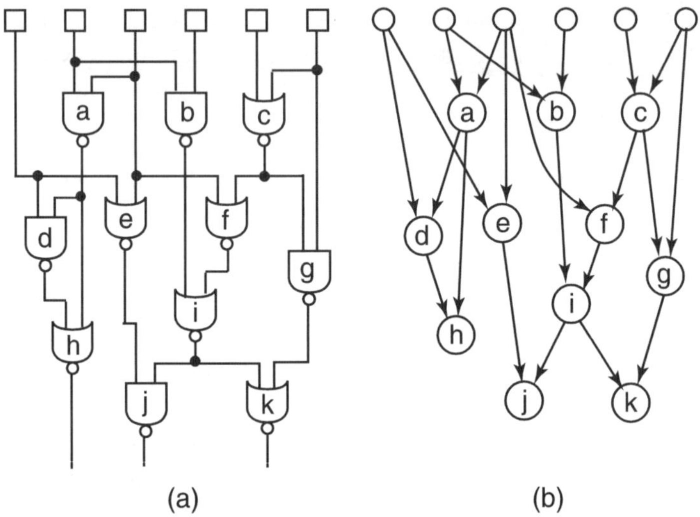

Figure 1.6. (**a**) 2-bounded Boolean network, (**b**) and its DAG. Note that we do not model the POs explicitly but treat h, j, and k as POs.

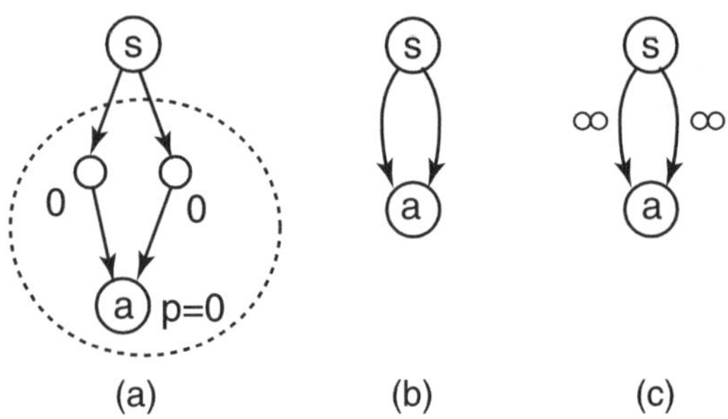

Figure 1.7. Visiting node a. (**a**) N_a, (**b**) N'_a, and (**c**) N''_a. The numbers next to the nodes denote their label.

(d) Node e: This is similar to node a, so $\overline{X}_e = \{e\}$, and $l(e) = 1$.

(e) Node f: This is similar to node d, so $\overline{X}_f = \{c, f\}$, and $l(f) = 1$.

(f) Node g: Figure 1.9 shows N_g, N'_g, and N''_g. There is only one cut possible in N''_g as shown on Figure 1.9(c). Thus, $\overline{X}_g = \{c, g\}$, and $l(g) = p = 1$.

(g) Node h: Figure 1.10 shows N_h, N'_h, and N''_h. In this case, N'_h, and N''_h are identical to those for node d. Thus, $\overline{X}_h = \{a, d, h\}$, and $l(h) = l(d) = 1$.

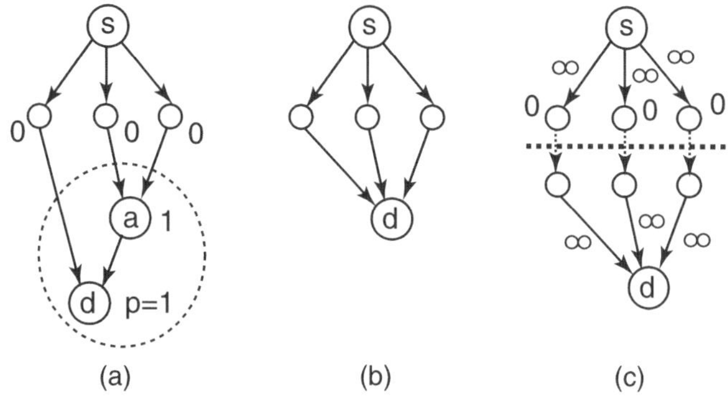

Figure 1.8. Visiting node d. (**a**) N_d, (**b**) N'_d, and (**c**) N''_d. Note that N''_d contains a K-feasible cut with the height of 0 as shown in dotted line. Bridging edges are shown in dotted arrows.

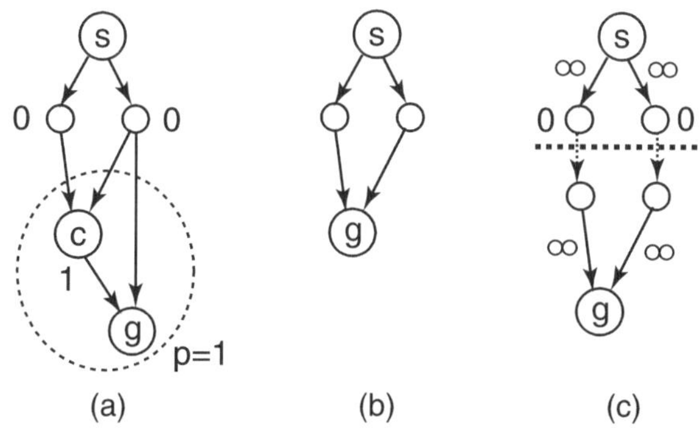

Figure 1.9. Visiting node g. (**a**) N_g, (**b**) N'_g, and (**c**) N''_g. N''_g contains a K-feasible cut with height of 0 shown in dotted line.

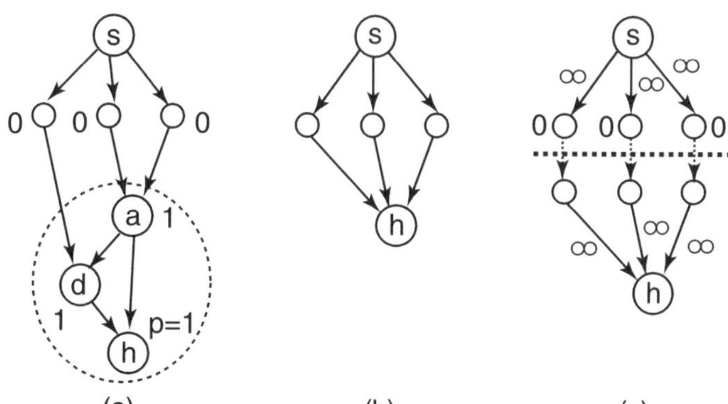

Figure 1.10. Visiting node h. (**a**) N_h, (**b**) N'_h, and (**c**) N''_h. N''_h contains a K-feasible cut with height of 0 shown in dotted line.

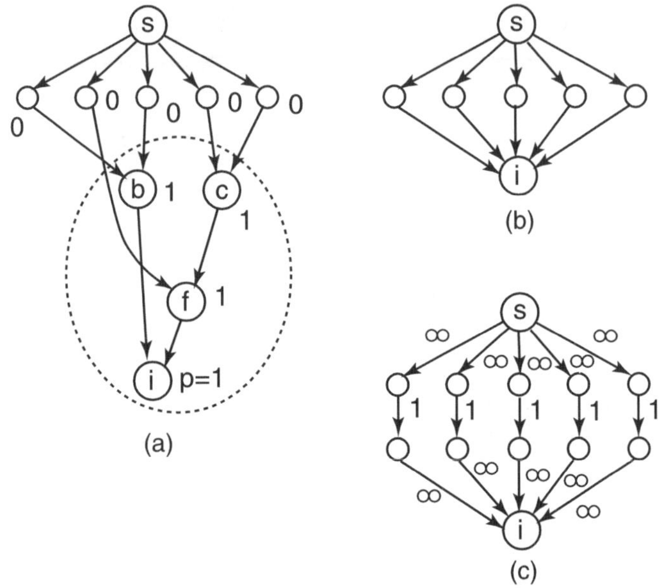

Figure 1.11. Visiting node i. (**a**) N_i, (**b**) N_i', and (**c**) N_i''. N_i'' does not contain a K-feasible cut.

(h) Node i: Figure 1.11 shows N_i, N_i', and N_i''. We see that $p = 1$. In this case, N_i'' does not contain a K-feasible cut. Thus, $\overline{X}_i = \{i\}$, and $l(i) = p + 1 = 2$.

(i) Node j: Figure 1.12 shows N_j, N_j', and N_j''. $p = 2$ in this case. There is only one K-feasible cut in N_j'', and its height is 1. Thus, $\overline{X}_j = \{i, j\}$, and $l(j) = p = 2$.

(j) Node k: Figure 1.13 shows N_k, N_k', and N_k''. $p = 2$ in this case. There is only one K-feasible cut in N_k'', and its height is 1. Thus, $\overline{X}_k = \{i, k\}$, and $l(k) = p = 2$.

Table 1.3 summarizes the labeling and clustering results.

2. Generate the LUTs for each node and draw the LUT-level network.

We initially set $L = \{h, j, k\}$. The following steps are based on Table 1.3 and Figure 1.6(b).

(a) Remove h from L. Then, h', the K-LUT implementation of h, contains $\{a, d, h\}$ according to Table 1.3. We note that $input(h')$ contains three PI nodes as shown in Figure 1.14(a). Since we do not add PI nodes into L, we have $L = \{j, k\}$.

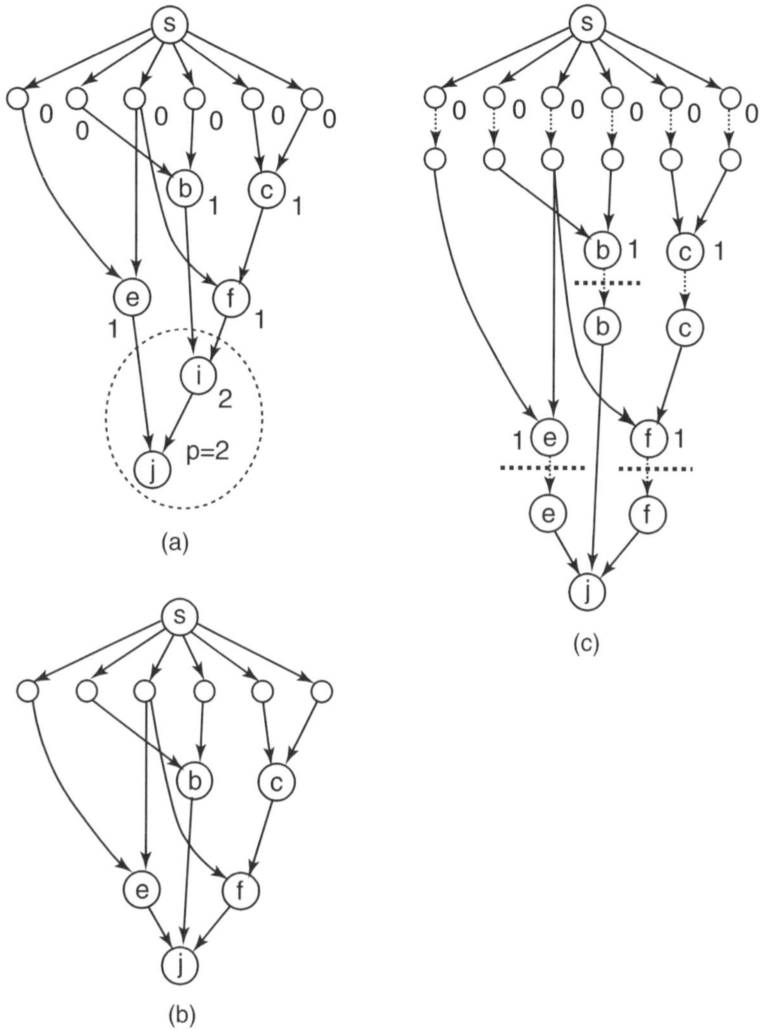

Figure 1.12. Visiting node j. (**a**) N_j, (**b**) N'_j, and (**c**) N''_j. N''_j contains a K-feasible cut with height of 1 shown in dotted line. Infinite capacities are not shown for simplicity.

(b) Remove j from L: $j' = \{i, j\}$ according to Table 1.3. We see that $input(j') = \{e, b, f\}$ as shown in Figure 1.14(b). Thus, $L = \{k\} \cup \{e, b, f\} = \{k, e, b, f\}$.

(c) Remove k from L: $k' = \{i, k\}$, and $input(k') = \{b, f, g\}$ as shown in Figure 1.14(c). Thus, $L = \{e, b, f\} \cup \{b, f, g\} = \{e, b, f, g\}$.

(d) Remove e from L: $e' = \{e\}$, and $input(e')$ contains two PI nodes. $L = \{b, f, g\}$.

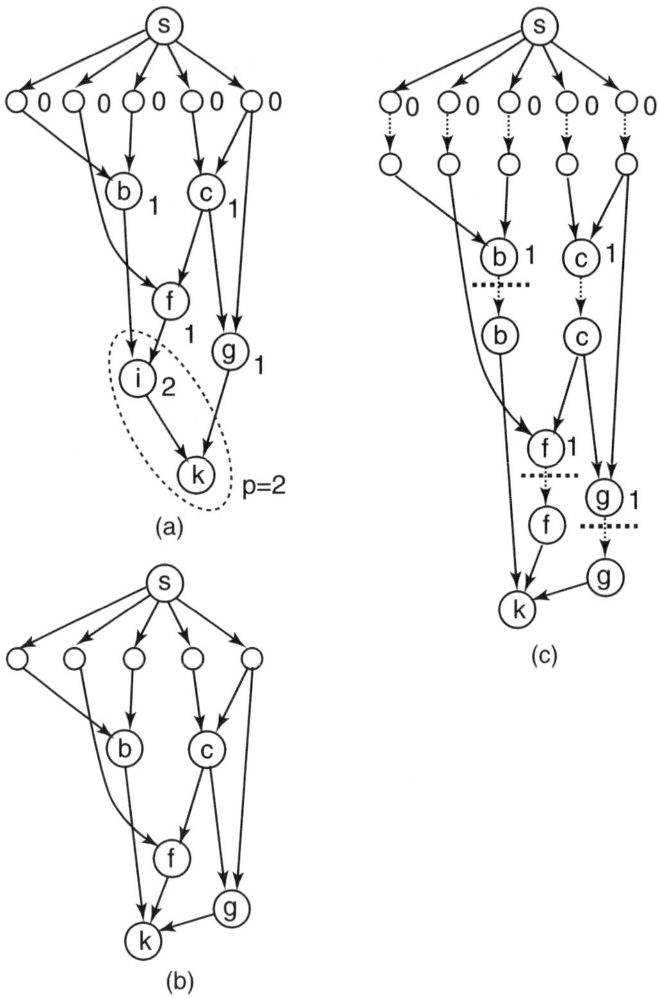

Figure 1.13. Visiting node k. (**a**) N_k, (**b**) N'_k, and (**c**) N''_k. N''_k contains a K-feasible cut with height of 1 shown in dotted line. Infinite capacities are not shown for simplicity.

(e) Remove b from L: $b' = \{b\}$, and $input(b')$ contains two PI nodes. $L = \{f, g\}$.

(f) Remove f from L: $f' = \{c, f\}$, and $input(f')$ contains three PI nodes. $L = \{g\}$.

(g) Remove g from L: $g' = \{c, g\}$, and $input(g')$ contains two PI nodes. $L = \emptyset$.

Table 1.4 shows the 7 LUTs generated. Figure 1.15 shows the LUT-level network. Nodes c and i are duplicated in the LUT network.

Table 1.3. Summary of the labeling and clustering process of the FlowMap algorithm.

Node	Label	Clustering
a	1	$\{a\}$
b	1	$\{b\}$
c	1	$\{c\}$
d	1	$\{a, d\}$
e	1	$\{e\}$
f	1	$\{c, f\}$
g	1	$\{c, g\}$
h	1	$\{a, d, h\}$
i	2	$\{i\}$
j	2	$\{i, j\}$
k	2	$\{i, k\}$

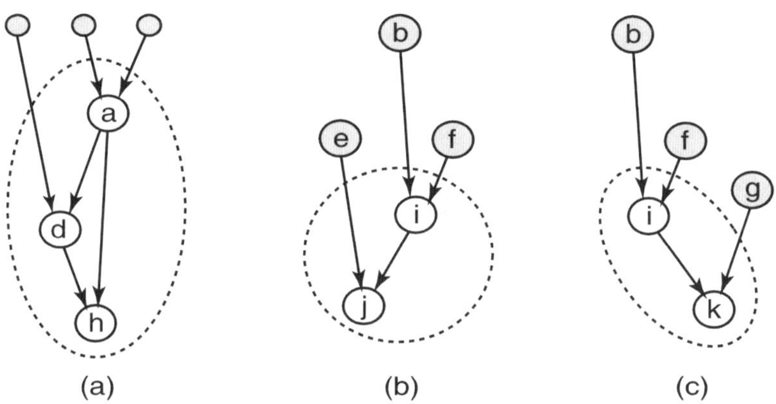

Figure 1.14. (**a**) Cluster rooted at h and its input nodes, (**b**) cluster rooted at j and its input nodes, (**c**) cluster rooted at k and its input nodes.

Table 1.4. List of LUTs generated by FlowMap algorithm.

Root	Elements
h	$\{a, d, h\}$
j	$\{i, j\}$
k	$\{i, k\}$
e	$\{e\}$
b	$\{b\}$
f	$\{c, f\}$
g	$\{c, g\}$

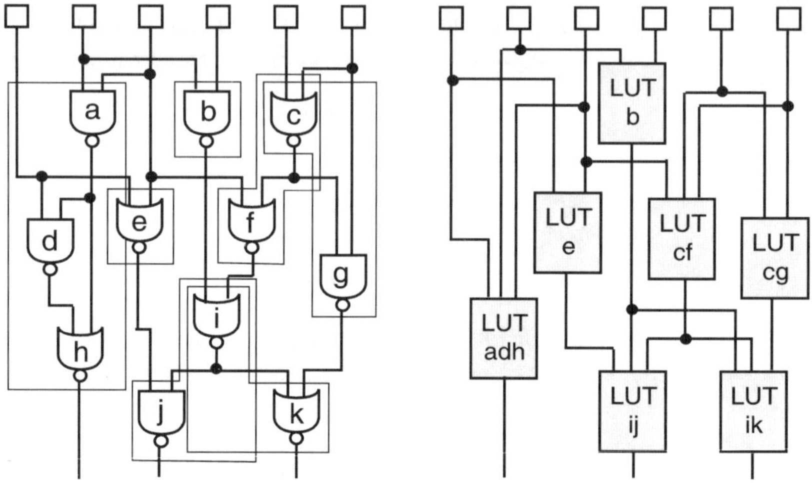

Figure 1.15. (**a**) Original network, (**b**) LUT-level network.

3. What is the maximum delay in the LUT-level network?

   According to Table 1.3, the maximum label is 2, which corresponds to the maximum delay value in the LUT-level network.

3. Multi-Level Coarsening Algorithm

hMetis algorithm [Karypis et al., 1997] solves the balanced bipartitioning problem, where the given circuit is divided into two roughly equal sized partitions. The objective is to minimize the number of inter-partition interconnects. hMetis consists of two phases, namely, clustering and partitioning. The basic principle behind the clustering (or coarsening as called in the paper) phase of hMetis is called "multi-level optimization" that requires multiple iterations of clustering process. During the first iteration, the gates in the original circuit are grouped to form level-1 clusters. At the end of this clustering process, we derive the network of these level-1 clusters. Note that the size (= number of nodes) of this network is smaller than that of the original circuit. Next, we group level-1 clusters together to form level-2 clusters and obtain their network. We repeat this process of "grouping clusters and reducing the size of their network" until we obtain the desired number of levels, say K, in the clustering hierarchy.

During the partitioning (or uncoarsening as called in the paper) phase of hMetis, we first perform bipartitioning among the level-K clusters using an existing partitioning algorithm such as FM [Fiduccia and Mattheyses, 1982]. After the partitioning solution reaches a local minima, these level-K clusters are decomposed (or uncoarsened) to reveal the level-K-1 clusters contained in them. We then further optimize the current partitioning solution using these level-K-1 clusters. This "decomposition and refinement" process is repeated until we obtain the partitioning of the gates in the original circuit. The clustering algorithms used in hMetis are shown to be effective in reducing the number of inter-cluster interconnects at higher levels, which in turn helps improve the partitioning solution quality.

Quick Overview

Given a hypergraph that models the gate-level circuit to be partitioned, hMetis algorithm utilizes three algorithms to compute the multi-level cluster hierarchy, namely, edge coarsening (EC), hyperedge coarsening (HEC), and modified hyperedge coarsening (MHEC).

- Edge coarsening (EC): The nodes in the hypergraph are first unmarked and visited in a random order. Given an unmarked node v, we first collect the "neighbors" of v, which is the set of nodes that are unmarked and are included in the hyperedges that contain v. For each neighbor n of v, we compute the weight of edge (v, n) by assigning a value $1/(|h| - 1)$, where h denotes the hyperedge that contains both n and v. After examining all neighbors of v, we select the neighbor with the maximum edge weight and merge v and n together. We mark both n and v so that these nodes are not clustered again later. This process completes when all nodes are visited.

- Hyperedge coarsening (HEC): Initially, all nodes are unmarked. The hyperedges are then sorted in an increasing order of their sizes. In case the hyperedges are weighted, we sort the hyperedges in a decreasing order of their weights and break ties in favor of smaller size. Next, we visit the hyperedges in the sorted order. Given a hyperedge, we examine if it contains any node that is already marked. If not, we group all nodes in the hyperedge to form a cluster. Otherwise, we skip to the next hyperedge. After visiting all hyperedges, each node that is not part of any cluster becomes a cluster of its own.

- Modified hyperedge coarsening (MHEC): This algorithm first applies HEC to the given hypergraph. After the hyperedges to be clustered have been selected, we visit the hyperedges again in the sorted order. For each hyperedge that has not yet been clustered (because it contains marked nodes), all the unmarked nodes in this hyperedge are clustered together. MHEC tends to further reduce the hyperedge counts after clustering and balance the size among the clusters.

Practice Problem

Consider the gate-level circuit shown in Figure 1.16. Assume that the weight of all nets is 1.

1. Perform Edge Coarsening and derive the cluster-level netlist. Visit the gates and break ties in alphabetical order. How many clusters were generated? How many hyperedges are included in the cluster-level circuit?

   (a) Visit a: Note that a is contained in n_1 only. So, $neighbor(a) = \{c, e\}$. The weight of $(a, c) = 1/(|n_1| - 1) = 0.5$. The weight of $(a, e) = 1/(|n_1| - 1) = 0.5$. Thus, we break the tie based on alphabetical order. So, a merges with c. We form $C_1 = \{a, c\}$ and mark a and c.

   (b) Visit b: Note that b is contained in n_2 only. So, $neighbor(b) = \{c, d\}$. Since c is already marked, b merges with d. We form $C_2 = \{b, d\}$ and mark b and d.

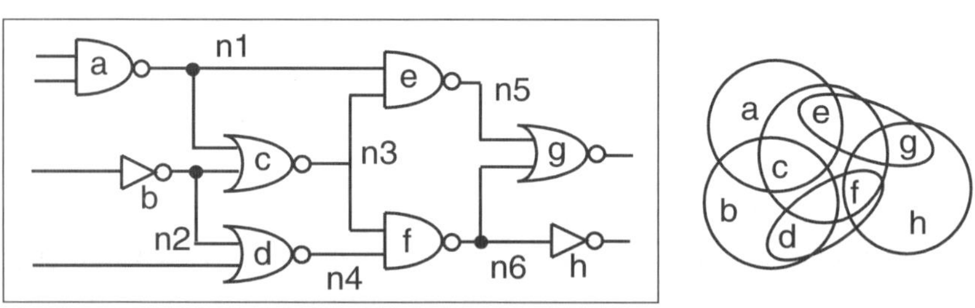

Figure 1.16. A gate-level circuit and its hypergraph representation.

(c) Since c and d are marked, we skip them.

(d) Visit e: the unmarked neighbors of e are g and f. We see that $w(e, g) = 1$ and $w(e, f) = 0.5$. So, e merges with g. We form $C_3 = \{e, g\}$ and mark e and g.

(e) Visit f: Node f is contained in n_3, n_4, and n_6. So, $neighbor(f) = \{c, d, e, g, h\}$. But, the only unmarked neighbor is h. So, f merges with h. We form $C_4 = \{f, h\}$ and mark f and h.

(f) Since g and h are marked, we skip them.

Table 1.5 shows the summary of the clustering result, and Table 1.6 shows how we obtain the cluster-level netlist. Note that (i) we removed duplicated entries in the cluster-level network, and (ii) the net that contains single cluster is deleted. Figure 1.17 shows the hypergraphs before and after EC. We generated 4 clusters that are connected by five hyperedges.

Table 1.5. Edge coarsening result.

Cluster	Nodes
C_1	$\{a, c\}$
C_2	$\{b, d\}$
C_3	$\{e, g\}$
C_4	$\{f, h\}$

Table 1.6. Netlist transformation based on EC result.

Net	Gate-level	Cluster-level	Final
n_1	$\{a, c, e\}$	$\{C_1, C_1, C_3\}$	$\{C_1, C_3\}$
n_2	$\{b, c, d\}$	$\{C_2, C_1, C_2\}$	$\{C_1, C_2\}$
n_3	$\{c, e, f\}$	$\{C_1, C_3, C_4\}$	$\{C_1, C_3, C_4\}$
n_4	$\{d, f\}$	$\{C_2, C_4\}$	$\{C_2, C_4\}$
n_5	$\{e, g\}$	$\{C_3, C_3\}$	$\emptyset$
n_6	$\{f, g, h\}$	$\{C_4, C_3, C_4\}$	$\{C_3, C_4\}$

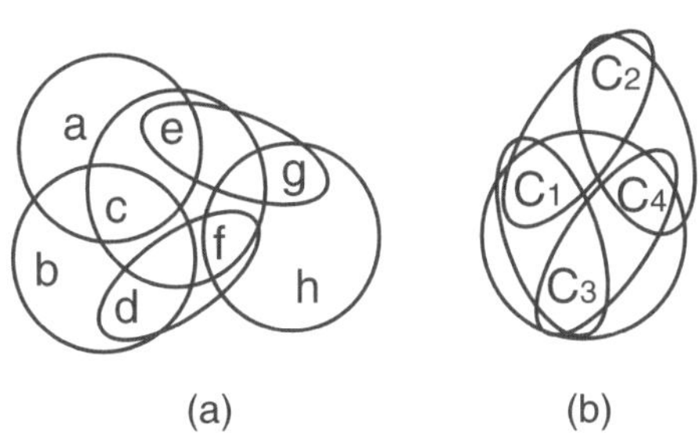

Figure 1.17. (**a**) Original hypergraph with six hyperedges, (**b**) hypergraph after edge coarsening, which has five hyperedges.

2. Perform Hyperedge Coarsening and derive the cluster-level netlist. Break ties in alphabetical order. How many clusters were generated? How many hyperedges are included in the cluster-level circuit?

   We first sort the hyperedges in an increasing order of their sizes: n_4, n_5, n_1, n_2, n_3, n_6. We unmark all nodes initially.

   (a) Visit $n_4 = \{d, f\}$: since d and f are not marked yet, we form $C_1 = \{d, f\}$ and mark d and f.

   (b) Visit $n_5 = \{e, g\}$: since e and g are not marked yet, we form $C_2 = \{e, g\}$ and mark e and g.

   (c) Visit $n_1 = \{a, c, e\}$: since e is already marked, we skip n_1.

   (d) Visit $n_2 = \{b, c, d\}$: since d is already marked, we skip n_2.

   (e) Visit $n_3 = \{c, e, f\}$: since e and f are already marked, we skip n_3.

   (f) Visit $n_6 = \{f, g, h\}$: since f and g are already marked, we skip n_6.

   At this point, each unmarked node becomes a cluster of its own: $C_3 = a$, $C_4 = b$, $C_5 = c$, and $C_6 = h$. Table 1.7 shows the summary of the HEC clustering result, and Table1.8 shows how we obtain the cluster-level netlist. Figure 1.18 shows the hypergraphs before and after HEC. We generated 6 clusters that are connected by four hyperedges.

3. Perform Modified Hyperedge Coarsening and derive the cluster-level netlist. Break ties in alphabetical order. How many clusters were generated? How many hyperedges are included in the cluster-level circuit?

Table 1.7. Hyperedge coarsening result.

Cluster	Nodes
C_1	$\{d, f\}$
C_2	$\{e, g\}$
C_3	$\{a\}$
C_4	$\{b\}$
C_5	$\{c\}$
C_6	$\{h\}$

Table 1.8. Netlist transformation based on HEC result.

Net	Gate-level	Cluster-level	Final
n_1	$\{a, c, e\}$	$\{C_3, C_5, C_2\}$	$\{C_3, C_5, C_2\}$
n_2	$\{b, c, d\}$	$\{C_4, C_5, C_1\}$	$\{C_4, C_5, C_1\}$
n_3	$\{c, e, f\}$	$\{C_5, C_2, C_1\}$	$\{C_5, C_2, C_1\}$
n_4	$\{d, f\}$	$\{C_1, C_1\}$	$\emptyset$
n_5	$\{e, g\}$	$\{C_2, C_2\}$	$\emptyset$
n_6	$\{f, g, h\}$	$\{C_1, C_2, C_6\}$	$\{C_1, C_2, C_6\}$

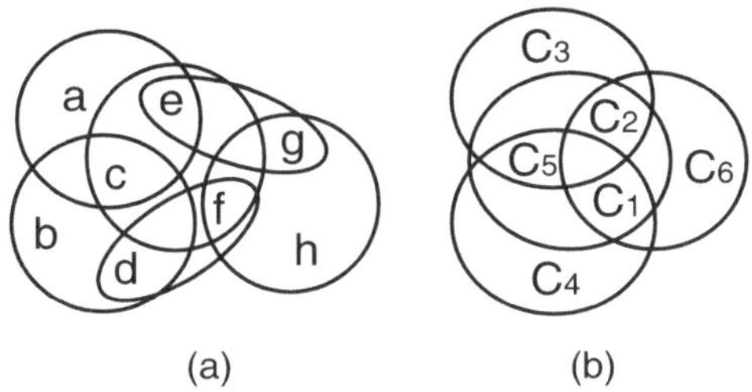

Figure 1.18. (a) Original hypergraph, (b) hypergraph after hyperedge coarsening, which has four hyperedges.

Table 1.9. Modified hyperedge coarsening result.

Cluster	Nodes
C_1	$\{d, f\}$
C_2	$\{e, g\}$
C_3	$\{a, c\}$
C_4	$\{b\}$
C_5	$\{h\}$

From the prior hyperedge coarsening, we note that we skipped n_1, n_2, n_3, and n_6. So we visit these again during MHEC:

(a) Visit $n_1 = \{a, c, e\}$: since e is already marked during HEC, we group the remaining unmarked nodes a and c. We form $C_3 = \{a, c\}$ and mark a and c.

(b) Visit $n_2 = \{b, c, d\}$: since d is marked during HEC and c during MHEC as above, we form $C_4 = \{b\}$ and mark b.

(c) Visit $n_3 = \{c, e, f\}$: since all of the nodes are already marked, we skip n_3.

(d) Visit $n_6 = \{f, g, h\}$: since f and g are already marked, we form $C_5 = \{h\}$ and mark h.

Table 1.9 shows the summary of the MHEC clustering result, and Table 1.10 shows how we obtain the cluster-level netlist. Figure 1.19 shows the hypergraphs before and after MHEC. We generated 5 clusters that are connected by four hyperedges.

Table 1.10. Netlist transformation based on MHEC result.

Net	Gate-level	Cluster-level	Final
n_1	$\{a, c, e\}$	$\{C_3, C_3, C_2\}$	$\{C_3, C_2\}$
n_2	$\{b, c, d\}$	$\{C_4, C_3, C_1\}$	$\{C_4, C_3, C_1\}$
n_3	$\{c, e, f\}$	$\{C_3, C_2, C_1\}$	$\{C_3, C_2, C_1\}$
n_4	$\{d, f\}$	$\{C_1, C_1\}$	$\emptyset$
n_5	$\{e, g\}$	$\{C_2, C_2\}$	$\emptyset$
n_6	$\{f, g, h\}$	$\{C_1, C_2, C_5\}$	$\{C_1, C_2, C_5\}$

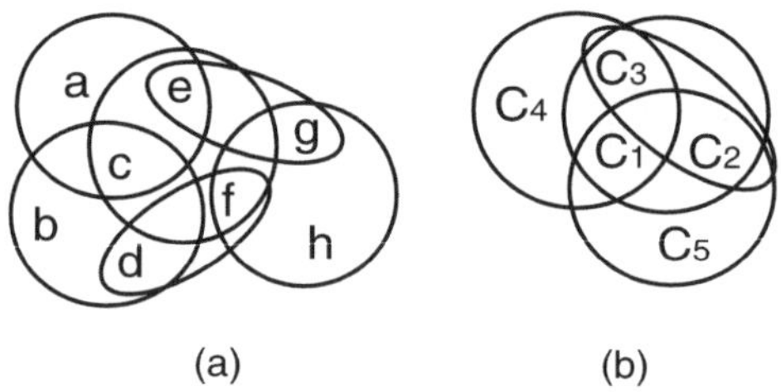

Figure 1.19. (**a**) Original hypergraph, (**b**) hypergraph after modified hyperedge coarsening, which has four hyperedges.

4. More Practice Problems

1. Perform Rajaraman and Wong algorithm on the graph shown in Figure 1.1 under the following assumptions:

   - Inter-cluster delay is 5.
   - Cluster size limit is set to 4.
   - The delay of nodes $\{a, b, c, k, l\}$ is 1. The delay of $\{d, f, h\}$ is 3, $\{e, g\}$ is 2, and $\{i, j\}$ is 4.

   Draw the cluster-level graph. How many clusters are generated? How many nodes are duplicated? What is the maximum delay in the clustered graph?

2. Perform Rajaraman and Wong algorithm on the following circuit:

   Assume that the inter-cluster delay is 3, cluster size limit is 3, and all gates have a unit delay.

3. Perform FlowMap algorithm to map the gate-level circuit shown in Figure 1.6 to 4-LUT, i.e., $K = 4$. Draw the LUT-level network. How many LUTs are generated? How many nodes are duplicated? What is the maximum delay in the LUT-level network?

4. Perform FlowMap algorithm to map the gate-level circuit shown in Figure 1.20 to 3-LUT. Draw the LUT-level network. How many LUTs are generated? How many nodes are duplicated? What is the maximum delay in the LUT-level network?

5. Perform Hyperedge Coarsening on the hypergraph shown in Figure 1.16. Assume that the weight of the nets is as follows: $w(n_1) = 5$, $w(n_2) = 2$, $w(n_3) = 3$, $w(n_4) = 3$, $w(n_5) = 1$, and $w(n_6) = 5$.

6. Perform Edge Coarsening on the hypergraph shown in Figure 1.17(b). How many clusters and hyperedges are included in the new clustered-level?

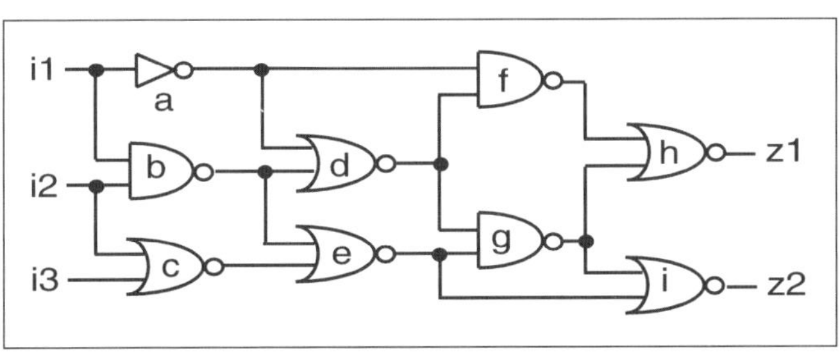

Figure 1.20. A gate-level circuit.

7. Perform Edge Coarsening, Hyperedge Coarsening, and Modified Hyperedge Coarsening on the following netlist and obtain the cluster-level netlists: $n_1 = \{a, b, d, e\}$, $n_2 = \{b, d, f\}$, $n_3 = \{a, e, g\}$, $n_4 = \{c, e, f, h\}$, $n_5 = \{c, g\}$, $n_6 = \{a, d, f, h\}$. Assume that the weight of all nets is 1. For EC, visit the gates and break ties in alphabetical order. For HEC/MHEC, break ties randomly.

5. Probing Further

Disclaimer: The list here is meant to be representative, not comprehensive. A comprehensive survey on LUT-based FPGA mapping algorithms is provided in [Cong and Ding, 1996].

Rajaraman and Wong Algorithm

The authors of [Yang and Wong, 1995] solve the min-cut replication problem by extending the Rajaraman and Wong algorithm [Rajaraman and Wong, 1995]. The problem is to determine min-cut replication sets for a k-way partitioning solution such that the cutsize of the partition is minimized after the replication. They optimally solve the min-area min-cut replication problem on directed graphs, which is to find min-cut replication sets with the minimum sizes. More importantly, they give an optimal solution to the hypergraph min-area min-cut replication problem using a much smaller flow network model.

Power minimization together with delay optimality was the goal in [Vaishnav and Pedram, 1995]. Their algorithm provides a way to implicitly enumerate alternate clustering solutions and selects a solution that has the same optimal delay but less power dissipation. The delay optimal clustering options are generated by the Rajaraman and Wong algorithm. For tree circuits, the proposed algorithm produces delay and power optimal partitioning, whereas for non-tree circuits it produces delay optimal clustering with significantly improved power dissipation.

The authors of [Yang and Wong, 1997] considered both area and pin constraints for delay optimal clustering. Note that [Rajaraman and Wong, 1995] considers area constraint only, and [Cong and Ding, 1992] considers pin constraint only. This work is used to divide the circuit into multiple FPGAs. They developed a repeated network cut technique for finding a cluster that is bounded by both area and pin constraints. Their algorithm achieves optimal delay under either the area constraint only or the pin constraint only. Under both area and pin constraints, this algorithm achieves optimal delay in most cases.

The authors of [Pan et al., 1998] perform delay optimal clustering for sequential circuits, whereas [Rajaraman and Wong, 1995] is targeting combinational circuits. In addition, retiming is performed simultaneously to determine the optimal locations of flip-flops (FF) for further delay reduction. This work models FFs as edge weights in its graph model of the netlist and performs retiming together with clustering. The algorithm produces clustering solutions with the optimal delay under the unit delay model, i.e., gates have delay, but interconnects do not. For the general delay model, i.e., both gate and interconnects have delay, it produces clustering solutions with a clock period provably close to optimal.

The basic delay optimal clustering problem is extended to two-level hierarchy in [Cong and Romesis, 2001], where clustering is applied twice recursively. This has an application to the latest field programmable gate array (FPGA) hierarchical architecture. The goal is to minimize the longest path delay in the two-level clustering solution. The authors showed that this problem, unlike the single-level clustering, is NP-hard. Their heuristic provides area-delay trade-off by controlling the amount of node duplication.

FlowMap Algorithm

The authors of [Yang and Wong, 1994] proposed EdgeMap, a delay optimal clustering algorithm under the "general delay model", where the gates and interconnect have delay. This is an extension of the "unit delay model" used in FlowMap algorithm [Cong and Ding, 1992], where only the gates have delay. EdgeMap is a non-trivial generalization of the FlowMap algorithm, where a key problem to solve is to compute a K-feasible network cut such that the circuit delay on every cut edge is upper-bounded by a specific value.

Area (= number of LUT clusters) minimization is another important objective for look-up table (LUT)-based FPGA mapping. The authors of FlowMap-r algorithm [Cong and Ding, 1994] studied the area and depth trade-off in LUT-based FPGA mapping. Starting from a depth-optimal mapping solution obtained by the FlowMap algorithm, they perform a sequence of depth relaxation operations and area-minimizing mapping procedures to produce a set of mapping solutions for a given design with smooth area and depth trade-off. A key part of FlowMap-r is a polynomial time optimal algorithm for computing an area-minimum mapping solution without node duplication for a K-bounded general Boolean network.

CutMap [Cong and Hwang, 1995] is another significant improvement of the FlowMap algorithm, which combines depth and area minimization during the LUT-based FPGA mapping process. CutMap computes min-cost min-height K-feasible cuts for critical nodes for depth minimization and computes min-cost K-feasible cuts for non-critical nodes for area minimization. CutMap guarantees depth-optimal mapping solutions in polynomial time as FlowMap but uses considerably fewer LUTs.

The authors of [Kukimoto et al., 1998] presented the first linear-time algorithm for the optimal delay technology mapping for standard library-based design instead of LUT-based FPGA. They showed that the basic dynamic programming approach in the FlowMap algorithm is not specific to FPGA mapping and can be easily adapted to library-based mapping. A key is to use the actual pin-to-pin delays of gates specified in a given library instead of unit delay as in the FlowMap algorithm.

Hermes [Teslenko and Dubrova, 2004] is a depth-optimal LUT based FPGA mapping algorithm, where a new strategy is presented to find depth-optimal

LUTs in a significantly shorter time compared to FlowMap. The quality of results is improved by enabling LUT re-implementation and by introducing a cost function which encourages input sharing among LUTs. In addition, Hermes performs flow computation directly on the original circuit graph instead of using the subgraphs as in FlowMap.

DAOmap [Chen and Cong, 2004] performs area minimization under delay constraints for LUT based FPGA mapping. The authors consider the potential node duplications during the cut enumeration/generation procedure so that the mapping costs encoded in the cuts drive the area-optimization objective more effectively. After the timing constraint is determined, they relax the non-critical paths by considering both local and global optimality information to minimize mapping area. DAOmap significant outperforms CutMap [Cong and Hwang, 1995] in terms of area and runtime.

The authors of [Mishchenko et al., 2007] avoid exhaustive cut enumeration for LUT mapping by computing only a small fixed number of so called "priority cuts" at each node. The criteria used to prioritize the cuts differ depending on the mapping goals. In case of delay minimization, the cuts are prioritized first by delay, then by the number of inputs, and finally by area. The authors showed that such prioritization gives a depth-optimum mapping for 95% of all benchmarks and LUT sizes, even if only one cut is stored at each node. The memory and runtime of the proposed algorithm are linear in circuit size.

Multi-Level Coarsening Algorithm

ML algorithm [Alpert et al., 1998] is another popular multi-level partitioning algorithm. Their multi-level clustering engine is similar to the Edge Coarsening algorithm used in the hMetis algorithm [Karypis et al., 1997], where a pair of nodes is matched and merged so that the area balance among the clusters is promoted. The authors enhanced their basic multi-level partitioning refinement engine by integrating last-in-first-out (LIFO) bucket structure [Hagen et al., 1997], and CLIP scheme [Dutt and Deng, 1996a] that basically promotes the moving of cells in local proximity.

hMetis-Kway algorithm [Karypis and Kumar, 1999] is an extension of the hMetis algorithm that solves the K-way partitioning problem. Given a circuit netlist, the K-way partitioning problem divides the circuit into K partitions so that the cutsize is minimized. The same set of coarsening algorithms used in the hMetis algorithm is used for the hMetis-Kway algorithm. A major difference, however, is that the cell move in their partitioning refinement engine is greedy, where we choose a cell to move randomly instead of based on its gain. Once a cell is chosen, the destination partition is determined by the gain.

Simultaneous cutsize and delay optimization is the goal of the multi-level partitioning algorithm named HPM [Cong et al., 2000]. The first-level of the cluster hierarchy is built first using the PRIME algorithm [Cong et al., 1999b] for performance optimization. The rest of the levels in the cluster hierarchy are built by the global connectivity-based algorithm named ESC [Cong and Lim, 2004] for cutsize minimization. This combination of delay and cutsize-oriented clustering in the multi-level optimization framework proves to be effective in simultaneous cutsize and delay optimization.

The authors of [Ababei and Bazargan, 2003] presented a statistical timing-driven hMetis-based partitioning. They adopt the "statistical timing criticality" concept to guide the partitioning process, which is done by extending the hyperedge coarsening scheme of the hMetis partitioner. Using this method, the most critical nets in the circuit are not cut so that the timing minimization is achieved. The use of the hMetis partitioning algorithm makes their partitioning algorithm run fast.

The authors of [Hwang and Pedram, 2005] presented a performance-oriented multi-level partitioning algorithm. Given a directed acyclic graph (DAG) representation of a sequential circuit, its bipartitioning solution (L, R), and its preferred direction (either L-to-R or R-to-L), the goal is to minimize the number of cut edges not in the preferred direction. This is useful because the delay of an IO path increases if it contains a "backward edge." In case of multi-level clustering, they use the Edge Coarsening scheme in hMetis and use the maximum depth or the maximum hop-count of any path containing the edge as a tie-breaker.

Chapter 2

PARTITIONING

Given a gate-level circuit, the goal of circuit partitioning problem is to divide the circuit into K roughly equal-sized partitions. The traditional objective is to minimize the number of nets connecting gates in multiple partitions, which is typically called the "cutsize" in the literature. Other objectives include critical path delay, total power consumption, etc. This chapter presents sample problems related to the following works:

- Kernighan and Lin algorithm [Kernighan and Lin, 1970]
- Fiduccia and Mattheyses algorithm [Fiduccia and Mattheyses, 1982]
- EIG algorithm [Hagen and Kahng, 1992]
- FBB algorithm [Yang and Wong, 1996]

The first two algorithms are so called move-based, where the given partitioning solution is improved by moving the gates across the partitions. The third algorithm minimizes so called the "ratio cut" metric based on eigenvector computation. The last algorithm adopts the maximum flow model to perform partitioning under cutsize minimization.

S.K. Lim, *Practical Problems in VLSI Physical Design Automation*

1. Kernighan and Lin Algorithm

Kernighan and Lin proposed an efficient heuristic algorithm [Kernighan and Lin, 1970] that solves the NP-hard Balanced Bipartitioning Problem, where the given gate-level circuit is divided into two equal-sized partitions. An important concept proposed by the authors is called the "gain-based cell swap." In this approach, a randomly generated initial partitioning solution is iteratively improved by swapping a pair of two gates (= cells) across the partitioning boundary. This cell swap is guided by the "gain" that represents how much the current cutsize will decrease after the swap. At every swap, the algorithm chooses the pair with the maximum gain.

Another important concept proposed by Kernighan and Lin is called the "pass", where all pairs are swapped exactly once during a single pass. When a pair is chosen to swap because it has the maximum gain, we swap it even when the gain value is negative. At the end of the current pass, we accept the first K swaps that lead to the minimum cutsize discovered during the entire pass, which may contain negative gain swaps. This concept of pass allows a limited degree of "uphill moves", where we accept worse moves during the exploration of the solution search in the hope of finding better local minima.

Kernighan and Lin motivated several successful follow-up algorithms such as [Fiduccia and Mattheyses, 1982; Karypis et al., 1997]. Instead of undirected graphs that Kernighan and Lin use to represent the circuits, the authors of [Fiduccia and Mattheyses, 1982] use hypergraphs, which are more natural representation of the circuits. In addition, they proposed to perform cell move instead of cell swap, where a cell is moved to the other partition instead of a pair of cells being swapped. Lastly, an efficient data structure is used to improve the time complexity significantly. More details of [Fiduccia and Mattheyses, 1982] are presented in Chapter 2, Section 2. In [Karypis et al., 1997], a multi-level bottom-up clustering is first performed, and an initial bipartitioning solution is obtained among the top-level clusters. This initial solution is refined by [Fiduccia and Mattheyses, 1982] in a "multi-level" fashion. More details of [Karypis et al., 1997] are presented in Chapter 1, Section 3.

Quick Overview

Given a gate-level circuit, the first step is to obtain an edge-weighted undirected graph G that represents the circuit. We typically use so called the k-clique model, where a net that contains k gates forms a k-clique in G, and each edge in the clique gets a weight of $1/(k-1)$. In case an edge (x, y) already exists from a prior net conversion, we just add the new weights instead of adding a parallel edge. We apply the KL algorithm on this graph, so the cutsize and gain are computed based on this graph, not the original circuit. Next, we obtain an initial balanced bipartitioning solution (P_1, P_2) of G, which is usually obtained randomly. For a cell $x \in P_1$, we define the *external cost of x* as follows:

$$E_x = \sum_{i \in P_2} c(x, i)$$

where $c(x, i)$ denotes the weight of the edge $e(x, i)$. This E_x denotes the sum of the weight of edges that connect x and its neighbors in the other partition, where the neighbors of x are defined to be the nodes that are connected with x via an edge. The *internal cost of x* is defined similarly as follows:

$$I_x = \sum_{i \in P_1} c(x, i)$$

This I_x denotes the sum of the weight of edges that connect x and its neighbors in the same partition. Finally, the gain of swapping x and y is defined as follows:

$$gain(x, y) = (E_x - I_x) + (E_y - I_y) - 2c(x, y)$$

We first unlock all cells before we start the first pass. Once the pass begins, we repeat the following at every swap until all cells are locked: (i) we compute the gain of all unlocked pairs, (ii) swap the pair with the maximum gain and lock the cells in the pair, and (iii) record the gain and the current cutsize. At the end of the pass, we identify and accept the first K swaps that lead to the minimum cutsize discovered during the entire pass. If the initial cutsize has reduced during the current pass, we attempt another pass using the best solution discovered from the current pass as the initial solution; otherwise we terminate the algorithm. Since we swap the cells, the area is always balanced between the two partitions. Note also that the entire Kernighan and Lin algorithm can be repeated with another random initial solution.

Practice Problem

Consider the gate-level circuit shown in Figure 2.1(a). Figure 2.1(b) shows an undirected graph model of this circuit.

1. Given an initial partition $\{abde, cfgh\}$, perform a single pass of KL algorithm. Break ties in lexicographical order.

   Figure 2.2(a) shows the initial partitioning. The initial cutsize is 5. Note that the cutsize is the sum of the weights of the cut edges, not the count of them. The gain values are computed based on the edge weights as well.

   (a) Swap 1: we first compute the gain of all unlocked pairs as shown in Table 2.1. Both pairs (d, c) and (e, c) have the maximum gain of 2, and we choose (d, c) based on lexicographical order. Figure 2.2(b) shows the resulting graph after swapping d and c.

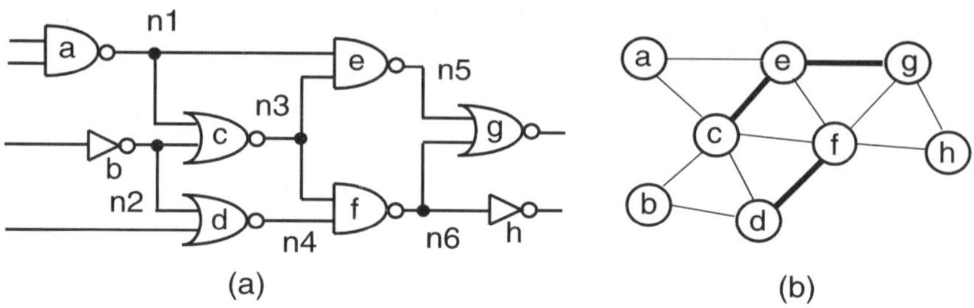

Figure 2.1. (**a**) Gate-level circuit, (**b**) its edge-weighted undirected graph representation. The thin and the thick lines indicate the weight of 0.5 and 1, respectively.

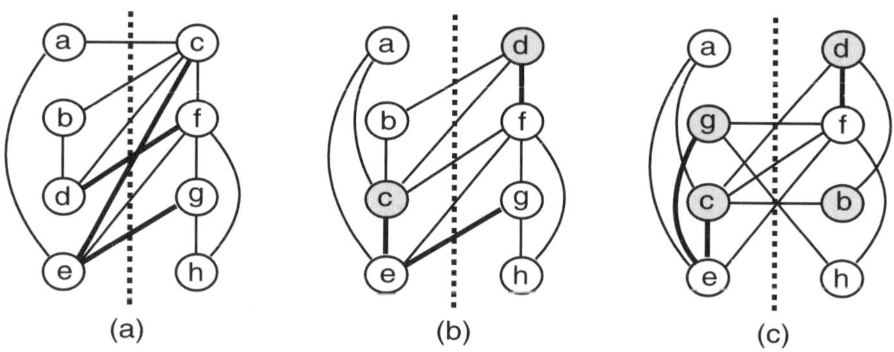

Figure 2.2. (**a**) Initial partitioning (cutsize = 5), (**b**) after swap 1, (**c**) after swap 2. The thin and the thick edges denote the weight of 0.5 and 1, respectively. The gray nodes are locked.

(b) Swap 2: we compute the gain of all unlocked pairs as shown in Table 2.2. (b, g) is the maximum gain pair. Figure 2.2(c) shows the resulting graph after the swap.

(c) Swap 3: we compute the gain of all unlocked pairs as shown in Table 2.3. (a, f) is the maximum gain pair (based on lexicographical order). Figure 2.3(a) shows the resulting graph after the swap.

(d) Swap 4: the only remaining unlocked pair is (e, h), and its gain is

$$(0.5 - 2.5) + (1 - 0) - 2 \cdot 0 = -1$$

Figure 2.3(b) shows the resulting graph after the swap.

2. Give the cutsize after each swap. What are the initial, the final, and the best cutsizes discovered during the pass?

   Table 2.4 shows the summary of the pass. The initial, the final, and the best cutsizes are 5, 5, and 3, respectively. We found two solutions with cutsize 3, and Figure 2.4 shows the first minimum cut.

Table 2.1. Gain computation for the first swap. The maximum gain swap chosen (due to lexicographic ordering) is shown in bold.

Pair	$E_x - I_x$	$E_y - I_y$	$c(x,y)$	Gain
(a,c)	$0.5-0.5$	$2.5-0.5$	0.5	1
(a,f)	$0.5-0.5$	$1.5-1.5$	0	0
(a,g)	$0.5-0.5$	$1-1$	0	0
(a,h)	$0.5-0.5$	$0-1$	0	-1
(b,c)	$0.5-0.5$	$2.5-0.5$	0.5	1
(b,f)	$0.5-0.5$	$1.5-1.5$	0	0
(b,g)	$0.5-0.5$	$1-1$	0	0
(b,h)	$0.5-0.5$	$0-1$	0	-1
$\mathbf{(d,c)}$	$\mathbf{1.5-0.5}$	$\mathbf{2.5-0.5}$	**0.5**	**2**
(d,f)	$1.5-0.5$	$1.5-1.5$	1	-1
(d,g)	$1.5-0.5$	$1-1$	0	1
(d,h)	$1.5-0.5$	$0-1$	0	0
(e,c)	$2.5-0.5$	$2.5-0.5$	1	2
(e,f)	$2.5-0.5$	$1.5-1.5$	0.5	1
(e,g)	$2.5-0.5$	$1-1$	1	0
(e,h)	$2.5-0.5$	$0-1$	0	1

Table 2.2. Gain computation for the second swap. The maximum gain swap is shown in bold.

Pair	$E_x - I_x$	$E_y - I_y$	$c(x,y)$	Gain
(a,f)	$0-1$	$1-2$	0	-2
(A,g)	$0-1$	$1-1$	0	-1
(a,h)	$0-1$	$0-1$	0	-2
(b,f)	$0.5-0.5$	$1-2$	0	-1
$\mathbf{(b,g)}$	$\mathbf{0.5-0.5}$	$\mathbf{1-1}$	**0**	**0**
(b,h)	$0.5-0.5$	$0-1$	0	-1
(e,f)	$1.5-1.5$	$1-2$	0.5	-2
(e,g)	$1.5-1.5$	$1-1$	1	-2
(e,h)	$1.5-1.5$	$0-1$	0	-1

Table 2.3. Gain computation for the third swap. The maximum gain swap is shown in bold.

Pair	$E_x - I_x$	$E_y - I_y$	$c(x,y)$	Gain
$\mathbf{(a,f)}$	$\mathbf{0-1}$	$\mathbf{1.5-1.5}$	**0**	$\mathbf{-1}$
(a,h)	$0-1$	$0.5-0.5$	0	-1
(e,f)	$0.5-2.5$	$1.5-1.5$	0.5	-3
(e,h)	$0.5-2.5$	$0.5-0.5$	0	-2

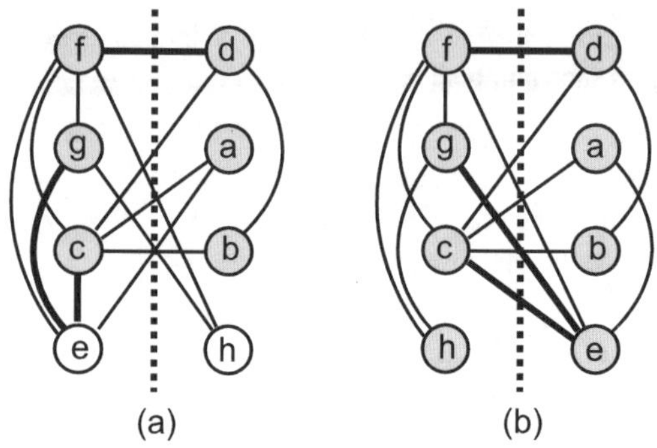

Figure 2.3. (**a**) After swap 3, (**b**) after swap 4. The thin and the thick lines denote the weight of 0.5 and 1, respectively. The gray nodes denote locked cells.

Table 2.4. A single pass of Kernighan and Lin algorithm. The minimum cutsize solutions are shown in bold.

i	Pair	$gain(i)$	$\sum gain(i)$	Cutsize
0	–	–	–	5
1	$(\boldsymbol{d}, \boldsymbol{c})$	**2**	**2**	**3**
2	$(\boldsymbol{b}, \boldsymbol{g})$	**0**	**2**	**3**
3	(a, f)	-1	1	4
4	(e, h)	-1	0	5

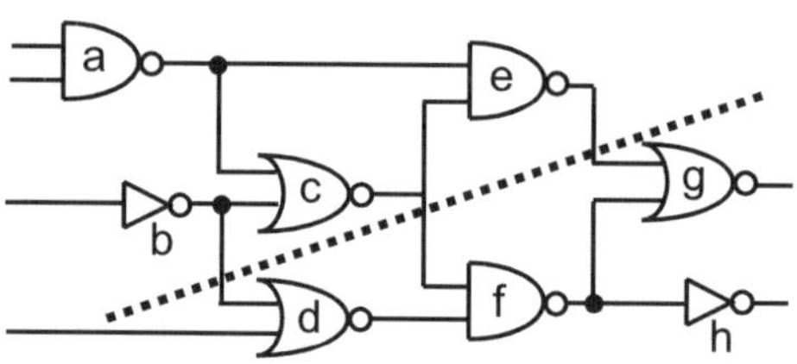

Figure 2.4. A bipartitioning solution of the circuit shown in Figure 2.1(a) with cutsize 3.

2. Fiduccia and Mattheyses Algorithm

Fiduccia and Mattheyses (FM) proposed a widely-used heuristic algorithm [Fiduccia and Mattheyses, 1982] for the Balanced Circuit Bipartitioning Problem. FM inherits several important concepts from Kernighan and Lin (KL) algorithm [Kernighan and Lin, 1970] such as gain computation and pass-based hill-climbing as explained in Chapter 2, Section 1. However, FM improves KL in the following three significant ways. First, FM applies directly on hypergraphs, a natural way to represent circuits. KL applies on an edge-weighted undirected graph instead. A study shows that it is not possible to assign weights to the edges in an undirected graph G so that any cut in G correctly represents the cutsize in the original circuit [Ihler et al., 1993]. On the other hand, the cutsize computed in the hypergraph exactly matches with the cutsize in the circuit.

Second, FM performs "cell moves" instead of "cell swaps" as in KL. At every move, the cell with the maximum gain is chosen to move to the other partition. Each cell move is constrained by the area balance requirement so that a cell move is legal only when the area constraint after the move is not violated. This cell move allows a significant time complexity improvement because we do not need $O(n^2)$ all-pair swap gain computation as in KL. Instead, we just need $O(n)$ gain computation.

Third, FM utilizes a special data structure called "bucket" to enable $O(1)$ search for the maximum gain cell, and $O(1)$ update of the gain values at each move. Before a pass begins in FM, we compute the gain values of all cells. Once the pass begins, we "update" the gain values of "affected cells" only instead of "computing" the gain values of "all cells" from scratch. This in turn means that the complexity of each cell move is $O(1)$ instead of $O(n^2)$ as in KL. Thus, the overall time complexity of FM is $O(n)$ compared to $O(n^3)$ in KL.

Quick Overview

The algorithm starts with an initial balanced bipartitioning solution (P_1, P_2) of the given hypergraph, which is usually obtained randomly. For a cell $x \in P_1$, we define $FS(x)$ as the number of nets that have x as the only cell in P_1. $TE(x)$ is defined as the number of nets that contain x and are entirely located in P_1, i.e., all cells in the net are partitioned in P_1. Finally, the gain of moving x from P_1 to P_2 is simply:

$$gain(x) = FS(x) - TE(x)$$

We perform the following three steps before the first pass begins: (i) unlock all cells, (ii) compute the gain of all cells based on the initial partitioning, and (iii) add the cells to the bucket structure. Once the pass begins, we repeat the following four steps at every move until all cells are locked: (i) we choose the cell with the maximum gain and is “legal”. A cell move is legal if moving it to the other partition does not violate the area constraint, (ii) move the chosen cell and lock it in the destination partition, (iii) update the gain values of the neighbors of the moved cell and update their positions in the bucket, and (iv) record the gain and the current cutsize. At the end of the pass, we identify and accept the first K moves that lead to the minimum cutsize discovered during the entire pass. If the initial cutsize has reduced during the current pass, we attempt another pass using the best solution discovered from the current pass as the initial solution; otherwise we terminate the algorithm. Note that the entire FM algorithm can be repeated with another random initial solution.

Practice Problem

Consider the gate-level circuit shown in Figure 2.5(a).

1. Model the circuit with a non-weighted hypergraph.

   See Figure 2.5(b).

2. Given an initial partition $\{acdg, befh\}$, calculate the initial gain of all cells and draw the bucket structure.

   Figure 2.6 shows the initial partitioning result. We compute the gain of the cells in the left partition as follows:

   (a) Cell a: a is contained in net $n_1 = \{a, c, e\}$. But a is not the only cell in n_1 that is located in the left partition, so $FS(a) = 0$. In addition, n_1 is not entirely located in the left partition. So, $TE(a) = 0$. Thus, $gain(a) = FS(a) - TE(a) = 0$.

   (b) Cell c: c is contained in net $n_1 = \{a, c, e\}$, $n_2 = \{b, c, d\}$, and $n_3 = \{c, f, e\}$. n_3 contains c as its only cell located in the left partition, so $FS(c) = 1$. In addition, none of these three nets are located entirely in the left partition. So, $TE(c) = 0$. Thus, $gain(c) = 1$.

   (c) Cell d: d is contained in net $n_2 = \{b, c, d\}$ and $n_5 = \{d, f\}$. n_5 contains d as its only cell located in the left partition, so $FS(d) =$

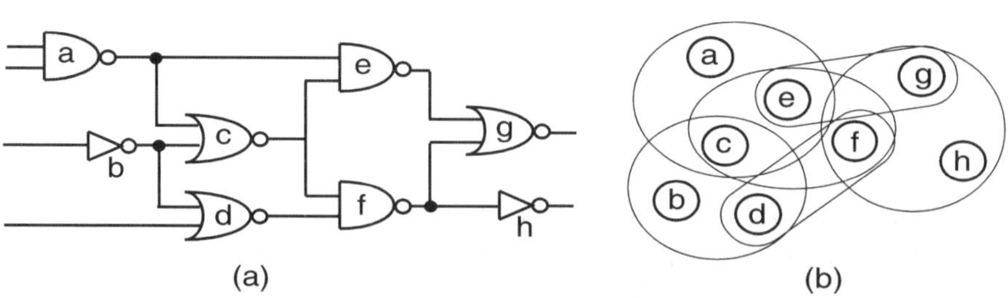

Figure 2.5. (**a**) Gate-level circuit, (**b**) hypergraph representation.

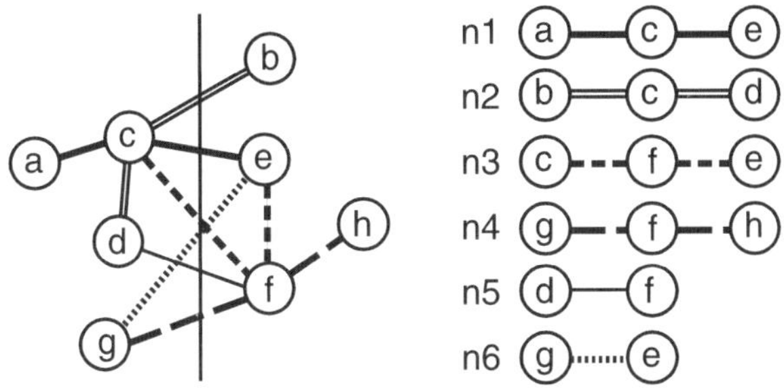

Figure 2.6. Netlist of the circuit in Figure 2.5 and its initial partitioning. Cutsize = 6.

1. In addition, none of these two nets are located entirely in the left partition. So, $TE(d) = 0$. Thus, $gain(d) = 1$.

(d) Cell g: g is contained in net $n_6 = \{g, e\}$ and $n_4 = \{g, f, h\}$. Both n_6 and n_4 contain g as their only cell located in the left partition, so $FS(g) = 2$. In addition, none of these two nets are located entirely in the left partition. So, $TE(g) = 0$. Thus, $gain(g) = 2$.

We compute the gain of the cells in the right partition as follows:

(a) Cell b: g is contained in net $n_2 = \{b, c, d\}$. n_2 contains b as its only cell located in the right partition, so $FS(b) = 1$. In addition, n_2 is not entirely located in the right partition, so, $TE(b) = 0$. Thus, $gain(b) = 1$.

(b) Cell e: e is contained in net $n_3 = \{c, f, e\}$, $n_6 = \{g, e\}$, and $n_1 = \{a, c, e\}$. Both n_6 and n_1 contain e as their only cell located in the right partition, so $FS(e) = 2$. In addition, none of these three nets are entirely located in the right partition, so, $TE(e) = 0$. Thus, $gain(e) = 2$.

(c) Cell f: f is contained in net $n_3 = \{c, f, e\}$, $n_5 = \{d, f\}$, and $n_4 = \{g, f, h\}$. n_5 contains f as its only cell located in the right partition, so $FS(f) = 1$. In addition, none of these three nets are entirely located in the right partition, so, $TE(f) = 0$. Thus, $gain(f) = 1$.

(d) Cell h: h is contained in net $n_4 = \{g, f, h\}$. But, h is not the only cell in n_4 that is located in the right partition, so $FS(h) = 0$. In addition, n_4 is not entirely located in the right partition. So, $TE(h) = 0$. Thus, $gain(h) = 0$.

From Figure 2.5(b) we see that the maximum net degree is $P_{max} = deg(c) = deg(e) = deg(f) = 3$, where $deg(x)$ denotes the number of hyperedges incident to x. Accordingly to [Fiduccia and Mattheyses, 1982],

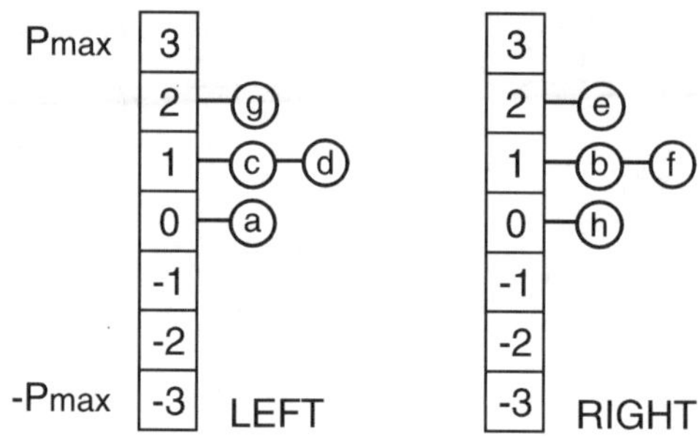

Figure 2.7. Bucket structure based on Figure 2.6.

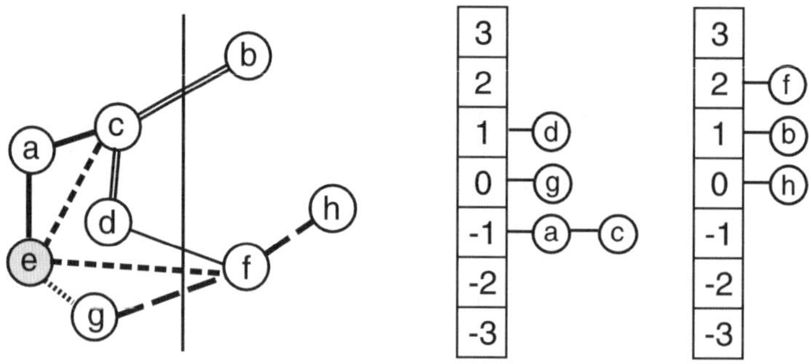

Figure 2.8. After moving e. Cutsize = 4.

P_{max} defines the range of the headers in the bucket structure. Figure 2.7 shows the bucket structure.

3. Perform a single pass of FM algorithm based on the area constraint $[3, 5]$. Ties should be broken in alphabetical order.

   (a) Move 1: From Figure 2.6, we see that both cell g and e have the maximum gain and can be moved without violating the area constraint. We move e based on alphabetical order. Figure 2.8 shows the resulting hypergraph. Next, we update the gain of the unlocked neighbors of e, $N(e) = \{a, c, g, f\}$, as follows: $gain(a) = FS(a) - TE(a) = 0 - 1 = -1$, $gain(c) = 0 - 1 = -1$, $gain(g) = 1 - 1 = 0$, $gain(f) = 2 - 0 = 2$. Figure 2.8 shows the updated bucket structure.

   (b) Move 2: f has the maximum gain, but moving f will violate the area constraint. So we move d. Figure 2.9 shows the resulting hypergraph. Next, we update the gain of the unlocked neighbors of d, $N(d) = \{b, c, f\}$, as follows: $gain(b) = 0 - 0 = 0$, $gain(c) = 1 - 1 = 0$, $gain(f) = 1 - 1 = 0$. Figure 2.9 shows the updated bucket structure.

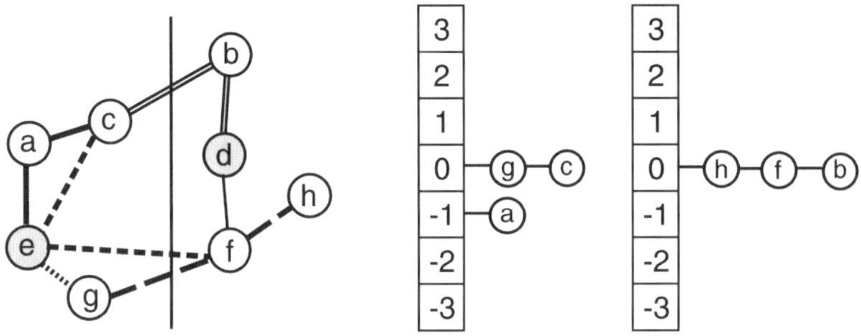

Figure 2.9. After moving d. Cutsize = 3.

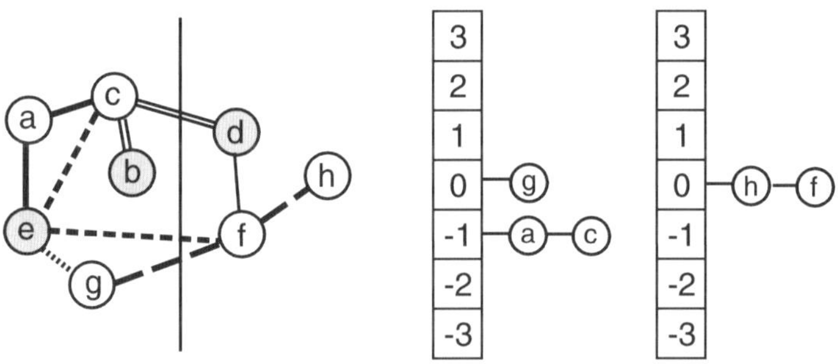

Figure 2.10. After moving b. Cutsize = 3.

(c) Move 3: Among the maximum gain cells $\{g, c, h, f, b\}$, we choose b based on alphabetical order. Figure 2.10 shows the resulting hypergraph. Next, we update the gain of the unlocked neighbors of b, $N(b) = \{c\}$ as follows: $gain(c) = 0 - 1 = -1$. Figure 2.10 shows the updated bucket structure.

(d) Move 4: Among the maximum gain cells $\{g, h, f\}$, we choose g based on the area constraint. Figure 2.11 shows the resulting hypergraph. Next, we update the gain of the unlocked neighbors of g, $N(g) = \{f, h\}$, as follows: $gain(f) = 1 - 2 = -1$, $gain(h) = 0 - 1 = -1$. Figure 2.11 shows the updated bucket structure.

(e) Move 5: We choose a based on alphabetical order. Figure 2.12 shows the resulting hypergraph. Next, we update the gain of the unlocked neighbors of a, $N(a) = \{c\}$, as follows: $gain(c) = 0 - 0 = 0$. Figure 2.12 shows the updated bucket structure.

(f) Move 6: We choose f based on the area constraint and alphabetical order. Figure 2.13 shows the resulting hypergraph. Next, we update

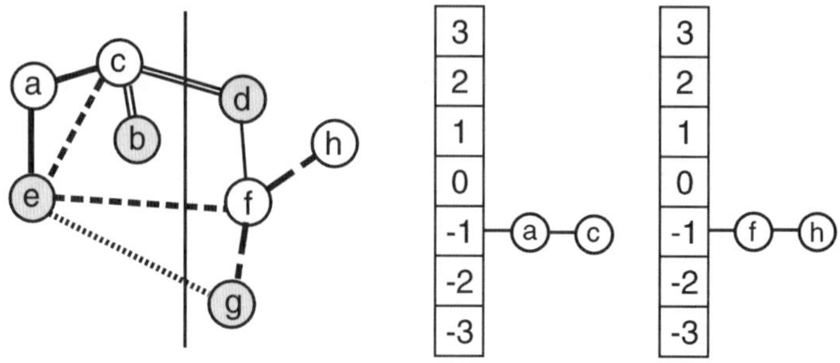

Figure 2.11. After moving g. Cutsize = 3.

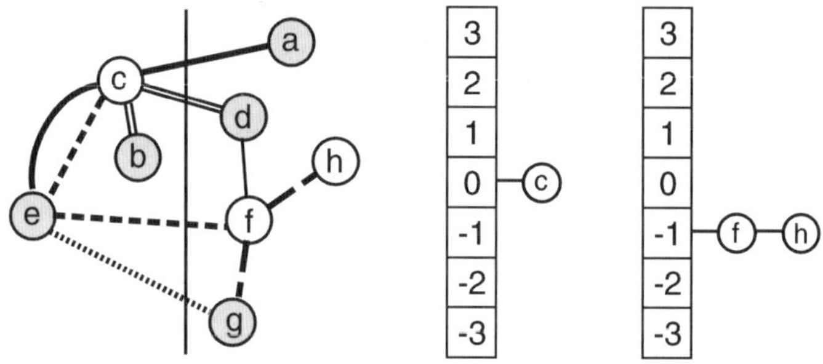

Figure 2.12. After moving a. Cutsize = 4.

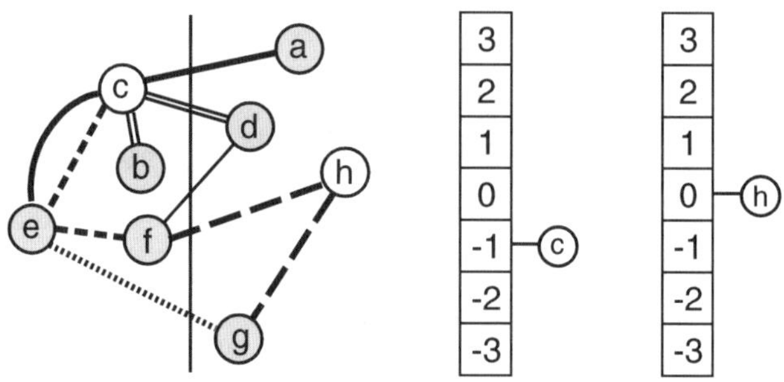

Figure 2.13. After moving f. Cutsize = 5.

the gain of the unlocked neighbors of f, $N(f) = \{h, c\}$, as follows: $gain(h) = 0 - 0 = 0$, $gain(c) = 0 - 1 = -1$. Figure 2.13 shows the updated bucket structure.

(g) Move 7: We move h. Figure 2.14 shows the resulting hypergraph and bucket structure. h has no unlocked neighbor.

(h) Move 8: We move c. Figure 2.15 shows the resulting hypergraph.

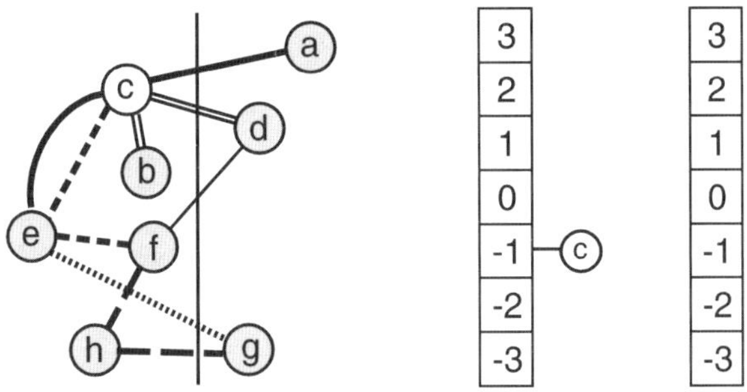

Figure 2.14. After moving h. Cutsize = 5.

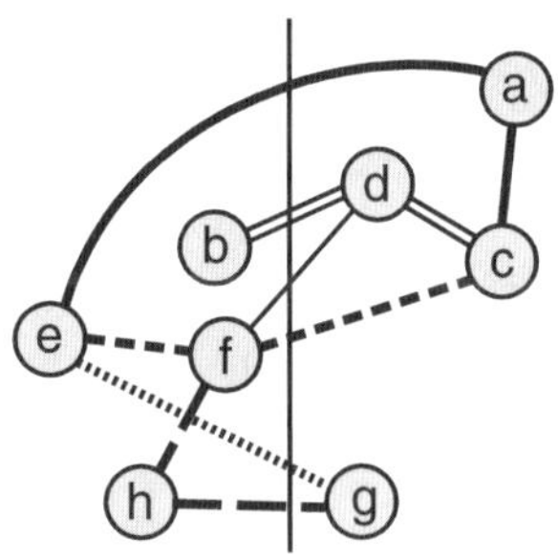

Figure 2.15. After moving c. Cutsize = 6.

Table 2.5. A single pass of FM. The minimum cutsize solutions are shown in bold.

i	Cell	$g(i)$	$\sum g(i)$	Cutsize
0	–	–	–	6
1	e	2	2	4
2	$\boldsymbol{d}$	**1**	**3**	**3**
3	$\boldsymbol{b}$	**0**	**3**	**3**
4	$\boldsymbol{g}$	**0**	**3**	**3**
5	a	–1	2	4
6	f	–1	1	5
7	h	0	1	5
8	c	–1	0	6

4. Give the cutsize after each move. What are the initial, final, and best cutsizes?

   Table 2.5 shows the summary of the FM pass. The initial, final, and best cutsizes are 6, 6, and 3, respectively. We found three solutions with cutsize 3 shown in Figures 2.9–2.11.

3. EIG Algorithm

Hagen and Kahng proposed a partitioning algorithm [Hagen and Kahng, 1992] named EIG that utilizes the second smallest eigenvalue and its eigenvector of a netlist matrix to optimize so called the "ratio cut" metric [Wei and Cheng, 1989]. This ratio cut metric, defined as $c(X,Y)/|X||Y|$, where $c(X,Y)$ denotes the cutsize between the two partitions X and Y, captures not only the cutsize minimization but also the area balancing that is important in circuit partitioning. Given an undirected graph G and its so called Laplacian matrix Q that basically shows the connectivity among the nodes in G, Hall [Hall, 1970] showed that the eigenvector of the second smallest eigenvalue of Q defines a one-dimensional placement of the nodes in G. In this placement solution, the "squared length" of the edges in G is minimized under the constraint $\sum_i x_i^2 = 1$, where x_i denotes the location of node i. Hagen and Kahng showed in [Hagen and Kahng, 1992] that the second smallest eigenvalue of Q is a tight lower bound of the ratio cut metric. Based on this theoretical result, they proposed a partitioning algorithm that utilizes this one-dimensional placement to derive bipartitioning solutions with minimal ratio cut metric.

Quick Overview

Given a gate-level circuit H, we first derive its undirected graph representation G based on the standard k-clique model. In this model, a net with k gates forms a k-clique, and each edge in the clique gets a weight of $1/(k-1)$. The remaining part of the algorithm proceeds as follows:

1. We build the $n \times n$ Laplacian matrix $Q = D - A$, where n is the number of nodes in G. A is the *adjacency matrix*, where each entry a_{ij} denotes the weight of edge $e(i,j)$ in G. D is the *degree matrix*, where each entry d_{ii} is the sum of the weights of all edges incident to node i in G.

2. We compute the second smallest eigenvalue and its eigenvector of Q using Lanczos method.

3. We sort the nodes in G based on their values in the eigenvector and obtain the node ordering $Z = \{v_1, v_2, \cdots, v_n\}$.

4. We use Z to derive and evaluate $n-1$ partitioning solutions. More specifically, we first obtain a bipartitioning solution $(\{v_1\}, \{v_2, \cdots, v_n\})$ and compute its ratio cut metric.[4] Next, we evaluate the ratio cut metric of $(\{v_1, v_2\}, \{v_3, \cdots, v_n\})$ from H, etc. Lastly, we choose the partitioning solution with the minimum ratio cut metric.

[4]Note that we can use H or G to compute cutsize. We use G in this section.

The authors presented another algorithm named EIG-IG. Instead of using G to compute eigenvectors, they use so called "intersection graph" (IG) instead. In IG, the nets in H become the nodes, and an edge (x, y) exists in IG if net x and y contain the same node.[5]

Practice Problem

Consider the gate-level circuit in Figure 2.16.

1. Build the k-clique based edge-weighted undirected graph of the circuit.

   See Figure 2.17.

2. Obtain the adjacency matrix A, degree matrix D, and Laplacian matrix Q.

   A is shown in Figure 2.18. D is shown in Figure 2.19. Lastly, Q is shown in Figure 2.20.

3. Find the second smallest eigenvalue and its eigenvector. Build the node ordering Z.

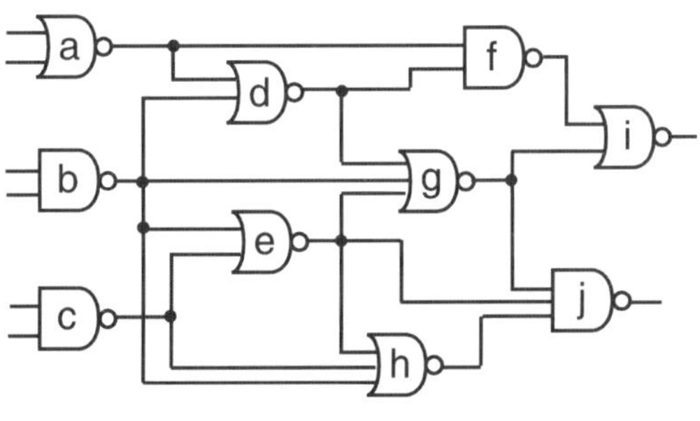

Figure 2.16. A gate-level circuit.

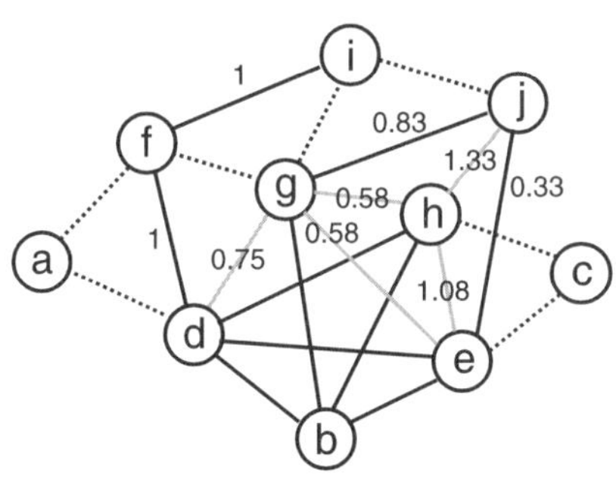

Figure 2.17. Clique-based edge-weighted undirected graph representation of the circuit in Figure 2.16. The dotted edges have the weight of 0.5, and the solid edges with no label have the weight of 0.25.

[5]See the related practice problem #5 on page 56.

	a	b	c	d	e	f	g	h	i	j
a	0	0	0	0.5	0	0.5	0	0	0	0
b	0	0	0	0.25	0.25	0	0.25	0.25	0	0
c	0	0	0	0	0.5	0	0	0.5	0	0
d	0.5	0.25	0	0	0.25	1.0	0.75	0.25	0	0
e	0	0.25	0.5	0.25	0	0	0.58	1.08	0	0.33
f	0.5	0	0	1.0	0	0	0.5	0	1.0	0
g	0	0.25	0	0.75	0.58	0.5	0	0.58	0.5	0.83
h	0	0.25	0.5	0.25	1.08	0	0.58	0	0	1.33
i	0	0	0	0	0	1.0	0.5	0	0	0.5
j	0	0	0	0	0.33	0	0.83	1.33	0.5	0

Figure 2.18. Adjacency matrix A.

	a	b	c	d	e	f	g	h	i	j
a	1.0	0	0	0	0	0	0	0	0	0
b	0	1.0	0	0	0	0	0	0	0	0
c	0	0	1.0	0	0	0	0	0	0	0
d	0	0	0	3.0	0	0	0	0	0	0
e	0	0	0	0	2.99	0	0	0	0	0
f	0	0	0	0	0	3.0	0	0	0	0
g	0	0	0	0	0	0	3.99	0	0	0
h	0	0	0	0	0	0	0	3.99	0	0
i	0	0	0	0	0	0	0	0	2.0	0
j	0	0	0	0	0	0	0	0	0	2.99

Figure 2.19. Degree matrix D.

	a	b	c	d	e	f	g	h	i	j
a	1.0	0	0	−0.5	0	−0.5	0	0	0	0
b	0	1.0	0	−0.25	−0.25	0	−0.25	−0.25	0	0
c	0	0	1.0	0	−0.5	0	0	−0.5	0	0
d	−0.5	−0.25	0	3.0	−0.25	−1.0	−0.75	−0.25	0	0
e	0	−0.25	−0.5	−0.25	2.99	0	−0.58	−1.08	0	−0.33
f	−0.5	0	0	−1.0	0	3.0	−0.5	0	−1.0	0
g	0	−0.25	0	−0.75	−0.58	−0.5	3.99	−0.58	−0.5	−0.83
h	0	−0.25	−0.5	−0.25	−1.08	0	−0.58	3.99	0	−1.33
i	0	0	0	0	0	−1.0	−0.5	0	2.0	−0.5
j	0	0	0	0	−0.33	0	−0.83	−1.33	−0.5	2.99

Figure 2.20. Laplacian matrix Q.

We use Matlab for the eigenvalue/eigenvector computation. The second smallest eigenvalue is 0.6281, and its eigenvector is: $[-0.6346, 0.1605, 0.5711, -0.1898, 0.2254, -0.2822, 0.0038, 0.1995, -0.1641, 0.1104]^T$. We observe the following:

- The squared sum of the values in the vector is 1 as shown by Hall [Hall, 1970].

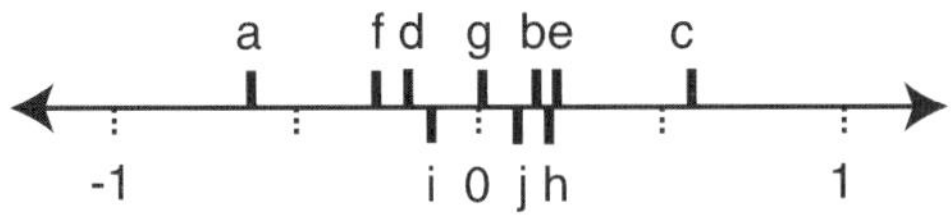

Figure 2.21. One-dimensional placement from the eigenvector.

- These values define a one-dimensional placement of the ten nodes within the range of $[-1, 1]$, where the sum of the squared length of all edges is minimized. Figure 2.21 shows this placement.
- These values define the following ordering among the nodes:

$$Z = \{a, f, d, i, g, j, b, h, e, c\}$$

4. Obtain the partitioning solution with the minimum ratio cut cost.

   (a) Partitioning $(\{a\}, \{f, d, i, g, j, b, h, e, c\})$:
   From Figure 2.17, we see that the cut edges are (a, f) and (a, d). Thus, the cutsize is $0.5 + 0.5 = 1.0$. Lastly, the ratio cut is $1.0/(1 \cdot 9) = 0.1111$.

   (b) Partitioning $(\{a, f\}, \{d, i, g, j, b, h, e, c\})$:
   From Figure 2.17, we see that the cut edges are (f, i), (f, g), (f, d) and (a, d). Thus, the cutsize is $1.0 + 0.5 + 1.0 + 0.5 = 3.0$. Lastly, the ratio cut is $3.0/(2 \cdot 8) = 0.1875$.

   (c) Partitioning $(\{a, f, d\}, \{i, g, j, b, h, e, c\})$:
   From Figure 2.17, we see that the cut edges are (f, i), (f, g), (d, g), (d, h), (d, e), and (d, b). Thus, the cutsize is $1.0 + 0.5 + 0.75 + 3 \cdot 0.25 = 3.0$. Lastly, the ratio cut is $3.0/(3 \cdot 7) = 0.1429$.

   (d) Partitioning $(\{a, f, d, i\}, \{g, j, b, h, e, c\})$:
   From Figure 2.17, we see that the cut edges are (i, j), (i, g), (f, g), (d, g), (d, h), (d, e), and (d, b). Thus, the cutsize is $0.5 \cdot 3 + 0.75 + 3 \cdot 0.25 = 3.0$. Lastly, the ratio cut is $3.0/(4 \cdot 6) = 0.125$.

   (e) Partitioning $(\{a, f, d, i, g\}, \{j, b, h, e, c\})$:
   From Figure 2.17, we see that the cut edges are (i, j), (g, j), (g, h), (g, e), (g, b), (d, h), (d, e), and (d, b). Thus, the cutsize is $0.5 + 0.83 + 0.58 \cdot 2 + 0.25 \cdot 4 = 3.49$. Lastly, the ratio cut is $3.49/(5 \cdot 5) = 0.1396$.

   (f) Partitioning $(\{a, f, d, i, g, j\}, \{b, h, e, c\})$:
   From Figure 2.17, we see that the cut edges are (j, e), (j, h), (g, h), (g, e), (g, b), (d, h), (d, e), and (d, b). Thus, the cutsize is $0.33 + 1.33 + 0.58 \cdot 2 + 0.25 \cdot 4 = 3.82$. Lastly, the ratio cut is $3.82/(6 \cdot 4) = 0.1592$.

   (g) Partitioning $(\{a, f, d, i, g, j, b\}, \{h, e, c\})$:
   From Figure 2.17, we see that the cut edges are (j, e), (j, h), (g, h),

Table 2.6. Summary of EIG algorithm.

P_A	P_B	Cutsize	Ratio cut
$\{a\}$	$\{f,d,i,g,j,b,h,e,c\}$	1.0	$1.0/(1 \cdot 9) = 0.1111$
$\{a,f\}$	$\{d,i,g,j,b,h,e,c\}$	3.0	$3.0/(2 \cdot 8) = 0.1875$
$\{a,f,d\}$	$\{i,g,j,b,h,e,c\}$	3.0	$3.0/(3 \cdot 7) = 0.1429$
$\{a,f,d,i\}$	$\{g,j,b,h,e,c\}$	3.0	$3.0/(4 \cdot 6) = 0.125$
$\{a,f,d,i,g\}$	$\{j,b,h,e,c\}$	3.49	$3.49/(5 \cdot 5) = 0.1396$
$\{a,f,d,i,g,j\}$	$\{b,h,e,c\}$	3.82	$3.82/(6 \cdot 4) = 0.1592$
$\{a,f,d,i,g,j,b\}$	$\{h,e,c\}$	3.82	$3.82/(7 \cdot 3) = 0.1819$
$\{a,f,d,i,g,j,b,h\}$	$\{e,c\}$	2.99	$2.99/(8 \cdot 2) = 0.1869$
$\{a,f,d,i,g,j,b,h,e\}$	$\{c\}$	1.0	$1.0/(9 \cdot 1) = 0.1111$

(g,e), (d,h), (d,e), (b,h), and (b,e). Thus, the cutsize is $0.33+1.33+0.58 \cdot 2+0.25 \cdot 4 = 3.82$. Lastly, the ratio cut is $3.82/(7 \cdot 3) = 0.1819$.

(h) Partitioning $(\{a,f,d,i,g,j,b,h\},\{e,c\})$:
From Figure 2.17, we see that the cut edges are (h,c), (h,e), (j,e), (g,e), (d,e), and (b,e). Thus, the cutsize is $0.5+1.08+0.33+0.58+0.25+0.25 = 2.99$. Lastly, the ratio cut is $2.99/(8 \cdot 2) = 0.1869$.

(i) Partitioning $(\{a,f,d,i,g,j,b,h,e\},\{c\})$:
From Figure 2.17, we see that the cut edges are (h,c) and (e,c). Thus, the cutsize is $0.5+0.5 = 1.0$. Lastly, the ratio cut is $1.0/(9 \cdot 1) = 0.1111$.

Table 2.6 shows the summary of EIG algorithm. We found two solutions with the minimum ratio cut cost value of 0.1111:

$$(\{a\},\{f,d,i,g,j,b,h,e,c\})$$
$$(\{a,f,d,i,g,j,b,h,e\},\{c\})$$

These solutions are very unbalanced in terms of area. But, the following solution is perfectly balanced and has the third lowest ratio cut cost of 0.1369:

$$(\{a,f,d,i,g\},\{j,b,h,e,c\})$$

5. Verify that the second smallest eigenvalue is a tight lower bound of the ratio cut metric.

The eigenvalue is $\lambda = 0.6281$. It is shown in [Hagen and Kahng, 1992] that $c \geq \lambda/n$, where c is the ratio cut cost, and n is the number of nodes in the graph. Since $n = 10$ in our case, we see that $\lambda/n = 0.06281$ is smaller than all of the ratio cut values shown in Table 2.6.

4. FBB Algorithm

Yang and Wong proposed a bipartitioning algorithm [Yang and Wong, 1996] named FBB that is based on the maximum flow computation. Given a flow network G and a pair of source and sink nodes (s, t), the Maximum Flow Minimum Cut Theorem [Ford and Fulkerson, 1962] states that the maximum flow from s to t defines a bipartitioning of G so that the weight of the cut is minimized among all cuts separating s and t. Since this so called *s-t mincut* may not be balanced in terms of the partition area, Yang and Wong proposed a method to repeat max-flow computations until a balanced partitioning is found. What is unique in their work is that the overall time complexity of this multi-iteration approach is asymptotically the same as a *single* max-flow computation. This is possible through the recycling of augmenting paths from the previous iterations. They also proposed a model to transform multi-terminal nets in the given circuit into a flow network G so that any cut in G preserves the correct cutsize information in the circuit. Lastly, they proposed an effective way to obtain the next cut if the current cut is not balanced, which is based on making a minor perturbation to the current flow network.

Quick Overview

Given a circuit NL, we first build the flow network G, where each net $n = \{v_1, v_2, \cdots, v_k\} \in NL$ is transformed as follows:

- We add node v_1, v_2, $\cdots$, v_k into G if not added yet.
- We add two auxiliary nodes n_1 and n_2 into G and connect them with so called "bridging edge" $e(n_1, n_2)$ that has a capacity of 1.
- We connect v_1, v_2, $\cdots$, v_k to n_1 with edges of ∞ capacity. We connect n_2 to v_1, v_2, $\cdots$, v_k with edges of ∞ capacity.

An illustration is shown in Figure 2.22. Next, we choose a pair of source and sink (s, t) randomly. We then repeat the following until we find a balanced bipartitioning solution:

- Step 1: We find the maximum flow from s to t using the augmenting path method.

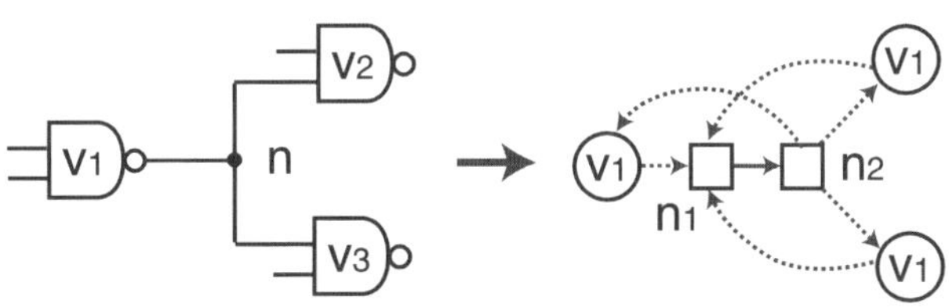

Figure 2.22. Modeling a net into a flow network.

- Step 2: We construct the partitioning solutions based on the max-flow computation, which is done by cutting some subset of the saturated nets. In the meantime, we choose the best solution, denoted $C(X, X')$, in terms of area balance. Note that all of these solutions have the same cutsize, which is equal to the max-flow value. If we find a solution that satisfies the area constraint, we terminate the algorithm.

- Step 3: If the solution above is still not area-balanced, we see if the area of X is smaller than the area lower bound. If yes, we choose a node $v \in Y$ that is contained in a cut net. Lastly, we merge all nodes in X and v into a single node, which becomes a new source s. We go back to step 1.

- Step 4: Otherwise, if the area of X is bigger than the area upper bound, we find a node $v \in X$ that is contained in a cut net. Lastly, we merge all nodes in X' and v into a single node, which becomes a new sink t. We go back to step 1.

During step 1 of some iteration, some of the augmenting paths are already found from the previous iterations. This simplifies the process of finding additional augmenting paths to find a new max-flow. In addition, while we are finding a node v to merge with X (step 3) or with X' (step 4), we choose v among the nodes in the cut net randomly. As the algorithm approaches closer to a balanced solution, we exhaustively search the best v among all the nodes in the cut nets.

Practice Problem

Consider the gate-level circuit shown in Figure 2.23. Assume that the area constraint is set to $[4, 5]$ for bipartitioning. Perform the node merging based on alphabetical order.

1. Model the circuit with a flow network.

   See Figure 2.24.

2. Find the maximum flow from node a to i.

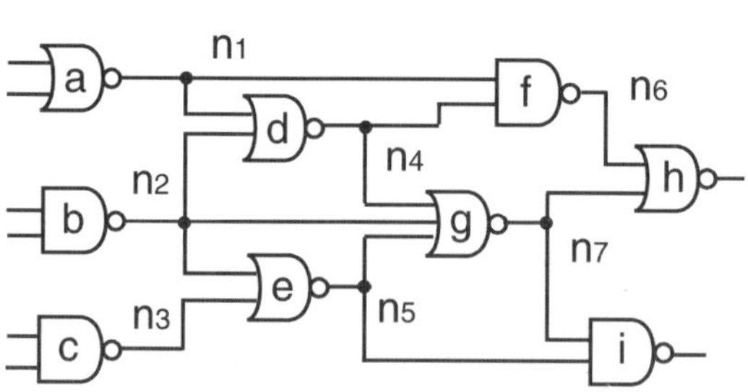

Figure 2.23. A gate-level circuit.

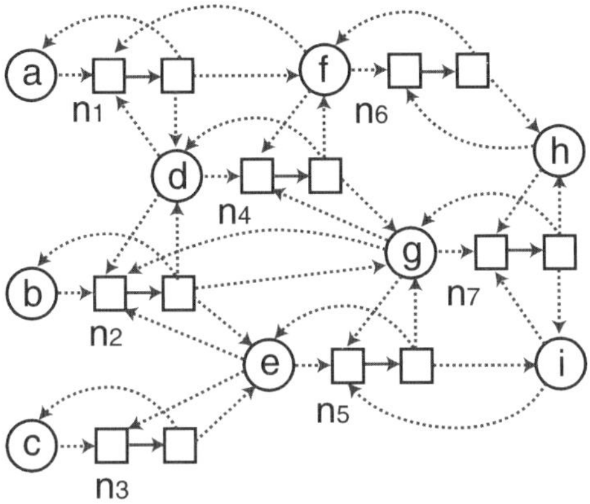

Figure 2.24. Flow network of the circuit in Figure 2.23. The capacity of dotted edges is infinity, while the solid edges have capacity of 1.

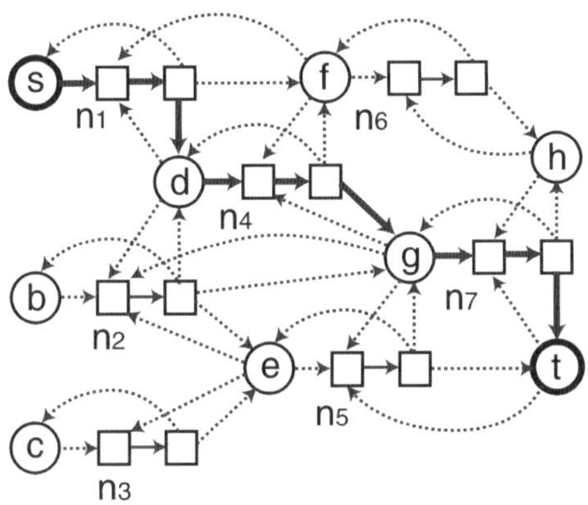

Figure 2.25. First maximum flow (value = 1) along with its augmenting path.

Table 2.7. Partitioning solutions derived from the first max-flow computation.

Cut net	Source partition	Sink partition
n_1	s	b, c, d, e, f, g, h, t
n_4	No cut	No cut
n_7	No cut	No cut

Figure 2.25 shows a maximum flow value of 1.[6] Nets n_1, n_4, and n_7 are saturated and define the partitioning solutions shown in Table 2.7. For example,

[6]Note that this maximum flow solution is not unique as there exist several other augmenting paths from s to t.

removal of n_1 from Figure 2.23 leads to a a-i mincut. But, removal of n_4 or n_7 does not lead to a a-i mincut. Thus, we cut n_1 and obtain the following solution:

$$P_s = \{s\}, P_t = \{b, c, d, e, f, g, h, t\}$$

3. Choose a node to merge with s.

   Since the area of P_s, the source-side partition, is smaller than the lower bound of 4, we choose a node from the sink side. In this case, the node should be contained in the cut net n_1. Since $n_1 = \{a, d, f\}$, we choose d based on alphabetical order.

4. Show the circuit and its flow network after the merging.

   See Figure 2.26. Note that the merged node now has multiple outputs attached to n_1 and n_4.

5. Perform the remaining part of FBB algorithm to obtain a balanced bipartitioning solution.

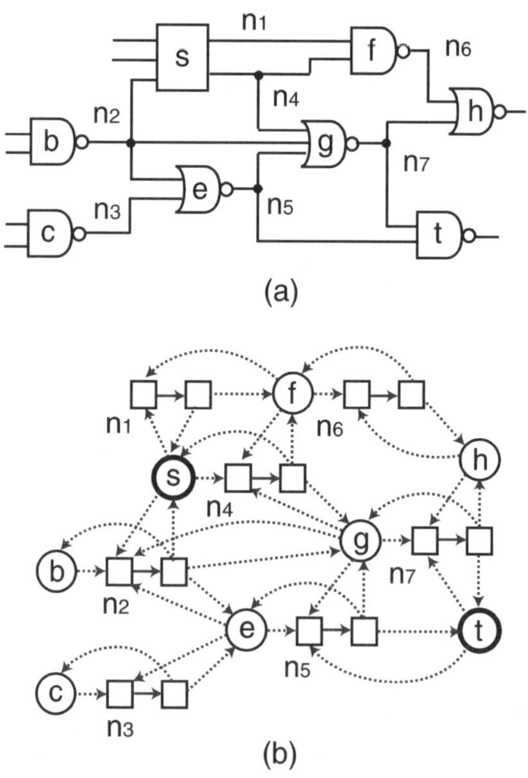

Figure 2.26. After merging s and d. (**a**) Circuit, (**b**) flow network.

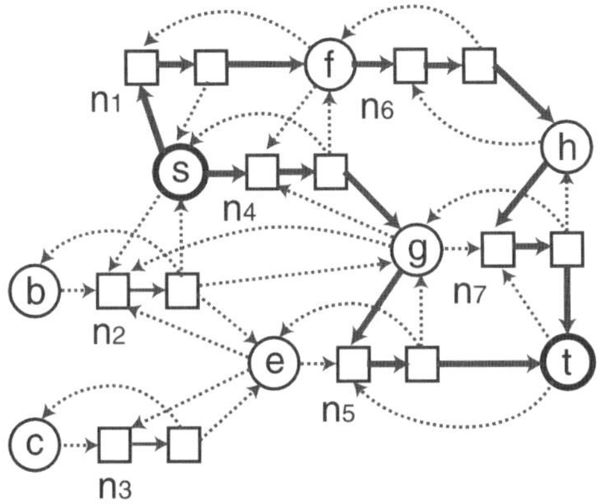

Figure 2.27. Second maximum flow (value = 2) along with its two augmenting paths.

Table 2.8. Partitioning solutions derived from the second max-flow computation.

Cut net	Source partition	Sink partition
n_1, n_4	No cut	No cut
n_1, n_5	No cut	No cut
n_6, n_4	No cut	No cut
n_6, n_5	No cut	No cut
n_7, n_4	No cut	No cut
n_7, n_5	s, b, c, e, f, g, h	t

(a) Second max-flow computation:

Figure 2.27 shows the augmenting paths, and the maximum flow (value = 2). Nets n_1, n_6, n_7, n_4, and n_5 are saturated and define the partitioning solutions shown in Table 2.8. Since the max-flow value is 2, our cutset will contain two nets. From Table 2.8 and Figure 2.26(b), we note that the cutset includes n_7 and n_5, which gives us the following partitioning solution:

$$P_s = \{s, b, c, e, f, g, h\}, P_t = \{t\}$$

Since the area of source partition is larger than the upper bound of 5 above, we choose a node from the source side. The set of nodes contained in n_7, n_5 and partitioned into the source side include $\{g, h, e\}$. Thus, we choose e to merge with t based on alphabetical order. Figure 2.28 shows the circuit and its flow network after the merging.

(b) Third max-flow computation:

Figure 2.29 shows the augmenting paths, and the maximum flow (value = 3). Nets n_1, n_6, n_7, n_4, n_5, and n_2 are saturated and define the

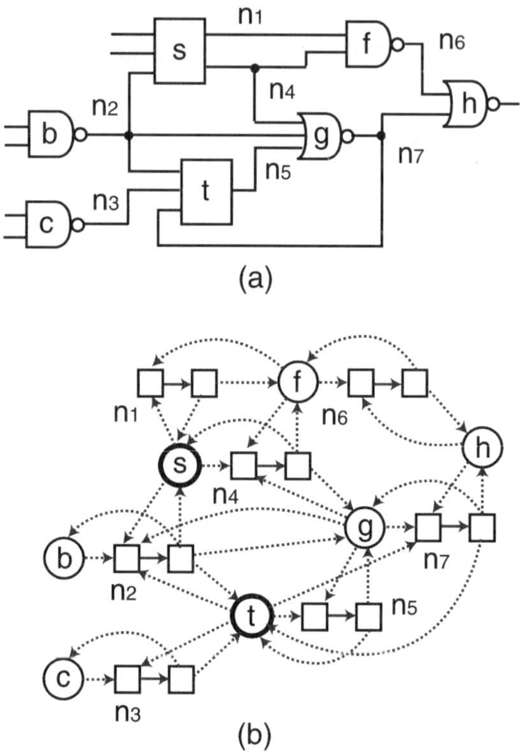

Figure 2.28. After merging t and e. (**a**) Circuit, (**b**) flow network.

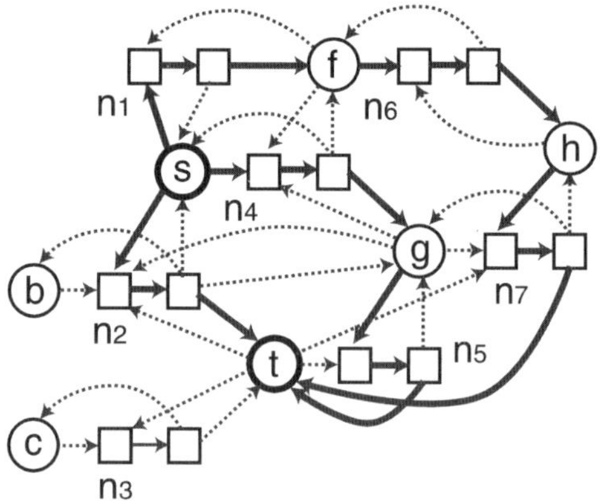

Figure 2.29. Third maximum flow (value = 3) along with its augmenting paths.

Table 2.9. Partitioning solutions derived from the third max-flow computation.

Cut net	Source partition	Sink partition
n_1, n_4, n_2	s, b	c, t, g, f, h
n_1, n_5, n_2	No cut	No cut
n_6, n_4, n_2	s, b, f	c, t, g, h
n_6, n_5, n_2	No cut	No cut
n_7, n_4, n_2	s, f, h, b	c, t, g
n_7, n_5, n_2	s, f, g, h	c, t, b

partitioning solutions shown in Table 2.9. Note that node b can be partitioned to either side because net n_2 will need to be always cut in the s-t mincut. In this case, node b is partitioned to improve the area balance in each cut. From Table 2.9 we found three balanced partitioning solutions with the cutsize of 3.

Thus, the final partitioning results are as follows:

$$(\{a, b, d, f\}, \{c, e, g, h, i\})$$
$$(\{a, b, d, f, h\}, \{c, e, g, i\})$$
$$(\{a, d, f, g, h\}, \{b, c, e, i\})$$

The cutsize in all of these solutions is 3.

5. More Practice Problems

1. Perform a single pass of Kernighan and Lin algorithm on the circuit in Figure 2.1(a) using $\{aceg, bdfh\}$ as the initial solution. Break ties in lexicographical order.

2. Consider the gate-level circuit in Figure 2.16. Perform a single pass of Kernighan and Lin algorithm using $\{acegi, bdfhj\}$ as the initial solution. Fix cells a and b in their initial partition, and do not move them.

3. Consider the clustered netlist shown in Table 1.9 and Table 1.10 based on Modified Hyperedge Coarsening algorithm. The area constraint for bipartitioning is set to $[2, 6]$.

   (a) Perform a single pass of Fiduccia and Mattheyses algorithm on the clustered netlist. The initial solution is $(\{C_1, C_4, C_5\}, \{C_2, C_3\})$.

   (b) Perform a single pass of Fiduccia and Mattheyses algorithm on the original netlist using the best solution obtained in part (a) as the initial solution.

4. Perform a single pass of Fiduccia and Mattheyses algorithm on the circuit shown in Figure 2.5(a) using $\{bcde, afgh\}$ as the initial solution. Break ties in alphabetical order. The area constraint is set to $[3, 5]$.

5. Perform EIG Algorithm on the circuit in Figure 2.5(a) and obtain bipartitioning solutions that minimize the following metrics:

   (a) Ratio cut under area constraint $[3, 5]$.

   (b) Cutsize under area constraint $[3, 5]$.

6. Consider the gate-level circuit shown in Figure 2.1(a).

   (a) Draw its "intersection graph" (see Figure 7 of [Hagen and Kahng, 1992]). Use the following correct definition of A'_{ab}:

$$A'_{ab} = \sum_{l=1}^{q} \frac{1}{d_l - 1} \left(\frac{1}{|s_a|} + \frac{1}{|s_b|} \right)$$

   (b) Obtain the second smallest eigenvalue and its eigenvector of the intersection graph.

   (c) Perform the module assignment heuristic shown in Figure 8 of [Hagen and Kahng, 1992] to obtain the best ratio-cut partition. What is the best ratio-cut value and its partitioning solution?

7. Perform FBB algorithm on the circuit in Figure 2.23 using (c, f) as the source/sink pair. The area constraint is set to $[4, 5]$. The node merging should be done in alphabetical order.

8. Perform FBB algorithm on the circuit in Figure 2.5(a) using (a, h) as the source/sink pair. The area constraint is set to $[3, 5]$. The node merging should be done in alphabetical order.

6. Probing Further

Disclaimer: The list here is meant to be representative, not comprehensive. A comprehensive survey on circuit partitioning algorithms is provided in [Alpert and Kahng, 1995b].

Kernighan and Lin Algorithm

Fiduccia and Mattheyses algorithm [Fiduccia and Mattheyses, 1982] is the most well-known extension of Kernighan and Lin algorithm [Kernighan and Lin, 1970]. As discussed earlier in Section 2, the three major extensions include handling of hypergraphs, replacing cell swaps with cell moves, and adopting bucket sorting.

Fiduccia and Mattheyses Algorithm

The author of [Krishnamurthy, 1984] extended the gain concept used in the Fiduccia and Mattheyses (FM) algorithm to quantify the impact of cell moves on future gains. Under this so called lookahead scheme, each cell has a vector of gains with k entries. The first entry is the same as the FM gain: immediate cutsize reduction. The second entry shows the change in the gain of neighboring cells if the cell is moved, i.e. a prediction of what will happen to other cells after the cell move. These gain vectors are then lexicographically sorted for cell move selection. This scheme provides an effective way to break ties when choosing the maximum gain cell.

The cell gain concept of FM is extended to handle multi-way partitioning problem in [Sanchis, 1989]. Under this so called K-way FM (KFM) algorithm, the entire netlist is first partitioned into K partitions. Each cell now has $K-1$ gain values to indicate the impact on the cutsize if the cell is moved to any of the $K-1$ target partitions. A follow-up study on KFM [Cong and Lim, 1998] showed that a recursive FM (RFM) significantly outperforms KFM for K-way partitioning problem and that a simple heuristic named KPM can improve KFM significantly KFM.

The authors of [Dutt and Deng, 1996b] extended FM to prevent densely connected sub-circuits (= clusters) from being cut. Under this so called CLIP scheme, the first cell move of each pass is chosen based on the initial gain. Once the first cell v is chosen, the gain values of all other cells are initialized to zero. We then move v, lock it, and update the gain of its neighboring cells. From this point on, FM finishes the remaining moves of the pass. The scheme pays more attention to the neighbors of moved cells and encourages the successive moves of closely connected cells (= clusters).

Buckets are the main data structure used in FM for managing cell gain values. The authors of [Hagen et al., 1997] showed that gain buckets maintained with last-in-first-out (LIFO) stacks lead to significantly better results than first-in-first-out (FIFO) queues or random management as in FM. They also showed that the LIFO selection scheme results in improvement over random schemes for KFM [Sanchis, 1989]. Under the LIFO scheme, most recently visited modules are placed near the beginning of the buckets, implicitly causing the neighborhoods or clusters of modules to be moved together. This has a similar "cluster removal" effect as the CLIP scheme [Dutt and Deng, 1996b].

The authors of [Hauck and Borriello, 1997] examined many of the existing techniques for FM and presented a methodology for determining the best mix of these approaches. These techniques include various clustering and unclustering schemes, initial partitioning creation, LIFO bucket management [Hagen et al., 1997], lookahead gain scheme [Krishnamurthy, 1984], and net partitioning. The result is a novel bipartitioning algorithm that includes both new and existing techniques.

EIG Algorithm

Spectral bipartitioning for ratio-cut minimization has been extended to multi-way partitioning by [Chan et al., 1994]. Their approach involves finding the k-smallest eigenvalue/eigenvector pairs of the Laplacian of the circuit graph. The eigenvectors provide an embedding of the graph's n vertices into a k-dimensional subspace. They also proposed a clustering heuristic to reduce the size of the problem before the eigenvalue/eigenvector computation.

The authors of [Riess et al., 1994] derived minimal ratio-cut partitioning solutions from the placement result obtained by the Gordian-L algorithm [Sigl et al., 1991a]. Gordian-L is an analytical placer, where the original quadratic placement formulation is modified to optimize linear objective under linear constraint. The 1-dimensional placement obtained by Gordian-L determines an ordering among the cells. Partitioning solutions are generated based on this ordering and evaluated in terms of ratio cut.

Spectral partitioning methods use the eigenvectors of the Laplacian matrix of the netlist graph to obtain partitioning solutions. Given d eigenvectors, the authors of [Alpert and Yao, 1995] map each vertex in the netlist graph to a vector in d-dimensional space such that these vectors constitute an instance of the vector partitioning problem. When all the eigenvectors are used, they showed that graph partitioning exactly reduces to vector partitioning. Based on this result, the authors presented an algorithm named MELO that is based on a simple vertex ordering scheme that can be used to yield high-quality 2-way and multi-way partitioning solutions.

The authors of [Alpert and Kahng, 1995a] presented a spectral partitioning algorithm that exploits both the geometric embedding and netlist topology information. The geometric embedding is done by the computation of d eigenvectors of the Laplacian matrix. This embedding is then partitioned based on the topological information from the netlist. They begin with a d-dimensional spectral embedding and construct a heuristic 1-dimensional ordering of the modules. Dynamic programming is then applied to efficiently compute the optimal k-way split of the ordering for a variety of objective functions. This integrated technique yields multi-way partitioning with lower cost than previous spectral approaches.

It has been shown that a linear objective yields better spectral placement in terms of wirelength than a quadratic objective. On the other hand, a quadratic objective tends to place nodes more sparsely than a linear objective, resulting in less overlap among the nodes. The authors of [Li et al., 1996] proposed so called the α-order objective function, which is linear$^\alpha$ ($1 \leq \alpha \leq 2$). Depending on how to tune α, the objective becomes closer to linear or quadratic. The goal is to capture the benefit of both the linear and quadratic objectives.

FBB Algorithm

Logic replication has been shown empirically to reduce pin count and partition size in partitioned networks. The authors of [Hwang and El Gamal, 1995] presented a network flow based algorithm for determining min-cut replication from a given partitioning solution. The algorithm is extended to hypergraphs, and replication heuristics are proposed to handle area balance constraints. When applied to the problem of partitioning a given design to multiple field-programmable gate arrays (FPGA), it is shown that min-cut replication provides substantial reductions in the numbers of FPGAs and pins required.

Given a flow network G with only two-pin nets, the authors of [Liu et al., 1995] first introduced an algorithm for optimum partitioning with replication under no area constraints. Compared with [Hwang and El Gamal, 1995] which requires an initial partitioning solution, this work computes both partitioning and replication simultaneously. Several heuristic extensions are then added to handle multi-pin nets and area balance constraints. The authors show that the area overhead from replication can be traded for cutsize reduction.

The authors of [Liu and Wong, 1998] extended the FBB algorithm to handle multi-way partitioning under both area and pin constraints. This problem occurs when a design needs to be partitioned into multiple FPGAs. Given a netlist graph $G(V, E)$, the basic approach is to find a subset of nodes $V_i \in V$ so that V_i satisfies both area and pin constraint. This is done by repeated computation of maximum flow. The authors try to maximize the area of V_i in order to reduce the number of FPGAs used.

Given a partitioning solution, the min-cut replication solution by [Hwang and El Gamal, 1995] is optimal only in terms of the cutsize but not in terms of the number of replicated nodes. Since the number of replicated nodes can be huge especially for large circuits, min-area replication is an important goal. The min-area min-cut replication problem is solved optimally by [Yang and

Wong, 1998] using network flow based algorithm. The authors also proposed a new compact flow network model to handle hypergraphs easily.

Dynamically reconfigurable FPGAs (DR-FPGA) have the potential to dramatically improve the logic density by time-sharing logic. To implement a design on a DR-FPGA, it has to be partitioned into multiple stages. The authors of [Mak and Young, 2003] proposed the idea of temporal logic replication in DR-FPGA to reduce the communication cost among the partitions. An optimal algorithm based on network flow model is presented to solve the min-area min-cut replication problem in the context of DR-FPGA partitioning.

Chapter 3

FLOORPLANNING

The input to the floorplanning problem is a set of blocks (soft or hard) and its netlist. The goal of floorplanning is to determine the location of the blocks so that the blocks do not overlap with each other. In case of soft blocks, we determine their dimensions as well. Traditional objectives include area and wirelength, and modern floorplanners address thermal hotspot, power supply noise, etc. Floorplanning problem is often solved under such design constraints as fixed outline, pre-placed blocks, alignment constraint, etc. This chapter presents sample problems related to the following works:

- Stockmeyer algorithm [Stockmeyer, 1983]
- Normalized polish expression [Wong and Liu, 1986]
- ILP-based floorplanning algorithm [Sutanthavibul et al., 1991]
- Sequence pair representation [Murata et al., 1995]

The first work determines the optimal orientation of the blocks in a given slicing floorplan. The second work presented an efficient way to represent slicing floorplans. The third work performs floorplanning for soft blocks with the integer linear programming (ILP) approach. The last work presented an efficient way to represent non-slicing floorplans. These works are targeting the traditional objectives such as area and wirelength.

S.K. Lim, *Practical Problems in VLSI Physical Design Automation*

1. Stockmeyer Algorithm

Given a slicing floorplan, Stockmeyer presented an optimal algorithm [Stockmeyer, 1983] that determines the orientation of the blocks in the floorplan so that the overall floorplan area is minimized. This algorithm is often used as a post-process of the given slicing floorplan to further optimize the area objective. Stockmeyer also proved in his paper that the optimal orientation problem for non-slicing floorplan is NP-complete.

Quick Overview

The algorithm starts with a tree that represents the given slicing floorplan. The goal is to traverse the internal nodes in this so called "slicing tree" in bottom-up fashion so that we compute the candidate dimensions of each internal node. When we obtain the dimension list of the top node in the tree, we choose the one with the minimum area. We then traverse the tree in top-down fashion to select the dimensions for the internal nodes as well as the orientation of the leaf nodes, i.e., the blocks themselves, based on the decision made for the parent node.

The core part of Stockmeyer algorithm is the computation of candidate dimensions of the given node during the bottom-up traversal. Given a vertical internal node for which we want to compute the dimension list, we begin by merging the first dimensions of the left (w_l, h_l) and the right child (w_r, h_r). In this case, the dimensions of the left and the right child are sorted so that the width is increasing and the height is decreasing. The resulting dimension after the merging is $(w_l + w_r, \max\{h_l, h_r\})$. If $h_l > h_r$, we merge the second dimension of the left child and the first dimension of the right child. If $h_l < h_r$, we merge the first dimension of the left child and the second dimension of the right child. In case $h_l = h_r$, we merge the second dimension of both the left and the right child. The merging (or joining as called in the paper) finishes when we reach the end of the dimension list for either the left or the right child.

In case we want to compute the dimensions for a horizontal internal node, the dimensions of the children are sorted so that the width is decreasing and the height is increasing. We merge the first dimensions of the left (w_l, h_l) and the right child (w_r, h_r). The resulting dimension after the merging is $(\max\{w_l, w_r\}, h_l + h_r)$. The way to choose the next merging candidate is the same as in vertical cut. An important observation made by Stockmeyer is that the total number of dimensions for any internal node is $O(L + R)$ instead of $O(L \cdot R)$, where L and R respectively denote the number of dimensions for the left and the right child. This helps reduce the runtime and space complexity of Stockmeyer algorithm significantly.

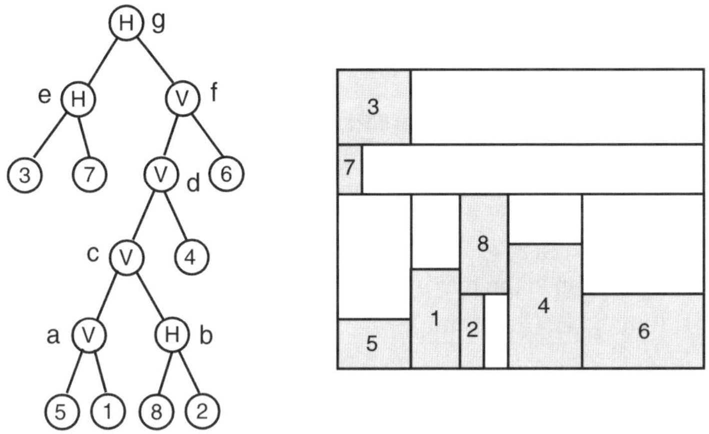

Figure 3.1. A slicing tree and its floorplan. Note that the lower left corner of each block is placed at the lower left corner of its room.

Practice Problem

Consider the slicing floorplan shown in Figure 3.1. The (width, height) of the blocks 1 through 8 are {(2,4), (1,3), (3,3), (3,5), (3,2), (5,3), (1,2), (2,4)}. Assume that xHy means x is top and y is bottom, and xVy means x is left and y is right. Place the lower left corner of each block to the lower left corner of its room.

1. Perform Stockmeyer algorithm to compute the minimum area of the floorplan.

   In case of vertical cut, the width is increasing and the height is decreasing in the dimension list. In case of horizontal cut, the width is decreasing and the height is increasing.

   (a) Visit node a: Since the cut orientation is vertical;

   $$L = \{(2,3),(3,2)\}$$
   $$R = \{(2,4),(4,2)\}$$

   (i) Join $l_1 = (2,3)$ and $r_1 = (2,4)$: we get $(2+2, \max\{3,4\}) = (4,4)$. Since the maximum is from R, we join l_1 and r_2 next.

   (ii) Join $l_1 = (2,3)$ and $r_2 = (4,2)$: we get $(2+4, \max\{3,2\}) = (6,3)$. Since the maximum is from L, we join l_2 and r_2 next.

   (iii) Join $l_2 = (3,2)$ and $r_2 = (4,2)$: we get $(3+4, \max\{2,2\}) = (7,2)$.

   Thus, the resulting dimensions are $\{(4,4),(6,3),(7,2)\}$.

(b) Visit node b: Since the cut orientation is horizontal;

$$L = \{(4,2),(2,4)\}$$
$$R = \{(3,1),(1,3)\}$$

(i) Join $l_1 = (4,2)$ and $r_1 = (3,1)$: we get $(\max\{4,3\}, 2+1) = (4,3)$. Since the maximum is from L, we join l_2 and r_1 next.
(ii) Join $l_2 = (2,4)$ and $r_1 = (3,1)$: we get $(\max\{2,3\}, 4+1) = (3,5)$. Since the maximum is from R, we join l_2 and r_2 next.
(iii) Join $l_2 = (2,4)$ and $r_2 = (1,3)$: we get $(\max\{2,1\}, 4+3) = (2,7)$.

Thus, the resulting dimensions are $\{(4,3),(3,5),(2,7)\}$.

(c) Visit node c: Since the cut orientation is vertical;

$$L = \{(4,4),(6,3),(7,2)\}$$
$$R = \{(2,7),(3,5),(4,3)\}$$

Note that we obtained R from reversing the order we computed for b earlier.

(i) Join $l_1 = (4,4)$ and $r_1 = (2,7)$: we get $(4+2, \max\{4,7\}) = (6,7)$. Since the maximum is from R, we join l_1 and r_2 next.
(ii) Join $l_1 = (4,4)$ and $r_2 = (3,5)$: we get $(4+3, \max\{4,5\}) = (7,5)$. Since the maximum is from R, we join l_1 and r_3 next.
(iii) Join $l_1 = (4,4)$ and $r_2 = (4,3)$: we get $(4+4, \max\{4,3\}) = (8,4)$. Since the maximum is from L, we join l_2 and r_3 next.
(iv) Join $l_2 = (6,3)$ and $r_3 = (4,3)$: we get $(6+4, \max\{3,3\}) = (10,3)$. Since the maximum is from R (and L), we reach the end of R and thus terminate.

Thus, the resulting dimensions are $\{(6,7),(7,5),(8,4),(10,3)\}$.

(d) Visit node d: Since the cut orientation is vertical;

$$L = \{(6,7),(7,5),(8,4),(10,3)\}$$
$$R = \{(3,5)(5,3)\}$$

(i) Join $l_1 = (6,7)$ and $r_1 = (3,5)$: we get $(6+3, \max\{7,5\}) = (9,7)$. Since the maximum is from L, we join l_2 and r_1 next.
(ii) Join $l_2 = (7,5)$ and $r_1 = (3,5)$: we get $(7+3, \max\{5,5\}) = (10,5)$. Since the maximum is from both L and R, we join l_3 and r_2 next.
(iii) Join $l_3 = (8,4)$ and $r_2 = (5,3)$: we get $(8+5, \max\{4,3\}) = (13,4)$. Since the maximum is from L, we join l_4 and r_2 next.

(iv) Join $l_4 = (10, 3)$ and $r_2 = (5, 3)$: we get $(10 + 5, \max\{3, 3\}) = (15, 3)$.

Thus, the resulting dimensions are $\{(9, 7), (10, 5), (13, 4), (15, 3)\}$.

(e) Visit node f: Since the cut orientation is vertical;

$$L = \{(9, 7), (10, 5), (13, 4), (15, 3)\}$$
$$R = \{(3, 5)(5, 3)\}$$

(i) Join $l_1 = (9, 7)$ and $r_1 = (3, 5)$: we get $(9 + 3, \max\{7, 5\}) = (12, 7)$. Since the maximum is from L, we join l_2 and r_1 next.

(ii) Join $l_2 = (10, 5)$ and $r_1 = (3, 5)$: we get $(10 + 3, \max\{5, 5\}) = (13, 5)$. Since the maximum is from both L and R, we join l_3 and r_2 next.

(iii) Join $l_3 = (13, 4)$ and $r_1 = (5, 3)$: we get $(13 + 5, \max\{4, 3\}) = (18, 4)$. Since the maximum is from L, we join l_4 and r_2 next.

(iv) Join $l_4 = (15, 3)$ and $r_1 = (5, 3)$: we get $(15 + 5, \max\{3, 3\}) = (20, 3)$.

Thus, the resulting dimensions are $\{(12, 7), (13, 5), (18, 4), (20, 3)\}$.

(f) Visit node e: Since the cut orientation is horizontal;

$$L = \{(3, 3)\}$$
$$R = \{(2, 1)(1, 2)\}$$

(i) Join $l_1 = (3, 3)$ and $r_1 = (2, 1)$: we get $(\max\{3, 2\}, 3 + 1) = (3, 4)$. Since the maximum is from L, we terminate.

Thus, the resulting dimension is $\{(3, 4)\}$.

(g) Visit node g: Since the cut orientation is horizontal;

$$L = \{(3, 4)\}$$
$$R = \{(20, 3), (18, 4), (13, 5), (12, 7)\}$$

(i) Join $l_1 = (3, 4)$ and $r_1 = (20, 3)$: we get $(\max\{3, 20\}, 4 + 3) = (20, 7)$. Since the maximum is from R, we join l_1 and r_2 next.

(ii) Join $l_1 = (3, 4)$ and $r_2 = (18, 4)$: we get $(\max\{3, 18\}, 4 + 4) = (18, 8)$. Since the maximum is from R, we join l_1 and r_3 next.

(iii) Join $l_1 = (3, 4)$ and $r_3 = (13, 5)$: we get $(\max\{3, 13\}, 4 + 5) = (13, 9)$. Since the maximum is from R, we join l_1 and r_4 next.

(iv) Join $l_1 = (3, 4)$ and $r_4 = (12, 7)$: we get $(\max\{3, 12\}, 4 + 7) = (12, 11)$.

Thus, the resulting dimensions are $\{(20, 7), (18, 8), (13, 9), (12, 11)\}$. The minimum area floorplan is $13 \times 9 = 117$.

Table 3.1. Summary of the bottom-up dimension computation in Stockmeyer algorithm. The minimum area floorplan is $13 \times 9 = 117$.

Node	Dir	Dimensions
a	ver	$L = \{(2,3),(3,2)\}$ $R = \{(2,4),(4,2)\}$ $D = \{(4,4),(6,3),(7,2)\}$
b	hor	$L = \{(4,2),(2,4)\}$ $R = \{(3,1),(1,3)\}$ $D = \{(4,3),(3,5),(2,7)\}$
c	ver	$L = \{(4,4),(6,3),(7,2)\}$ $R = \{(2,7),(3,5),(4,3)\}$ $D = \{(6,7),(7,5),(8,4),(10,3)\}$
d	ver	$L = \{(6,7),(7,5),(8,4),(10,3)\}$ $R = \{(3,5)(5,3)\}$ $D = \{(9,7),(10,5),(13,4),(15,3)\}$
f	ver	$L = \{(9,7),(10,5),(13,4),(15,3)\}$ $R = \{(3,5)(5,3)\}$ $D = \{(12,7),(13,5),(18,4),(20,3)\}$
e	hor	$L = \{(3,3)\}$ $R = \{(2,1)(1,2)\}$ $D = \{(3,4)\}$
g	hor	$L = \{(3,4)\}$ $R = \{(20,3),(18,4),(13,5),(12,7)\}$ $D = \{(20,7),(18,8),(13,9),(12,11)\}$

Table 3.1 shows the summary of the bottom-up dimension computation.

2. Find the optimal orientation and draw the floorplan.

   We now visit the nodes in the slicing tree in top-down fashion to compute the dimension and location of the internal nodes.

   (a) Node g: we choose $(13,9)$ for this root node. This is from joining $(3,4)$ and $(13,5)$. Thus, node e is $(3,4)$, and node f is $(13,5)$.

   (b) Node e: we choose $(3,4)$, which is from joining $(3,3)$ and $(2,1)$. Thus, the optimal orientation of block 3 is $(3,3)$, and block 7 is $(2,1)$.

   (c) Node f: we choose $(13,5)$, which is from joining $(10,5)$ and $(3,5)$. Thus, node d is $(10,5)$, and the optimal orientation of block 6 is $(3,5)$.

   (d) Node d: we choose $(10,5)$, which is from joining $(7,5)$ and $(3,5)$. Thus, node c is $(7,5)$, and the optimal orientation of block 4 is $(3,5)$.

   (e) Node c: we choose $(7,5)$, which is from joining $(4,4)$ and $(3,5)$. Thus, node a is $(4,4)$, and node b is $(3,5)$.

   (f) Node a: we choose $(4,4)$, which is from joining $(2,3)$ and $(2,4)$. The optimal orientation of block 5 is $(2,3)$, and block 1 is $(2,4)$.

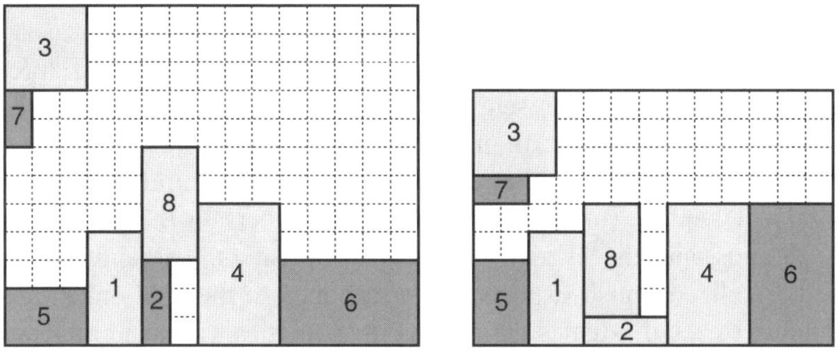

Figure 3.2. Slicing floorplan before and after the optimal rotation. The darker blocks are rotated.

(g) Node b: we choose $(3, 5)$, which is from joining $(2, 4)$ and $(3, 1)$. The optimal orientation of block 8 is $(2, 4)$, and block 2 is $(3, 1)$.

Figure 3.2 shows the floorplans before and after the rotation.

2. Normalized Polish Expression

Wong and Liu presented a method named Normalized Polish Expression [Wong and Liu, 1986] to represent slicing floorplans. Given a binary tree that represents a slicing floorplan of n blocks, the polish expression of this tree is a string of length $2n - 1$ that consists of the block numbers and H (for horizontal cut) and V (for vertical cut). The numbers in the expression are called operands, and the H and V are called operators. The authors showed that the normalized polish expression corresponds to the post-order traversal of the slicing tree and satisfies the following properties: (i) each block appears exactly once in the string, (ii) the number of operands is larger than the number of operators at all positions in the string, which is called the "balloting property" in the paper, and (iii) there are no consecutive operators of the same type in the string, which is called the "normality property." This normalized polish expression has 1-to-1 correspondence to a slicing floorplan so that we can obtain a unique slicing floorplan from a normalized polish expression, vice versa.

The main advantage of normalized polish expression is twofold. First, we can easily perturb the current slicing floorplanning solution to obtain a new neighboring solution. The authors provided three kinds of "moves" to perturb the normalized polish expression so that the resulting expression remains normalized and satisfies the balloting property. This is helpful when we utilize an iterative improvement type of optimization method such as Simulated Annealing [Kirkpatrick et al., 1983]. Second, we can quickly evaluate the quality of the given polish expression, which is done by computing the location of the blocks in the floorplan with a $O(n \log n)$ bottom-up traversal of the corresponding slicing tree. We can then obtain the area of the floorplan as well as the total wirelength. This is again important for Simulated Annealing-based optimization because this quick evaluation allows us to explore more solutions and increase the chance of finding high quality solutions.

Quick Overview

We first obtain a random initial polish expression PE_0. Next, we determine the following parameters used in Simulated Annealing: initial and final temperature, cooling rate, and number of moves at each temperature ($= M_t$). We compute C_0, the cost of PE_0, by $C_0 = A_0 + \lambda \cdot W_0$, where A_0 and W_0 respectively denote the area and wirelength of PE_0, and λ is the user-defined parameter. We set $Z = PE_0$, the best solution we return at the end of annealing. Once the annealing process begins, we make M_t moves at the initial temperature level. There are three types of moves: M1 is swapping two adjacent operands, M2 is complementing some chain, and M3 is swapping a pair of adjacent operand and operator. A chain is a set of consecutive operators in a polish expression, and its complementation involves swapping H and V in the chain. At each move we randomly select one of these three types and randomly choose a pair or chain. In case of M3, we examine if the balloting and normality properties are violated.

Next, we perform the chosen move and obtain a new neighboring polish expression. If the cost of this new solution is lower than the current, we accept the move; otherwise, we accept the move based on a probability function that is temperature-dependent. This function offers a high probability of accepting "bad moves" during the high temperature period and a low probability during the low temperature period. At the end of the move, we update Z, the best solution to be returned. After making all the moves at the current temperature level, we reduce the temperature using a cooling ratio $r < 1$ and repeat the moves. When the temperature reaches the final temperature, or the number of moves accepted is sufficiently low, we stop the annealing process and return Z.

Practice Problem

Consider the following polish expression:

$$PE_1 = 25V1H374VH6V8VH$$

The (width, height) of the modules 1 through 8 are {(2,4), (1,3), (3,3), (3,5), (3,2), (5,3), (1,2), (2,4)}.

1. Draw the corresponding slicing tree.

   See Figure 3.3.

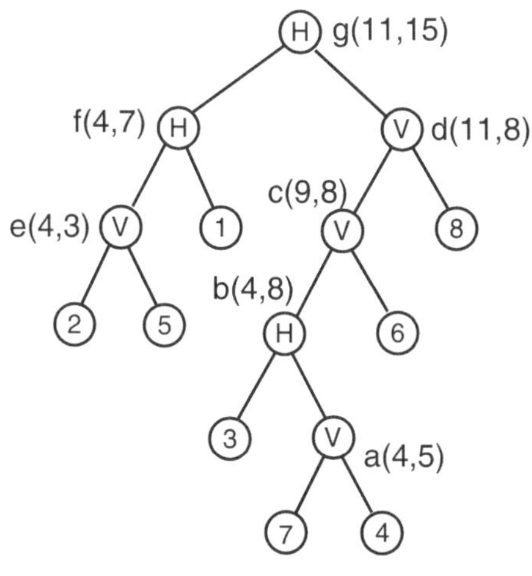

Figure 3.3. Slicing tree of PE_1.

2. What is the minimum area of the slicing floorplan? Rotation is not allowed.

    (a) Visit node a: vertical merging of $(1, 2)$ and $(3, 5)$ results in

    $$(1 + 3, \max\{2, 5\}) = (4, 5)$$

    (b) Visit node b: horizontal merging of $(3, 3)$ and $(4, 5)$ results in

    $$(\max\{3, 4\}, 3 + 5) = (4, 8)$$

    (c) Visit node c: vertical merging of $(4, 8)$ and $(5, 3)$ results in $(9, 8)$.
    (d) Visit node d: vertical merging of $(9, 8)$ and $(2, 4)$ results in $(11, 8)$.
    (e) Visit node e: vertical merging of $(1, 3)$ and $(3, 2)$ results in $(4, 3)$.
    (f) Visit node f: horizontal merging of $(4, 3)$ and $(2, 4)$ results in $(4, 7)$.
    (g) Visit node g: horizontal merging of $(4, 7)$ and $(11, 8)$ results in $(11, 15)$.

3. Draw the corresponding slicing floorplan. Place the lower left corner of each block to the lower left corner of its room.

    See Figure 3.7(a).

4. Consider an M1 move that swaps module 3 and 7 in PE_1. Draw the new slicing tree and perform incremental area computation.

    The new polish expression is

    $$PE_2 = 25V1H734VH6V8VH$$

    The slicing tree is shown in Figure 3.4. We only need to update the dimension of nodes a, b, c, d, and g.

    (a) Visit node a: vertical merging of $(3, 3)$ and $(3, 5)$ results in $(6, 5)$.
    (b) Visit node b: horizontal merging of $(1, 2)$ and $(6, 5)$ results in $(6, 7)$.
    (c) Visit node c: vertical merging of $(6, 7)$ and $(5, 3)$ results in $(11, 7)$.
    (d) Visit node d: vertical merging of $(11, 7)$ and $(2, 4)$ results in $(13, 7)$.
    (e) Visit node g: horizontal merging of $(4, 7)$ and $(13, 7)$ results in $(13, 14)$.

    Figure 3.7(b) shows the change on the floorplan.

5. Consider an M2 move that complements the last chain in PE_2. Draw the new slicing tree and perform incremental area computation.

    The new polish expression is

    $$PE_3 = 25V1H734VH6V8HV$$

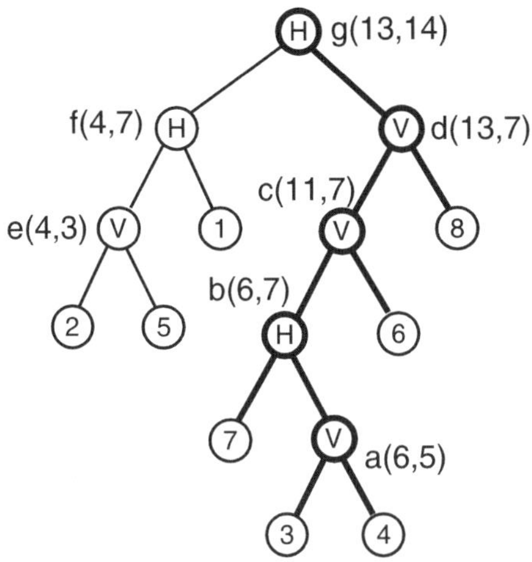

Figure 3.4. Slicing tree after swapping blocks 3 and 7 in PE_1. The bold part of the tree was updated.

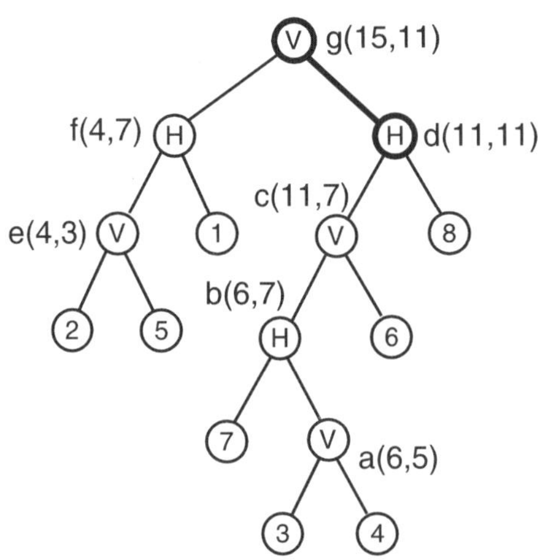

Figure 3.5. Slicing tree after complementing the last chain (= the orientation of nodes d and g) in PE_2. The bold part of the tree was updated.

The slicing tree is shown in Figure 3.5. We only need to update the dimension of nodes d and g.

(a) Visit node d: horizontal merging of $(11, 7)$ and $(2, 4)$ results in $(11, 11)$.

(b) Visit node g: vertical merging of $(4, 7)$ and $(11, 11)$ results in $(15, 11)$.

Figure 3.7(c) shows the change on the floorplan.

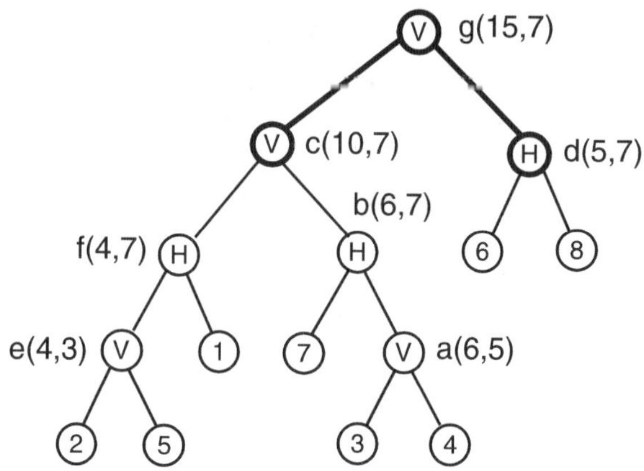

Figure 3.6. Slicing tree after swapping block 6 and V in PE_3. The bold part of the tree was updated.

6. Consider an M3 move that swaps 6 and V in PE_3. Draw the new slicing tree and perform incremental area computation.

   The new polish expression is

$$PE_4 = 25V1H734VHV68HV$$

   The slicing tree is shown in Figure 3.6. We only need to update the dimension of nodes c, d and g.

   (a) Visit node c: vertical merging of $(4, 7)$ and $(6, 7)$ results in $(10, 7)$.
   (b) Visit node d: horizontal merging of $(5, 3)$ and $(2, 4)$ results in $(5, 7)$.
   (c) Visit node g: vertical merging of $(10, 7)$ and $(5, 7)$ results in $(15, 7)$.

   Figure 3.7(d) shows the change on the floorplan.

7. What is the average change on the area among the three moves M1, M2, and M3? Compute the initial annealing temperature based on this average and the acceptance probability of 0.9.

   The area changed from 11×15 to 13×14 to 15×11 to 15×7. Thus, the average area change is

$$\Delta_{ave} = \frac{|165 - 182| + |182 - 165| + |165 - 105|}{3} = 31.33$$

   Thus,

$$T_0 = \frac{-\Delta_{ave}}{ln(0.9)} = 297.39$$

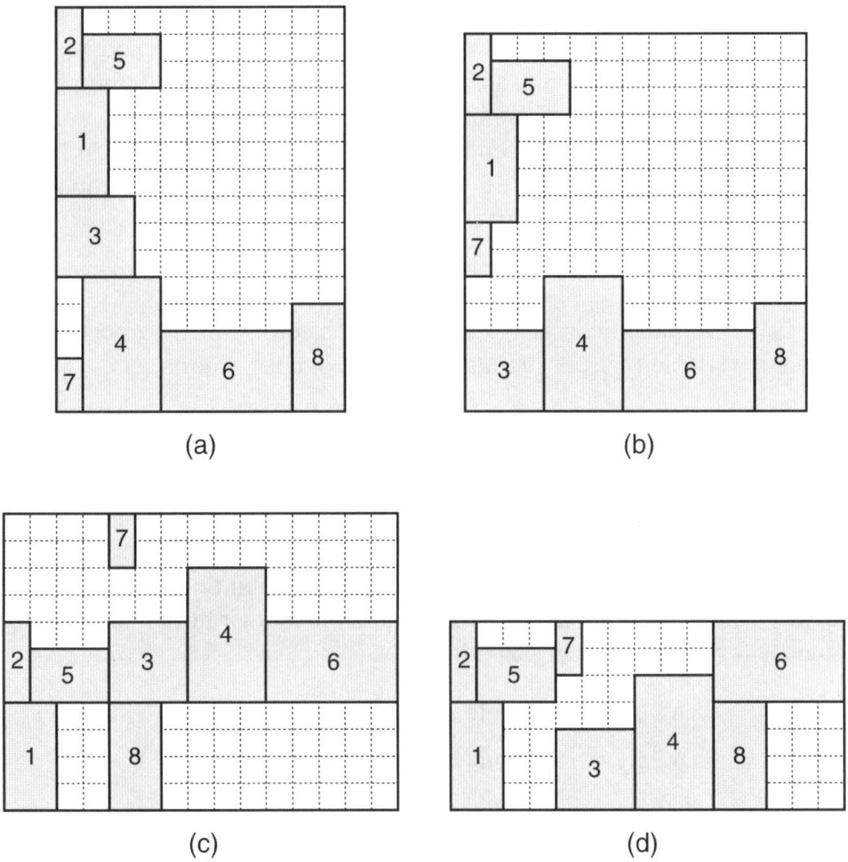

Figure 3.7. Changes on the floorplan based on the M1, M2, and M3 moves. (**a**) Initial floorplan, (**b**) after M1, (**c**) after M2, (**d**) after M3.

3. ILP Floorplanning Algorithm

An analytical method for floorplanning optimization is presented in [Sutanthavibul et al., 1991]. This method is based on mixed integer linear programming (ILP) formulation that considers various floorplanning objectives including area, wirelength, and routability. This work can handle both fixed and flexible modules and considers rotation of the fixed modules. Various techniques were utilized to convert non-linear objectives and constraints into linear equations. In order to handle large-scale problems, the authors presented a heuristic called "successive augmentation". In this method, a subset of modules is first floorplanned, and these modules are merged to form bigger modules. We then floorplan these bigger modules together with the next subset of modules. The goal is to keep the number of modules to be floorplanned (= number of variables in the ILP formulation) low so that the computation time remains reasonable.

Quick Overview

The ILP formulation consists of four parts: objective function, non-overlap constraints, variable type constraints, and chip boundary constraints.

- Objective function: The primary objective is the floorplan area minimization. Since the area objective is non-linear (width $\times$ height), we set the width as a constraint and minimize the height.

- Non-overlap constraints: Given a pair of modules, we assign only one "relative position relation" to this pair out of four possibilities: right of, left of, below, and above. This can be done by utilizing all-pair binary variables x_{ij} and y_{ij} for modules i and j as follows:

  - $x_{ij} = 0$ and $y_{ij} = 0$: module i is to the left of module j.
  - $x_{ij} = 0$ and $y_{ij} = 1$: module i is below module j.
  - $x_{ij} = 1$ and $y_{ij} = 0$: module i is to the right of module j.
  - $x_{ij} = 1$ and $y_{ij} = 1$: module i is above module j.

  For each pair of blocks, we set up these four equations so that only one of them becomes non-trivial based on the actual relative position. In case the module rotation is desired, an integer variable z_i for module i is used so that if $z_i = 1$, the non-overlap constraint equations utilize the width in place of the height, vice versa. If we have a flexible module with fixed area and bounded aspect ratio, we set the width as a continuous variable and obtain the linear equation for the height. Since the width and height relation is non-linear, i.e., $w_i \cdot h_i = A_i$, we utilize Taylor series to approximate the width-height relation.

- Variable type constraints: We define the type and range of the continuous and integer variables.

- Chip boundary constraints: We make sure that the modules are located within the chip boundary.

Practice Problem

Formulate the ILP floorplanning for the following problem instances. Assume that the dimension of the fixed modules is given as (width, height). The desired aspect ratio (= width/height) is 1.

1. Four fixed modules: $m_1(4,5)$, $m_2(3,7)$, $m_3(6,4)$, and $m_4(7,7)$. Rotation is not allowed.

   First, we obtain the list of continuous and integer variables as follows:

   - 8 continuous variables: the coordinate variables (x_1, x_2, x_3, x_4, y_1, y_2, y_3, y_4)
   - 12 integer variables: the all-pair relative position variables (x_{12}, x_{13}, x_{14}, x_{23}, x_{24}, x_{34}, y_{12}, y_{13}, y_{14}, y_{23}, y_{24}, y_{34}).

   The upper bound of chip boundary is computed as follows:

$$W = \sum w_i = 4 + 3 + 6 + 7 = 20$$
$$H = \sum h_i = 5 + 7 + 4 + 7 = 23$$

   Lastly, we construct the ILP formulation as follows:

   Minimize y^*
   Subject to
   non-overlap constraints:

$$\begin{aligned}
x_1 + w_1 &\leq x_2 + 20(x_{12} + y_{12}) \\
x_1 - w_2 &\geq x_2 - 20(1 - x_{12} + y_{12}) \\
y_1 + h_1 &\leq y_2 + 23(1 + x_{12} - y_{12}) \\
y_1 - h_2 &\geq y_2 - 23(2 - x_{12} - y_{12})
\end{aligned}$$

$$\begin{aligned}
x_1 + w_1 &\leq x_3 + 20(x_{13} + y_{13}) \\
x_1 - w_3 &\geq x_3 - 20(1 - x_{13} + y_{13}) \\
y_1 + h_1 &\leq y_3 + 23(1 + x_{13} - y_{13}) \\
y_1 - h_3 &\geq y_3 - 23(2 - x_{13} - y_{13})
\end{aligned}$$

$$x_1 + w_1 \le x_4 + 20(x_{14} + y_{14})$$
$$x_1 - w_4 \ge x_4 - 20(1 - x_{14} + y_{14})$$
$$y_1 + h_1 \le y_4 + 23(1 + x_{14} - y_{14})$$
$$y_1 - h_4 \ge y_4 - 23(2 - x_{14} - y_{14})$$

$$x_2 + w_2 \le x_3 + 20(x_{23} + y_{23})$$
$$x_2 - w_3 \ge x_3 - 20(1 - x_{23} + y_{23})$$
$$y_2 + h_2 \le y_3 + 23(1 + x_{23} - y_{23})$$
$$y_2 - h_3 \ge y_3 - 23(2 - x_{23} - y_{23})$$

$$x_2 + w_2 \le x_4 + 20(x_{24} + y_{24})$$
$$x_2 - w_4 \ge x_4 - 20(1 - x_{24} + y_{24})$$
$$y_2 + h_2 \le y_4 + 23(1 + x_{24} - y_{24})$$
$$y_2 - h_4 \ge y_4 - 23(2 - x_{24} - y_{24})$$

$$x_3 + w_3 \le x_4 + 20(x_{34} + y_{34})$$
$$x_3 - w_4 \ge x_4 - 20(1 - x_{34} + y_{34})$$
$$y_3 + h_3 \le y_4 + 23(1 + x_{34} - y_{34})$$
$$y_3 - h_4 \ge y_4 - 23(2 - x_{34} - y_{34})$$

variable type constraints:

$$x_1 \ge 0,\ x_2 \ge 0,\ x_3 \ge 0,\ x_4 \ge 0$$
$$y_1 \ge 0,\ y_2 \ge 0,\ y_3 \ge 0,\ y_4 \ge 0$$
$$x_{12},\ x_{13},\ x_{14},\ x_{23},\ x_{24},\ x_{34} \in \{0, 1\}$$
$$y_{12},\ y_{13},\ y_{14},\ y_{23},\ y_{24},\ y_{34} \in \{0, 1\}$$

chip width constraints:

$$x_1 + w_1 \le y^*$$
$$x_2 + w_2 \le y^*$$
$$x_3 + w_3 \le y^*$$
$$x_4 + w_4 \le y^*$$

chip height constraints:

$$y_1 + h_1 \le y^*$$
$$y_2 + h_2 \le y^*$$
$$y_3 + h_3 \le y^*$$
$$y_4 + h_4 \le y^*$$

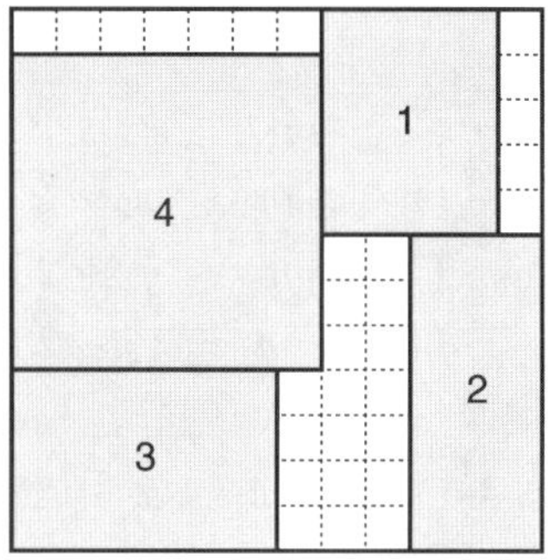

Figure 3.8. ILP floorplanning with fixed modules. The chip dimension is 12×12.

We solve this ILP formulation using GLPK [FSF, 2006] and obtain the following solutions:[7]

$$
\begin{aligned}
& y^* = 12 \\
& (x_1, y_1) = (7,7), (x_2, y_2) = (9,0), (x_3, y_3) = (0,0), (x_4, y_4) = (0,4) \\
& (x_{12}, y_{12}) = (1,1) : (1 \text{ is above } 2) \\
& (x_{13}, y_{13}) = (1,1) : (1 \text{ is above } 3) \\
& (x_{14}, y_{14}) = (1,0) : (1 \text{ is to the right of } 4) \\
& (x_{23}, y_{23}) = (1,0) : (2 \text{ is to the right of } 3) \\
& (x_{24}, y_{24}) = (1,0) : (2 \text{ is to the right of } 4) \\
& (x_{34}, y_{34}) = (0,1) : (3 \text{ is below } 4)
\end{aligned}
$$

Figure 3.8 shows the floorplanning result. The chip dimension is 12×12. Note that this floorplanning is sub-optimal because module 2 can be shifted to the left by 1 to reduce the chip width to 11. The source of this sub-optimality is twofold: linear approximation of the area objective $(= y^*)$ and the chip boundary constraint $(= x_i + w_i \leq y^*)$. In fact, $y^* = 12$ is an optimal solution, and the solutions satisfy all of the constraints specified.

2. The same modules as in problem 1, but rotation is allowed for all modules.

   In addition to the variables used in problem 1, we need four more integer variables for rotation: z_1, z_2, z_3, and z_4. In addition, we need $M = \max\{W, H\} = 23$. The ILP is given as follows:

   Minimize y^*
   Subject to

[7]The source files for all of the integer linear programming formulations presented in this section are available at: `http://users.ece.gatech.edu/limsk/book`.

non-overlap constraints:

$$
\begin{aligned}
x_1 + z_1 h_1 + (1 - z_1) w_1 &\leq x_2 + 23(x_{12} + y_{12}) \\
x_1 - z_2 h_2 - (1 - z_2) w_2 &\geq x_2 - 23(1 - x_{12} + y_{12}) \\
y_1 + z_1 w_1 + (1 - z_1) h_1 &\leq y_2 + 23(1 + x_{12} - y_{12}) \\
y_1 - z_2 w_2 - (1 - z_2) h_2 &\geq y_2 - 23(2 - x_{12} - y_{12})
\end{aligned}
$$

$$
\begin{aligned}
x_1 + z_1 h_1 + (1 - z_1) w_1 &\leq x_3 + 23(x_{13} + y_{13}) \\
x_1 - z_3 h_3 - (1 - z_3) w_3 &\geq x_3 - 23(1 - x_{13} + y_{13}) \\
y_1 + z_1 w_1 + (1 - z_1) h_1 &\leq y_3 + 23(1 + x_{13} - y_{13}) \\
y_1 - z_3 w_3 - (1 - z_3) h_3 &\geq y_3 - 23(2 - x_{13} - y_{13})
\end{aligned}
$$

$$
\begin{aligned}
x_1 + z_1 h_1 + (1 - z_1) w_1 &\leq x_4 + 23(x_{14} + y_{14}) \\
x_1 - z_4 h_4 - (1 - z_4) w_4 &\geq x_4 - 23(1 - x_{14} + y_{14}) \\
y_1 + z_1 w_1 + (1 - z_1) h_1 &\leq y_4 + 23(1 + x_{14} - y_{14}) \\
y_1 - z_4 w_4 - (1 - z_4) h_4 &\geq y_4 - 23(2 - x_{14} - y_{14})
\end{aligned}
$$

$$
\begin{aligned}
x_2 + z_2 h_2 + (1 - z_2) w_2 &\leq x_3 + 23(x_{23} + y_{23}) \\
x_2 - z_3 h_3 - (1 - z_3) w_3 &\geq x_3 - 23(1 - x_{23} + y_{23}) \\
y_2 + z_2 w_2 + (1 - z_2) h_2 &\leq y_3 + 23(1 + x_{23} - y_{23}) \\
y_2 - z_3 w_3 - (1 - z_3) h_3 &\geq y_3 - 23(2 - x_{23} - y_{23})
\end{aligned}
$$

$$
\begin{aligned}
x_2 + z_2 h_2 + (1 - z_2) w_2 &\leq x_4 + 23(x_{24} + y_{24}) \\
x_2 - z_4 h_4 - (1 - z_4) w_4 &\geq x_4 - 23(1 - x_{24} + y_{24}) \\
y_2 + z_2 w_2 + (1 - z_2) h_2 &\leq y_4 + 23(1 + x_{24} - y_{24}) \\
y_2 - z_4 w_4 - (1 - z_4) h_4 &\geq y_4 - 23(2 - x_{24} - y_{24})
\end{aligned}
$$

$$
\begin{aligned}
x_3 + z_3 h_3 + (1 - z_3) w_3 &\leq x_4 + 23(x_{34} + y_{34}) \\
x_3 - z_4 h_4 - (1 - z_4) w_4 &\geq x_4 - 23(1 - x_{34} + y_{34}) \\
y_3 + z_3 w_3 + (1 - z_3) h_3 &\leq y_4 + 23(1 + x_{34} - y_{34}) \\
y_3 - z_4 w_4 - (1 - z_4) h_4 &\geq y_4 - 23(2 - x_{34} - y_{34})
\end{aligned}
$$

variable type constraints:

$$\begin{aligned}
&x_1 \geq 0,\ x_2 \geq 0,\ x_3 \geq 0,\ x_4 \geq 0\\
&y_1 \geq 0,\ y_2 \geq 0,\ y_3 \geq 0,\ y_4 \geq 0\\
&x_{12},\ x_{13},\ x_{14},\ x_{23},\ x_{24},\ x_{34} \in \{0, 1\}\\
&y_{12},\ y_{13},\ y_{14},\ y_{23},\ y_{24},\ y_{34} \in \{0, 1\}\\
&z_1,\ z_2,\ z_3,\ z_4 \in \{0, 1\}
\end{aligned}$$

chip width constraints:

$$\begin{aligned}
x_1 + (1 - z_1)w_1 + z_1 h_1 \leq y^*\\
x_2 + (1 - z_2)w_2 + z_2 h_2 \leq y^*\\
x_3 + (1 - z_3)w_3 + z_3 h_3 \leq y^*\\
x_4 + (1 - z_4)w_4 + z_4 h_4 \leq y^*
\end{aligned}$$

chip height constraints:

$$\begin{aligned}
y_1 + (1 - z_1)h_1 + z_1 w_1 \leq y^*\\
y_2 + (1 - z_2)h_2 + z_2 w_2 \leq y^*\\
y_3 + (1 - z_3)h_3 + z_3 w_3 \leq y^*\\
y_4 + (1 - z_4)h_4 + z_4 w_4 \leq y^*
\end{aligned}$$

We solve this ILP formulation and obtain the following solutions:

$y^* = 11$

$(x_1, y_1) = (7, 6), (x_2, y_2) = (0, 0), (x_3, y_3) = (7, 0), (x_4, y_4) = (0, 3)$

$z_1 = 0, z_2 = 1, z_3 = 1, z_4 = 0$: (2 and 3 are rotated.)

$(x_{12}, y_{12}) = (1, 1)$: (1 is above 2)

$(x_{13}, y_{13}) = (1, 1)$: (1 is above 3)

$(x_{14}, y_{14}) = (1, 0)$: (1 is to the right of 4)

$(x_{23}, y_{23}) = (0, 0)$: (2 is to the left of 3)

$(x_{24}, y_{24}) = (0, 1)$: (2 is below 4)

$(x_{34}, y_{34}) = (1, 0)$: (3 is to the right of 4)

Figure 3.9 shows the floorplanning result. The chip dimension is 11×11.

3. Two fixed modules: $m_1(4, 5)$, $m_2(3, 7)$, and two flexible modules: m_3 (area is 24, and aspect ratio range is $[0.5, 2]$), and m_4 (area is 49, and aspect ratio range is $[0.3, 2.5]$). Rotation is allowed for the fixed modules.

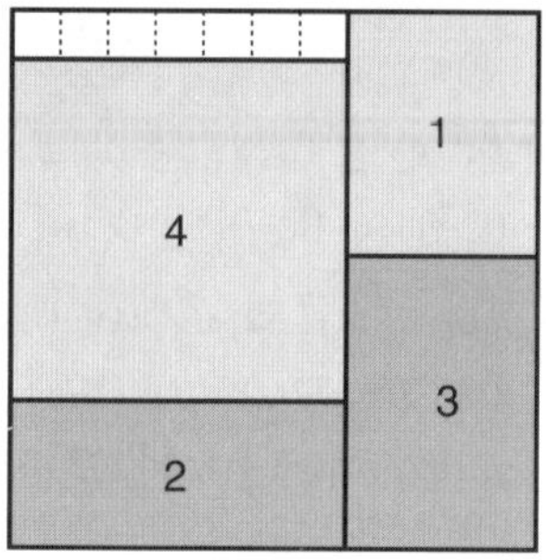

Figure 3.9. ILP floorplanning with fixed modules and rotation. Rotated modules are shown darker. The chip dimension is 11×11.

First, we obtain the list of continuous and integer variables as follows:

- Ten continuous variables: the coordinate variables (x_1, x_2, x_3, x_4, y_1, y_2, y_3, y_4), and the sizing variables (w_3, w_4) for the flexible modules.
- Fourteen integer variables: the all-pair relative position variables (x_{12}, x_{13}, x_{14}, x_{23}, x_{24}, x_{34}, y_{12}, y_{13}, y_{14}, y_{23}, y_{24}, y_{34}), and the rotation variables (z_1, z_2) for the fixed modules.

Next, we obtain a linear approximation of the height of flexible modules in terms of the width. The goal is to avoid using the non-linear relation $h_i = S_i/w_i$. Based on Taylor expansion, we have:

$$h_i = \frac{S_i}{w_{i,max}} + (w_{i,max} - w_i)\frac{S_i}{w_{i,max}^2}$$

where S_i denotes the area of module i. From the aspect ratio constraints, we get

$$w_i \cdot h_i \geq S_i,\ l_i \leq \frac{w_i}{h_i} \leq u_i$$

From the above, we get:

$$w_{i,min} = \sqrt{S_i \cdot l_i},\ w_{i,max} = \sqrt{S_i \cdot u_i}$$

which gives us the following ranges for the module width:

$$3.46 \leq w_3 \leq 6.93 \quad (3.1)$$

$$3.83 \leq w_4 \leq 11.07 \quad (3.2)$$

Thus, we obtain the following linear approximation for h_3 and h_4:

$$h_3 = \frac{24}{\sqrt{24 \cdot 2}} + (\sqrt{24 \cdot 2} - w_3)\frac{24}{24 \cdot 2} = -0.5w_3 + 6.93 \quad (3.3)$$

$$h_4 = \frac{49}{\sqrt{49 \cdot 2.5}} + (\sqrt{49 \cdot 2.5} - w_4)\frac{49}{49 \cdot 2.5} = -0.4w_4 + 8.85 \quad (3.4)$$

Based on Equation (3.1), (3.2), (3.3), and (3.4) we get

$$3.47 \leq h_3 \leq 5.20 \quad (3.5)$$
$$4.42 \leq h_4 \leq 7.32 \quad (3.6)$$

Lastly, we compute the upper bound of chip width $(= W)$ and chip height $(= H)$ as follows:

$$W = \sum w_i = \max\{4, 5\} + \max\{3, 7\} + 6.93 + 11.07 = 30.00$$
$$H = \sum h_i = \max\{4, 5\} + \max\{3, 7\} + 5.20 + 7.32 = 24.52$$

Thus, $M = \max\{W, H\} = 30.00$. We construct the ILP formulation as follows:

Minimize y^*
Subject to
non-overlap constraints:

$$x_1 + z_1h_1 + (1 - z_1)w_1 \leq x_2 + 30.00(x_{12} + y_{12})$$
$$x_1 - z_2h_2 - (1 - z_2)w_2 \geq x_2 - 30.00(1 - x_{12} + y_{12})$$
$$y_1 + z_1w_1 + (1 - z_1)h_1 \leq y_2 + 30.00(1 + x_{12} - y_{12})$$
$$y_1 - z_2w_2 - (1 - z_2)h_2 \geq y_2 - 30.00(2 - x_{12} - y_{12})$$

$$x_1 + z_1h_1 + (1 - z_1)w_1 \leq x_3 + 30.00(x_{13} + y_{13})$$
$$x_1 - w_3 \geq x_3 - 30.00(1 - x_{13} + y_{13})$$
$$y_1 + z_1w_1 + (1 - z_1)h_1 \leq y_3 + 30.00(1 + x_{13} - y_{13})$$
$$y_1 - (-0.5w_3 + 6.93) \geq y_3 - 30.00(2 - x_{13} - y_{13})$$

$$x_1 + z_1h_1 + (1 - z_1)w_1 \leq x_4 + 30.00(x_{14} + y_{14})$$
$$x_1 - w_4 \geq x_4 - 30.00(1 - x_{14} + y_{14})$$
$$y_1 + z_1w_1 + (1 - z_1)h_1 \leq y_4 + 30.00(1 + x_{14} - y_{14})$$
$$y_1 - (-0.4w_4 + 8.85) \geq y_4 - 30.00(2 - x_{14} - y_{14})$$

$$x_2 + z_2h_2 + (1 - z_2)w_2 \leq x_3 + 30.00(x_{23} + y_{23})$$
$$x_2 - w_3 \geq x_3 - 30.00(1 - x_{23} + y_{23})$$

$$y_2 + z_2w_2 + (1 - z_2)h_2 \le y_3 + 30.00(1 + x_{23} - y_{23})$$
$$y_2 - (-0.5w_3 + 6.93) \ge y_3 - 30.00(2 - x_{23} - y_{23})$$

$$x_2 + z_2h_2 + (1 - z_2)w_2 \le x_4 + 30.00(x_{24} + y_{24})$$
$$x_2 - w_4 \ge x_4 - 30.00(1 - x_{24} + y_{24})$$
$$y_2 + z_2w_2 + (1 - z_2)h_2 \le y_4 + 30.00(1 + x_{24} - y_{24})$$
$$y_2 - (-0.4w_4 + 8.85) \ge y_4 - 30.00(2 - x_{24} - y_{24})$$

$$x_3 + w_3 \le x_4 + 30.00(x_{34} + y_{34})$$
$$x_3 - w_4 \ge x_4 - 30.00(1 - x_{34} + y_{34})$$
$$y_3 + (-0.5w_3 + 6.93) \le y_4 + 30.00(1 + x_{34} - y_{34})$$
$$y_3 - (-0.4w_4 + 8.85) \ge y_4 - 30.00(2 - x_{34} - y_{34})$$

variable type constraints:

$$x_1 \ge 0,\ x_2 \ge 0,\ x_3 \ge 0,\ x_4 \ge 0$$
$$y_1 \ge 0,\ y_2 \ge 0,\ y_3 \ge 0,\ y_4 \ge 0$$
$$3.46 \le w_3 \le 6.93$$
$$3.83 \le w_4 \le 11.07$$
$$x_{12},\ x_{13},\ x_{14},\ x_{23},\ x_{24},\ x_{34} \in \{0, 1\}$$
$$y_{12},\ y_{13},\ y_{14},\ y_{23},\ y_{24},\ y_{34} \in \{0, 1\}$$
$$z_1,\ z_2 \in \{0, 1\}$$

chip width constraints:

$$x_1 + (1 - z_1)w_1 + z_1h_1 \le y^*$$
$$x_2 + (1 - z_2)w_2 + z_2h_2 \le y^*$$
$$x_3 + w_3 \le y^*$$
$$x_4 + w_4 \le y^*$$

chip height constraints:

$$y_1 + (1 - z_1)h_1 + z_1w_1 \le y^*$$
$$y_2 + (1 - z_2)h_2 + z_2w_2 \le y^*$$
$$y_3 + (-0.5w_3 + 6.93) \le y^*$$
$$y_4 + (-0.4w_4 + 8.85) \le y^*$$

We solve this ILP formulation and obtain the following solutions:

$$
\begin{aligned}
&y^* = 10.46\\
&(x_1, y_1) = (5.46, 5.20, (x_2, y_2) = (0, 0), (x_3, y_3) = (7, 0),\\
&(x_4, y_4) = (0, 3)\\
&z_1 = 1, z_2 = 1\text{: (1 and 2 are rotated)}\\
&w_3 = 3.46,\ \ w_4 = 3.83\\
&(x_{12}, y_{12}) = (1, 1) : \text{(1 is above 2)}\\
&(x_{13}, y_{13}) = (1, 1) : \text{(1 is above 3)}\\
&(x_{14}, y_{14}) = (1, 0) : \text{(1 is to the right of 4)}\\
&(x_{23}, y_{23}) = (0, 0) : \text{(2 is to the left of 3)}\\
&(x_{24}, y_{24}) = (0, 1) : \text{(2 is below 4)}\\
&(x_{34}, y_{34}) = (1, 0) : \text{(3 is to the right of 4)}
\end{aligned}
$$

Based on our linear approximation, i.e., Equation (3.3) and (3.4), we get

$$
\begin{aligned}
h_3 &= -0.5w_3 + 6.93 = 5.20\\
h_4 &= -0.4w_4 + 8.85 = 7.32
\end{aligned}
$$

Figure 3.10 shows the floorplanning result. The chip dimension is $(7 + 3.46) \times (3 + 7.32) = 10.46 \times 10.32$.

One important thing to note here is that the area of module 3 based on the linear approximation is $3.46 \times 5.20 = 17.99$, which is smaller than the actual area 24. A similar error is found for module 4, where $3.83 \times 7.32 = 28.04$ vs 49. Thus, the correct dimension for module 3 is $(3.46, 24/3.46 = 6.94)$ and module 4 is $(3.83, 49/3.83 = 12.79)$. Figure 3.11(a) shows the floorplan based on the correct module dimensions. We note that the

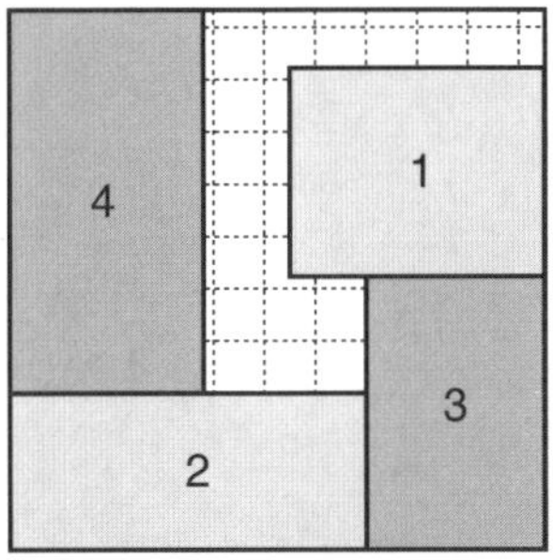

Figure 3.10. ILP floorplanning with fixed (1 and 2) and flexible (3 and 4) modules. The dimension of flexible modules is based on linear approximation. Modules 1 and 2 are rotated. The chip dimension is 10.46×10.32.

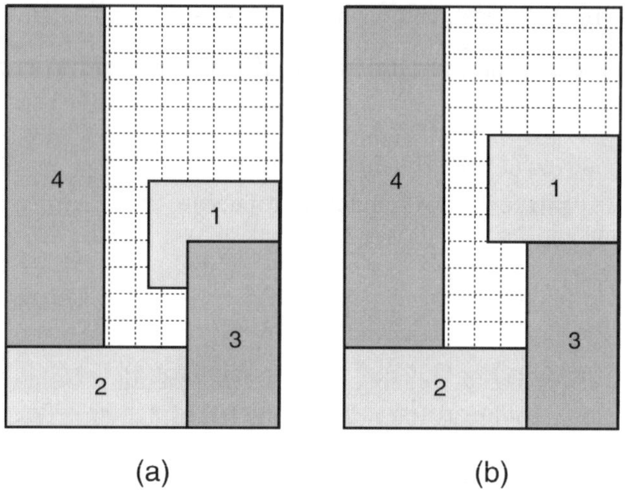

Figure 3.11. (**a**) Modified floorplan from the one shown in Figure 3.10 by using actual module dimension, (**b**) after removing the overlap. The chip dimension is 10.46×15.79.

chip height increased considerably, and overlap is introduced. The floorplan dimension after removing the overlaps is $(7 + 3.46) \times (3 + 12.79) = 10.46 \times 15.79$ as shown in Figure 3.11(b). The aspect ratio is 0.66. Note that the module overlap can be avoided by using a linear approximation that over-estimates the area instead of under-estimating it.[8]

[8]A sample problem included at the end of this chapter (problem 3, page 95) discusses this point.

4. Sequence Pair Representation

Murata et al. presented a method named Sequence Pair [Murata et al., 1995] to encode non-slicing floorplans. Given a non-slicing floorplan of n modules, the sequence pair is a pair of module name sequences that contains all the information about which subset of modules is above, below, to the right of, and to the left of a given module. There is 1-to-1 correspondence between a sequence pair and its non-slicing floorplan so that we can obtain a unique non-slicing floorplan from a sequence pair, vice versa.

Quick Overview

To go from a floorplan to its sequence-pair, we first draw so called "up-right step-line" for each module as follows: starting from the upper right corner of a module until we reach the upper right corner of the floorplan, we draw vertical lines (going up) and horizontal lines (going right) in an alternate fashion so that we do not cross any module boundary in the floorplan. We also draw "down-left step-lines" in a similar way. A pair of up-right step-line and down-left step-line for a given module forms the "positive step-line." Likewise, a pair of left-up and right-down step-lines of a module forms the "negative step-line." The authors showed that the positive step-lines of the modules do not cross each other, and the same is true for the negative step-lines. Finally, the order among the positive step-lines from the left to right forms the first sequence in a sequence pair. Likewise, the order among the negative step-lines from the bottom to top forms the second sequence in a sequence pair.

To go from a sequence pair to its floorplan, we do the following: given a module x in a sequence pair $SP(S_1, S_2)$, we obtain the list of modules that appear before x in both S_1 and S_2. These modules are located to the left of x in the floorplan. The set of modules that appear after x in both S_1 and S_2 are located to the right of x in the floorplan. The set of modules that appear after x in S_1 and before x in S_2 are located below x in the floorplan. Lastly, the set of modules that appear before x in S_1 and after x in S_2 are located above x in the floorplan. Next, we build a directed graph named Horizontal Constraint Graph (HCG) based on the "right-of" and "left-of" relation, where a directed edge $e(a, b)$ means module a is to the left of b. We add a source node and connect it to all nodes in HCG. We also add a sink node to HCG and connect all nodes to this sink. A longest path length from the source to each node in HCG denotes the x coordinate of the module in the floorplan. The longest source-sink path length is the width of the floorplan. Likewise, we construct a Vertical Constraint Graph (VCG) using the "above" and "below" relation and compute the y coordinates of the modules and the height of the floorplan in a similar way.

Sequence pair offers similar kinds of advantages as the Polish Expression [Wong and Liu, 1986] when used in conjunction with Simulated Annealing: (i) efficient exploration of the solution space via local search, and (ii) polynomial time evaluation of a candidate solution. In addition, the authors showed that the solution space defined by sequence pair is so called "P-admissible" in that there always exists an optimal solution to the floorplanning problem that can be encoded using sequence pair. The authors provide three types of moves that are used to perturb the current sequence pair: M1 is swapping a random pair of modules in the first sequence, M2 is swapping a random pair of modules in both sequences, and M3 is rotating a randomly selected module by 90-degree.

Practice Problem

Consider the following sequence pair $SP_1 = (17452638, 84725361)$. The (width, height) of the modules 1 through 8 are {(2,4), (1,3), (3,3), (3,5), (3,2), (5,3), (1,2), (2,4)}.

1. Draw the horizontal and vertical constraint graphs.

   Table 3.2 shows the relative positions derived from SP_1. The corresponding horizontal and vertical constraint graphs are shown in Figures 3.12 and 3.13, respectively.

2. What is the minimum area of the non-slicing floorplan?

   Figure 3.14 shows the constraint graphs with the longest s-t paths. The width of the floorplan is 11 from the HCG, and the height is 15 from the VCG. Thus, the floorplan area is $11 \times 15 = 165$.

3. Draw the corresponding non-slicing floorplan. Find the location of the lower left corner of each module.

   Table 3.3 shows the longest path length from the source to each module in HCG and VCG. Note that the weight of the module itself is not included, which result in the (x, y) location of the lower left corner of the modules. Figure 3.15 shows the floorplan.

Table 3.2. Relative positions among the modules in SP_1.

Module	Right-of	Left-of	Above	Below
1	$\emptyset$	$\emptyset$	$\emptyset$	$\{2,3,4,5,6,7,8\}$
2	$\{3,6\}$	$\{4,7\}$	$\{1,5\}$	$\{8\}$
3	$\emptyset$	$\{2,4,5,7\}$	$\{1,6\}$	$\{8\}$
4	$\{2,3,5,6\}$	$\emptyset$	$\{1,7\}$	$\{8\}$
5	$\{3,6\}$	$\{4,7\}$	$\{1\}$	$\{2,8\}$
6	$\emptyset$	$\{2,4,5,7\}$	$\{1\}$	$\{3,8\}$
7	$\{2,3,5,6\}$	$\emptyset$	$\{1\}$	$\{4,8\}$
8	$\emptyset$	$\emptyset$	$\{1,2,3,4,5,6,7\}$	$\emptyset$

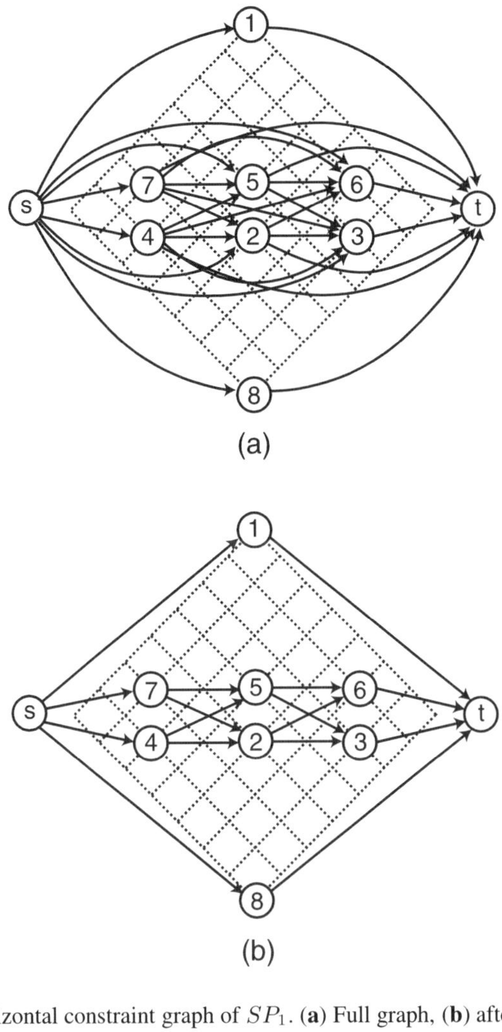

Figure 3.12. Horizontal constraint graph of SP_1. (**a**) Full graph, (**b**) after removing transitive edges for simplicity.

4. Consider a new sequence pair SP_2, where the modules 1 and 3 in the positive sequence of SP_1 are swapped. Draw the constraint graphs and the floorplan.

   The resulting sequence pair is $(37452618, 84725361)$. Table 3.4 shows the relative positions derived from the sequence pair. Figure 3.16 shows the corresponding HCG and VCG along with their s-t longest paths. The chip dimension is 13×14. Table 3.5 shows the longest path length from the source to each module in HCG and VCG, which corresponds to the (x, y) location of the lower left corner of each module. Figure 3.17 shows the floorplan.

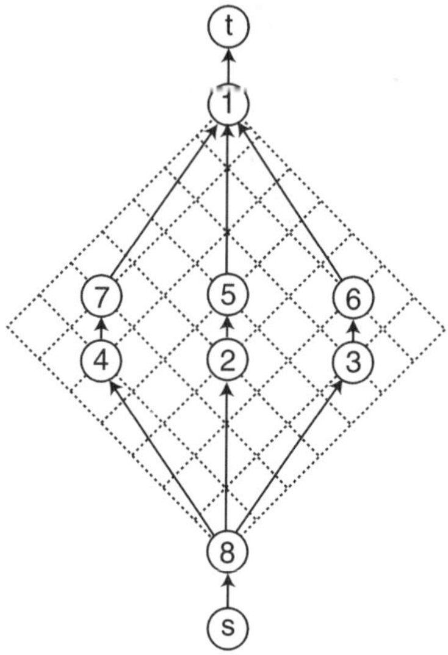

Figure 3.13. Vertical constraint graph of SP_1 with transitive edges removed.

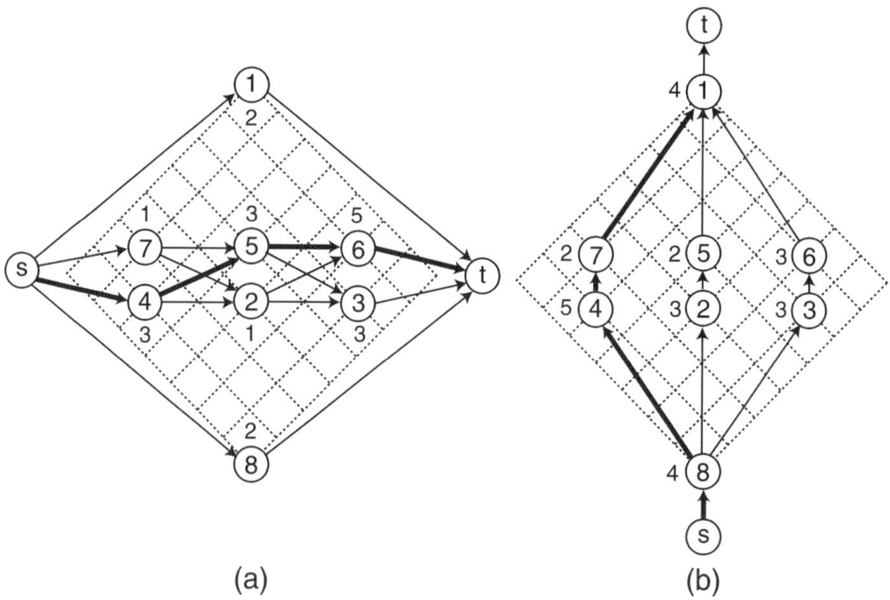

Figure 3.14. Sequence pair = SP_1. (**a**) HCG with longest s-t path length 11, (**b**) VCG with longest s-t path length 15. The numbers next to each node denotes its width (in HCG) or height (in VCG).

Table 3.3. Longest path lengths for the modules in HCG and VCG for SP_1. These values correspond to the location of the lower left corner of each module.

Module	HCV	VCG
1	0	11
2	3	4
3	6	4
4	0	4
5	3	7
6	6	7
7	0	9
8	0	0

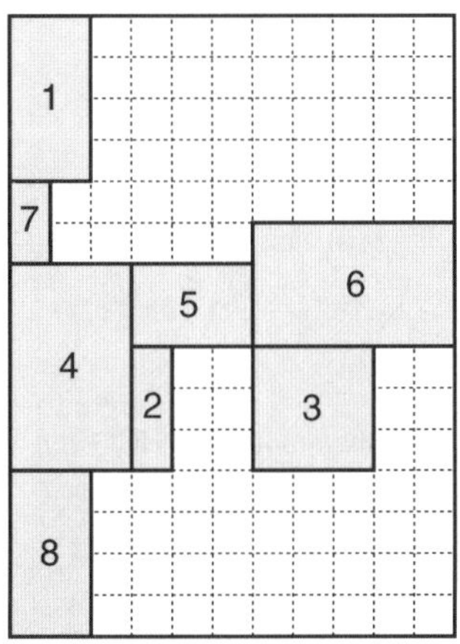

Figure 3.15. Non-slicing floorplan based on SP_1.

Table 3.4. Relative positions among the modules in SP_2.

Module	Right-of	Left-of	Above	Below
1	$\emptyset$	$\{2,3,4,5,6,7\}$	$\emptyset$	$\{8\}$
2	$\{1,6\}$	$\{4,7\}$	$\{3,5\}$	$\{8\}$
3	$\{1,6\}$	$\emptyset$	$\emptyset$	$\{2,4,5,7,8\}$
4	$\{1,2,5,6\}$	$\emptyset$	$\{3,7\}$	$\{8\}$
5	$\{1,6\}$	$\{4,7\}$	$\{3\}$	$\{2,8\}$
6	$\{1\}$	$\{2,3,4,5,7\}$	$\emptyset$	$\{8\}$
7	$\{1,2,5,6\}$	$\emptyset$	$\{3\}$	$\{4,8\}$
8	$\emptyset$	$\emptyset$	$\{1,2,3,4,5,6,7\}$	$\emptyset$

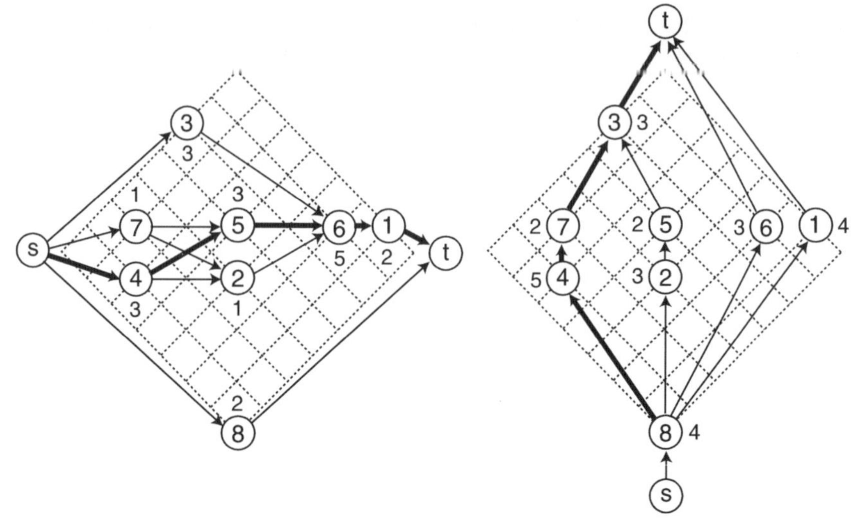

Figure 3.16. Sequence pair = SP_2. (**a**) HCG with longest s-t path length 13, (**b**) VCG with longest s-t path length 14.

Table 3.5. Longest path lengths for the modules in HCG and VCG for SP_2.

Module	HCV	VCG
1	11	4
2	3	4
3	0	11
4	0	4
5	3	7
6	6	4
7	0	9
8	0	0

5. Consider a new sequence pair SP_3, where the modules 4 and 6 in both sequences of SP_2 are swapped. Draw the constraint graphs and the floorplan.

   The resulting sequence pair is $(37652418, 86725341)$. Table 3.6 shows the relative positions derived from the sequence pair. Figure 3.18 shows the corresponding HCG and VCG along with their s-t longest paths.[9] The chip dimension is 13×12. Table 3.7 shows the longest path length from the source to each module in HCG and VCG, which corresponds to the (x, y)

[9]The path s-8-6-7-3-t is another longest path.

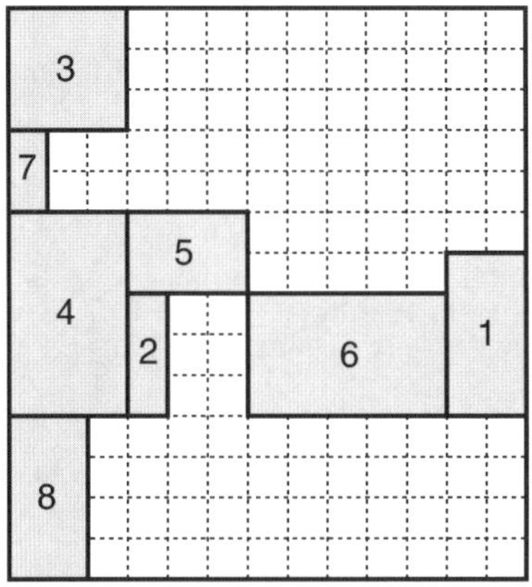

Figure 3.17. Non-slicing floorplan based on SP_2.

Table 3.6. Relative positions among the modules in SP_3.

Module	Right-of	Left-of	Above	Below
1	$\emptyset$	$\{2, 3, 4, 5, 6, 7\}$	$\emptyset$	$\{8\}$
2	$\{1, 4\}$	$\{6, 7\}$	$\{3, 5\}$	$\{8\}$
3	$\{1, 4\}$	$\emptyset$	$\emptyset$	$\{2, 5, 6, 7, 8\}$
4	$\{1\}$	$\{2, 3, 5, 6, 7\}$	$\emptyset$	$\{8\}$
5	$\{1, 4\}$	$\{6, 7\}$	$\{3\}$	$\{2, 8\}$
6	$\{1, 2, 4, 5\}$	$\emptyset$	$\{3, 7\}$	$\{8\}$
7	$\{1, 2, 4, 5\}$	$\emptyset$	$\{3\}$	$\{6, 8\}$
8	$\emptyset$	$\emptyset$	$\{1, 2, 3, 4, 5, 6, 7\}$	$\emptyset$

location of the lower left corner of each module. Figure 3.19 shows the floorplan.

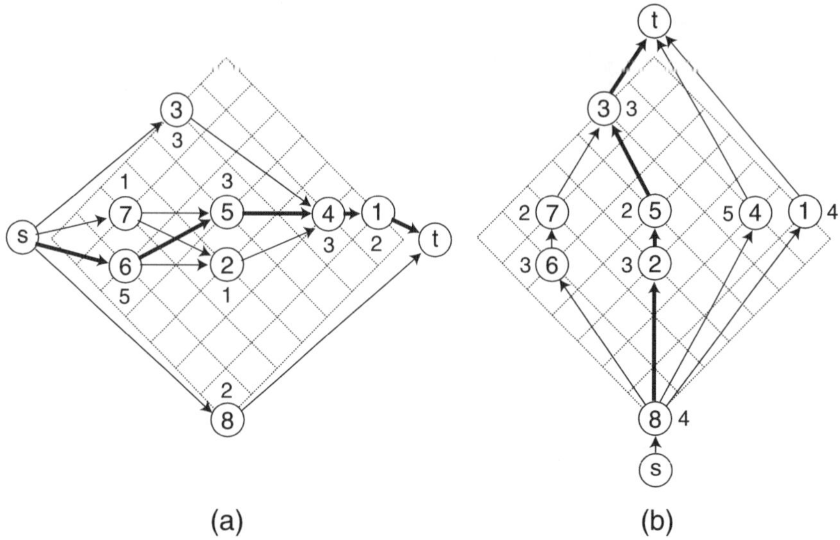

Figure 3.18. Sequence pair = SP_3. (**a**) HCG with longest s-t path length 13, (**b**) VCG with longest s-t path length 12. The numbers next to the nodes denote the weights.

Table 3.7. Longest path lengths for the modules in HCG and VCG for SP_3.

Module	HCV	VCG
1	11	4
2	5	4
3	0	9
4	8	4
5	5	7
6	0	4
7	0	7
8	0	0

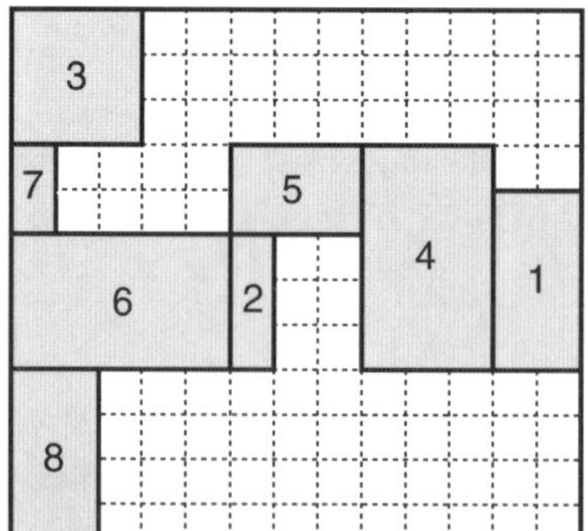

Figure 3.19. Non-slicing floorpl an based on SP_3.

5. More Practice Problems

For problems #1 and #2, assume that xHy means x is top and y is bottom, and xVy means x is left and y is right in the polish expression. The module dimension is given as (width, height). Place the lower left corner of each block to the lower left corner of its room.

1. Perform Stockmeyer algorithm on the slicing floorplan that is represented by $435H1V67VH2V8HV$. The dimension of the modules 1 through 8 are $\{(5,3), (2,3), (6,3), (2,5), (6,2), (5,1), (3,8), (6,3)\}$. Draw the floorplans before and after the orientation optimization.

2. Consider the following polish expression:

$$PE_1 = 435H1V67VH2V8HV$$

   The (width, height) of the modules 1 through 8 are $\{(5,3), (2,3), (6,3), (2,5), (6,2), (5,1), (3,8), (6,3)\}$. Assume that rotation is not allowed.

   (a) Draw the slicing tree and its floorplan for PE_1.

   (b) Consider an M1 move that swaps module 2 and 6 in PE_1. Name this new polish expression PE_2. Draw the new slicing tree and its floorplan.

   (c) Consider an M2 move that complements the third chain from the left in PE_2. Name this new polish expression PE_3. Draw the new slicing tree and its floorplan.

   (d) Consider an M3 move that swaps 5 and H in PE_3. Draw the new slicing tree and its floorplan.

3. Consider the ILP floorplanning problem with flexible modules shown in page 81. The authors of [Sutanthavibul et al., 1991] used a linear approximation that under-estimates the area (line 1 in Figure 3.20).

   (a) Obtain the linear equations for the module height h_3 and h_4 based on the over-estimation (line 2 in Figure 3.20).

   (b) Formulate the ILP formulation using the over-estimation in part (a) and obtain the floorplan.

   (c) What is the drawback of the over-estimation? Which method results in smaller floorplan area between the under- and over-estimation?

4. Formulate the ILP floorplanning for the following problem instances. The desired aspect ratio (= width/height) is 1.

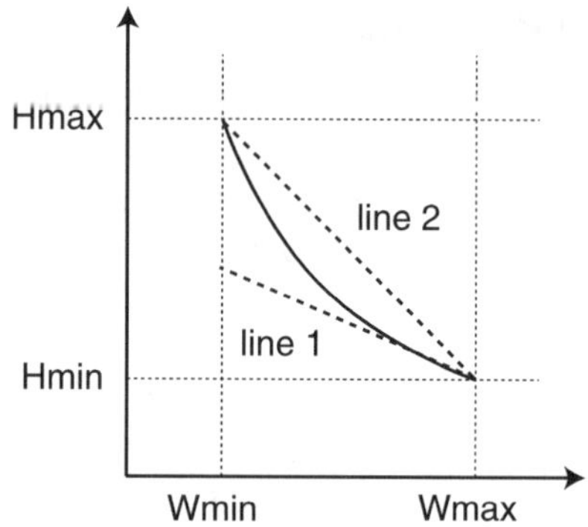

Figure 3.20. Linear approximation of area.

(a) Four fixed modules: $m_1(2,5)$, $m_2(3,6)$, $m_3(7,4)$, and $m_4(6,6)$. Rotation is not allowed.

(b) The same modules as in problem (a), but rotation is allowed for all modules.

(c) Two fixed modules: $m_1(2,5)$, $m_2(3,6)$, and two flexible modules: m_3 (area is 28, and aspect ratio range is $[0.3, 2.5]$), and m_4 (area is 36, and aspect ratio range is $[0.5, 2]$). Rotation is allowed for the fixed modules. Use the under-estimating linear area approximation shown in [Sutanthavibul et al., 1991].

5. Consider the following sequence pair:

$$SP_1 = (63528417, 17452836)$$

The dimension of the modules 1 through 8 are {(5,3), (2,3), (6,3), (2,5), (6,2), (5,1), (3,8), (6,3)}.

(a) Draw the constraint graphs and the floorplan for SP_1.

(b) Consider a new sequence pair SP_2, where the modules 3 and 4 in the positive sequence of SP_1 are swapped. Draw the constraint graphs and the floorplan.

(c) Consider a new sequence pair SP_3, where the modules 1 and 6 in both sequences of SP_2 are swapped. Draw the constraint graphs and the floorplan.

6. Probing Further

Disclaimer: The list here is meant to be representative, not comprehensive. A comprehensive survey on non-slicing floorplan representation methods is provided in [Chen and Chang, 2007].

Stockmeyer Algorithm

The authors of [Pan and Liu, 1992] presented a way to generalize Stockmeyer's algorithm [Stockmeyer, 1983] to handle non-slicing floorplans. Their method handles any non-slicing floorplans but works more effectively with the ones that are "approximately" slicing. Given a non-slicing floorplan, they identify a group of contiguous lines (instead of single line) that separates the blocks into two partitions. This so called "generalized cutline" can be viewed as a generalization of a slicing line. Using this concept together with Stockmeyer's bottom-up recursive merging process, they minimize the area of the final non-slicing floorplan.

Stockmeyer's algorithm has $O(n \cdot d)$ time/space complexity, where n is the number of blocks, and d is the depth of the slicing tree. Thus, the worst-case complexity is $O(n^2)$ if the tree is not balanced, and the best-case is $O(n \log n)$ if the tree is balanced. The author of [Shi, 1995] presented an algorithm that improves the worst-case complexity to $O(n \log n)$ regardless of the tree structure. This is done by a new data structure called "realization tree", which is utilized during the bottom-up merging process.

Normalized Polish Expression

The authors of [Young and Wong, 1998] extended the normalized polish expression [Wong and Liu, 1986] to handle pre-placed modules. They proposed a shape curve computation procedure that can take the positions of the pre-placed modules into consideration. The shape curve computation procedure is used repeatedly during the floorplanning process to fully exploit the shape flexibility of the modules.

Floorplanning with uncertainty is the problem of obtaining a good floorplan when the information about module dimensions is not complete. In this formulation introduced by [Bazargan et al., 1999], the width and the height of the modules are random variables instead of fixed values. The authors extended the normalized polish expression to handle these random variables and used statistical addition and maximum operations to compute the distributions of the final floorplan area.

Bus routability is another important issue in floorplanning because the blocks are usually connected with very wide buses, causing congestion and routing failure. The authors of [Rafiq et al., 2003] considered bus routability, area, and timing objectives in their slicing floorplanning formulation. In order to improve the timing, they use channels in between the blocks for buffer insertion, which is done by exploiting the slicing structure and its normalized polish expression. The bus routing is done using a mixed integer linear programming (ILP).

Slicing floorplans are commonly believed to suffer from poor utilization of space when all modules are hard. For this reason, a large body of literature has recently been devoted to various new representations of non-slicing floorplans to improve space utilization. The authors of [Lai and Wong, 2001] showed that a simple compaction procedure extends the capability of normalized polish expression to represent non-slicing floorplans. They conclude that slicing tree is a complete floorplan representation for all non-slicing floorplans as well.

ILP-based Floorplanning Algorithm

In [Moh et al., 1996], floorplanning with soft modules is formulated as a non-linear geometric programming. It is well-known that the geometric programming problem, after a simple transformation, can be converted into a convex optimization problem, which in turn can be solved efficiently by a non-linear solver. The advantage of this approach over the mixed integer linear programming by [Sutanthavibul et al., 1991] is that it does not suffer from the solution degradation caused by various schemes to linearize non-linear objectives and constraints.

The number of variables and constraints used in [Moh et al., 1996] is significantly reduced in a new convex programming formulation presented in [Chen and Fan, 1998]. Since the complexity of solving a convex programming problem typically increases dramatically with the numbers of variables and constraints, this new formulation leads to a significant reduction of computational effort in solving the floorplan area minimization problem.

The works by [Moh et al., 1996] and [Chen and Fan, 1998] are not scalable because they suffer from the high complexity to solve a convex programming problem. The authors of [Chen and Kuh, 2000] presented a new linear programming (LP) based formulation that handles soft, hard, and pre-placed modules. They solve a set of LP problems in an iterative fashion to obtain global minimum solution. They do not use integer variables and constraints, thereby further speeding up the runtime compared to [Sutanthavibul et al., 1991].

The authors of [Ekpanyapong et al., 2004] adopted the mixed integer linear programming formulation of [Sutanthavibul et al., 1991] to perform floorplanning with micro-architectural modules. A major difference between floorplanning with circuit modules and micro-architectural modules is that an architectural simulation can be performed for a given software application to collect various kinds of dynamic runtime profiling information. They exploited the frequency of block-to-block interconnect usage to obtain floorplans that improve the performance of the target micro-architecture.

Sequence Pair Representation

Since the introduction of sequence pair representation [Murata et al., 1995], VLSI CAD community has seen a huge volume of works on efficient representation of non-slicing floorplans including Bounded Slice-line Grid (BSG), O-tree, B*-tree, Corner Block List (CBL), Transitive Closure Graph (TCG), T-tree, Adjacent Constraint Graph (ACG), Q sequence, MP-tree, etc. In addition, many studies have been done on how to exploit these representation methods to satisfy various geometric constraints in the floorplan such as adjacency, boundary, alignment, etc. Comprehensive reviews of these methods and comparisons are provided in [Yao et al., 2001; Cong et al., 2004a; Chen and Chang, 2007]. The following works are strictly related to Sequence Pair extension and application.

The original paper on sequence pair [Murata et al., 1995] did not discuss how to handle soft modules or pre-placed modules. A sequence pair based non-slicing floorplanning problem with a mixture of soft, hard, and pre-placed modules is solved by the authors of [Murata and Kuh, 1998]. Given a sequence pair, they first formulated a convex programming to determine the width and height of the soft modules under the area constraint. The solution to this subproblem is then used in Simulated Annealing framework to obtain the sequence pair that results in the best possible floorplan area. They also perform a feasibility test to see if the given sequence pair satisfies the pre-placed module constraint.

The time complexity of obtaining a block placement from a given sequence pair is significantly improved by the authors of [Tang et al., 2001]. Since this process is repeated many times during Simulated Annealing, this runtime saving translates to significantly more sequence pairs we can explore. The main idea is to compute the longest common subsequence in a pair of weighted sequences. With the help of sophisticated data structure, the complexity of placement construction from a given sequence pair is reduced from $O(n^2)$ to $O(n \log \log n)$.

The authors of [Adya and Markov, 2003] studied the fixed-outline floorplan problem, where the goal is to floorplan the modules inside a given rectangular "outline" so that the wirelength is minimized. Compared with the classical formulation where no target outline is given, this new formulation is shown to be more relevant to hierarchical design style for very large-scale ASIC designs [Kahng, 2000]. The authors empirically showed that instances of the fixed-outline floorplan problem are significantly harder than the related instances of classical floorplan problems. New objective functions and new types of moves are presented in the context of sequence pair based non-slicing floorplanner named Parquet to handle the fixed outline constraint efficiently.

The authors of [Lin and Chang, 2004] presented a new non-slicing floorplan representation scheme named TCG-S, which is a combination of transitive constraint graph (TCG) representation [Lin and Chang, 2001] and sequence pair (SP). The goal is to combine the advantages of SP and TCG and eliminate their disadvantages. The benefits of SP include faster packing and perturbation schemes. The benefits of TCG include faster convergence to a desired solution, easy handling of geometric constraints, and incremental update for cost evaluation. These nice properties make TCG-S a superior representation than TCG and SP themselves.

Traditional floorplanning problem deals with module sizing and placement in a single 2D plane. In 3D floorplanning (or alternatively called 2.5D), which is applicable in 3D stacked IC technology, the modules are placed in a stack of multiple 2D planes. The authors of [Cong et al., 2004b] presented a thermal-driven 3D floorplanning algorithm that is based on transitive constraint graph (TCG) [Lin and Chang, 2001]. Their new floorplan representation scheme called Combined Bucket and 2D Array (CBA) is an extension of TCG, where the placement information among the modules across the planes is encoded efficiently. CBA is used in its Simulated Annealing-based optimization [Kirkpatrick et al., 1983] to obtain 3D floorplans with high thermal and area quality results.

Most of the modern floorplanners rely on efficient floorplan representation schemes such as the normalized polish expression [Wong and Liu, 1986] or the sequence pair [Murata et al., 1995] that are optimized with Simulated Annealing [Kirkpatrick et al., 1983] under various objectives and constraints. The runtime and parameter tuning effort involved with such approaches, however, are a major drawback. The authors of [Cong et al., 2006] presented a fast floorplanner named PATOMA that does not rely on Simulated Annealing. PATOMA utilizes top-down recursive partitioning and legalization to obtain high quality floorplans within a fraction of runtime. PATOMA also outperforms other existing fast floorplanners in the literature [Ranjan et al., 2001; Adya and Markov, 2003; Adya et al., 2004; Sassone and Lim, 2006].

Chapter 4

PLACEMENT

Circuit placement is the process of determining the location of each gate (or block in some cases) in the netlist. The traditional objectives include wirelength, timing, and congestion. Recently, thermal hotspot, power consumption, and power supply noise issues drew much attention because the location of gates has a non-negligible impact on these reliability concerns. In order to handle large-scale circuits, placement is usually done in two steps: global placement and detailed placement as in the case with routing. Global placement is mainly concerned with "rough" location of the gates, e.g., which region of the chip a gate is located. Some gates may be overlapping with each other in a global placement solution. These overlaps are then removed during the detailed placement. This chapter presents sample problems related to the following works:

- Mincut placement [Breuer, 1977] and terminal propagation [Dunlop and Kernighan, 1985]
- Gordian placement [Kleinhans et al., 1991]
- TimberWolf algorithm [Sun and Sechen, 1995]

The first work is based on utilizing partitioning to perform placement. The second work is based on quadratic placement. The third work is based on Simulated Annealing [Kirkpatrick et al., 1983]. The first two are global placers, while the third does both. These works are targeting wirelength reduction.

S.K. Lim, *Practical Problems in VLSI Physical Design Automation*

1. Mincut Placement

The mincut placement method pioneered by Breuer [Breuer, 1977] utilizes partitioning to perform placement. In this method the given circuit is repeatedly partitioned into two sub-circuits. Correspondingly, the layout region occupied by the circuit is divided either horizontally or vertically to accommodate the sub-circuits. This partitioning process repeats until each partition is occupied by a single gate or by a sub-circuit that is small enough. In case of the former, we have a legal placement with no overlap among the gates. In case of the latter, a post-process so called "placement legalization" is performed to find unique locations for the gates in each sub-circuit. Breuer suggested that the cutsize be minimized during the partitioning, which tends to minimize the overall wirelength of the final placement. Breuer presented several "cut orientation sequences" to decide how to partition the placement region. The one that is presented in this section is called the quadrature placement, where the placement region is divided into four partitions by a pair of vertical and horizontal cut. This process can be recursively performed to obtain the desired number of partitions.

Dunlop and Kernighan presented an important enhancement to mincut placement called "terminal propagation" [Dunlop and Kernighan, 1985]. They noted that during the partitioning process of mincut placement, some gates in the partition to be divided are connected to the gates located outside the partition. They suggested that this kind of "external connection" should be used to bias the bipartitioning process. In order to facilitate this biasing, they add terminals that represent the external connections to the both sides of the placement region to be partitioned and use these terminals as anchor points to "pull the gates" to certain partitions. They proposed a way to limit the number of external connections to be considered in order to balance the "internal" vs "external" connections. In order to best benefit from terminal propagation, a cut orientation sequence called "breadth-first recursive bipartitioning" is used, where an alternate sequence of horizontal and vertical cuts are recursively inserted to divide the placement region in a breadth-first fashion.

Quick Overview

Here we discuss the recursive bipartitioning based mincut placement along with terminal propagation. We assume that all gates are located at the center of the placement region (= alternatively called block) before the partitioning starts. In addition, if a bipartitioning cutline is inserted to divide the given block, the gates are located at the center of the two sub-blocks. Assuming that the first cutline is vertical, the gates are now located at the center of the left and the right sub-block. Next, we add a horizontal cutline to divide the left block into the left top and the left bottom blocks. The gates are now located at three distinct locations. We then insert the third cutline to divide the right block into the right top and the right bottom blocks. This breadth-first cutline insertion

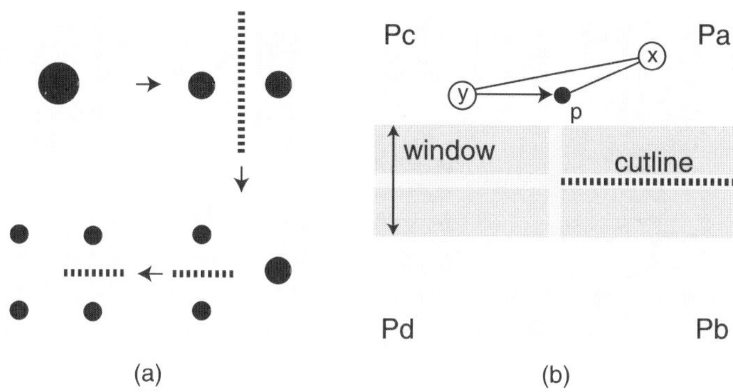

Figure 4.1. (**a**) Breadth-first recursive bipartitioning, (**b**) terminal propagation, where the right-half is being cut by a horizontal cut (shown in dotted line). The left-half is already partitioned into P_c and P_d, and $y \in P_C$ is located at the center of P_c. We propagate y to p because it is located outside the window.

continues until we obtain the desired number of partitions. An illustration is shown in Figure 4.1(a).

Given a block to be bipartitioned, we perform terminal propagation as illustrated in Figure 4.1(b). Assume that the left half of the chip L is already partitioned into $\{P_c, P_d\}$. Also assume that $y \in P_c$ is located at the center of P_c. We are now partitioning the right half R into $\{P_a, P_b\}$. For each gate $x \in R$ that is connected to another gate y located *outside* R, we check to see if y is too closely located to the cutline. In this case, y is considered too close if y is located inside the "window" created by two parallel lines to the cutline. If y is outside the window, we "propagate" y to p, a point on the boundary of R. In this case, p becomes the "propagated terminal" for y. We then connect x and p and ignore the x-p connection. During the partitioning refinement, p is fixed at this initial location and acts like an "anchor" that pulls x into P_a that p is locked in. This is due to the fact that if x is not partitioned together with p, the x-p connection increases the cutsize by one. If an external neighbor is located inside the window, we do not propagate it.

Practice Problem

Consider the gate-level netlist shown in Table 4.1. Figure 4.2 shows the undirected graph model of the netlist, where the thick and the thin edges have weights of 1 and 0.5, respectively.[10] The primary inputs and outputs do not need to be placed.

[10]This undirected graph representation is suitable for KL partitioning algorithm [Kernighan and Lin, 1970]. However, a hypergraph representation can be used if FM partitioning algorithm [Fiduccia and Mattheyses, 1982] is desired.

Table 4.1. Gate-level netlist used for mincut placement.

$n_1 = \{e, f\}$
$n_2 = \{a, e, i\}$
$n_3 = \{b, f, g\}$
$n_4 = \{c, g, l\}$
$n_5 = \{d, l, h\}$
$n_6 = \{e, i, j\}$
$n_7 = \{f, j\}$
$n_8 = \{g, j, k\}$
$n_9 = \{l, o, p\}$
$n_{10} = \{h, p\}$
$n_{11} = \{i, m\}$
$n_{12} = \{j, m, n\}$
$n_{13} = \{k, n, o\}$

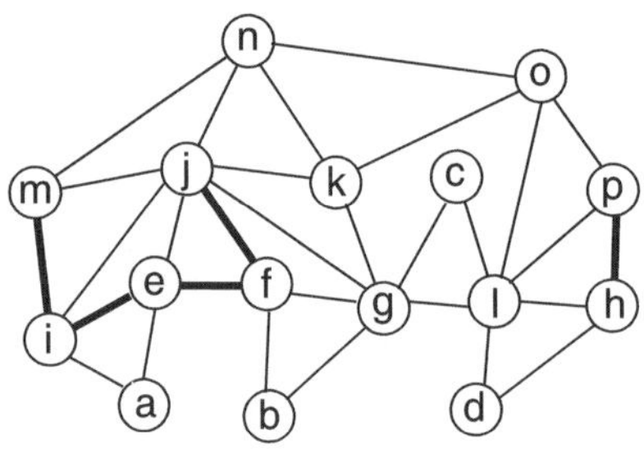

Figure 4.2. Clique-based graph model of the netlist shown in Table 4.1. The thick and the thin edges have weights of 1 and 0.5, respectively.

1. Perform the quadrature mincut starting with vertical cut first, and place the 16 gates into 4×4 grid. Show the placement after each cut. The area skew is set to zero, i.e., we desire perfectly balanced bipartitioning (= bisection) at every mincut.

   The following six cuts are added sequentially. Note that any partitioning algorithm can be used for the bisection such as KL [Kernighan and Lin, 1970] or FM [Fiduccia and Mattheyses, 1982].

   - Cut 1: Figure 4.3(a) shows the placement after the first cut (= vertical) with cutsize 3. Note that there exist other bisections with the same cutsize.
   - Cut 2: Figure 4.3(b) shows the placement after the second cut (= horizontal) with cutsize 5. At this point the first-level quadrants are formed.
   - Cut 3: Figure 4.3(c) shows the placement after the third cut (= vertical) with cutsize 5.

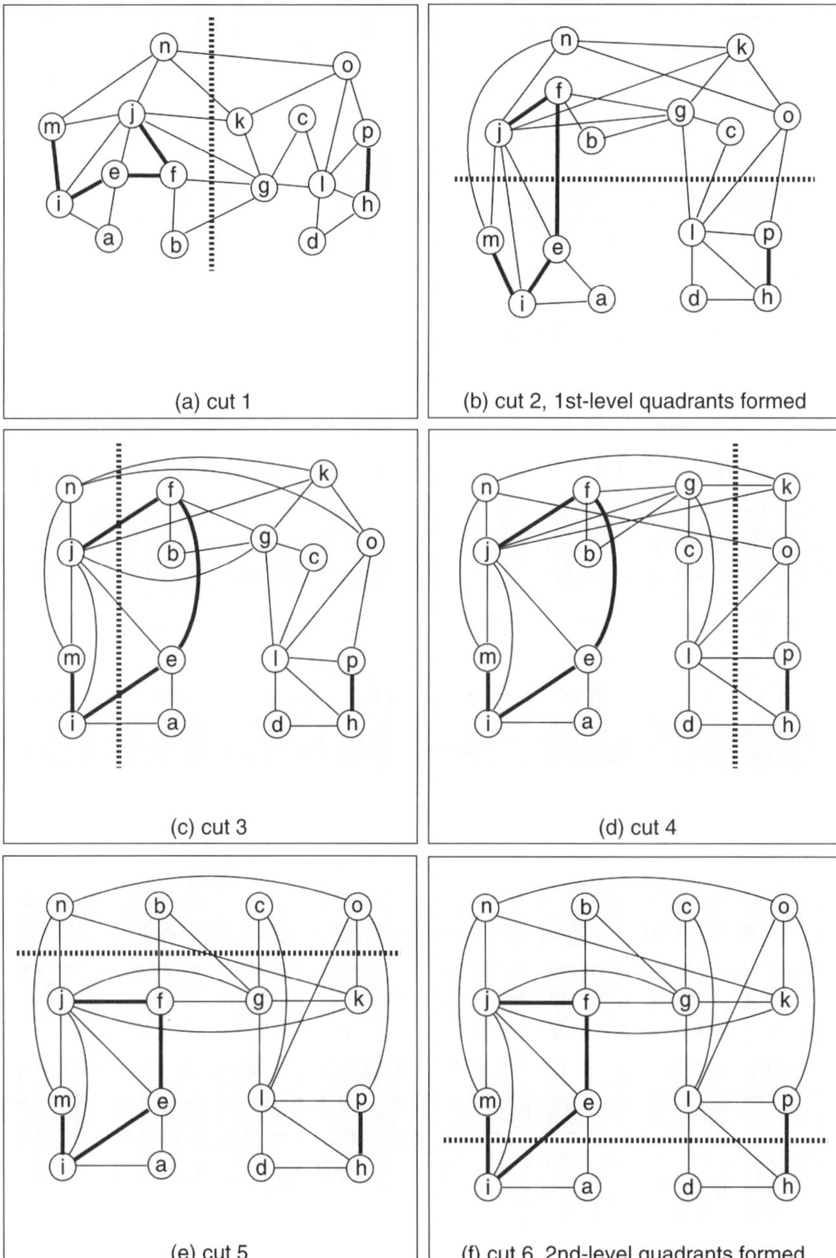

Figure 4.3. Quadrature mincut placement. The thick edges have a weight of 1, and the thin edges have 0.5. The dotted lines show the current partitioning.

- Cut 4: Figure 4.3(d) shows the placement after the fourth cut (= vertical) with cutsize 4.
- Cut 5: Figure 4.3(e) shows the placement after the fifth cut (= horizontal) with cutsize 5.
- Cut 6: Figure 4.3(f) shows the placement after the sixth cut (= horizontal) with cutsize 5. At this point the second-level quadrants are formed.

From Figure 4.3(f), we compute the half-perimeter of the bounding box of the nets as follows: $w(n_1) = 1$, $w(n_2) = 2$, $w(n_3) = 2$, $w(n_4) = 2$, $w(n_5) = 2$, $w(n_6) = 3$, $w(n_7) = 1$, $w(n_8) = 3$, $w(n_9) = 3$, $w(n_{10}) = 1$, $w(n_{11}) = 1$, $w(n_{12}) = 2$, $w(n_{13}) = 4$. Thus, the total wirelength cost is 27.

2. Perform the recursive bipartitioning mincut starting with a vertical cut first, and place the gates into 4×4 grid. Use terminal propagation, where the terminals located within the mid-third window should be ignored. Show the placement after each cut. The area skew is set to zero.

   The following 15 cuts are added sequentially in a breadth-first fashion.

   - Cut 1: Figure 4.9(a) shows the placement after the first cut. No terminal propagation is possible.
   - Cut 2: Figure 4.9(b) shows the placement after the second cut. No terminal propagation is possible.
   - Cut 3: Figure 4.9(c) shows the placement after the third cut. Figure 4.4 shows the terminal propagation performed for this cut. Nodes k and o have external connections to n and j. Assuming n and j are placed at the center of the left top partition, we propagate their terminal p_1. Note that p_1 is located outside the mid-third window. In addition, node g has external connections to j, f, and b. Assuming that f and b are placed at the center of the left bottom partition, p_2 represents their propagated terminal. Note again that p_2 is located outside the mid-third window. Based on these fixed terminals, the nodes k and o are partitioned to the top, and g to the bottom to minimize the cutsize across the partition boundary.
   - Cut 4: Figure 4.9(d) shows the placement after the fourth cut. Figure 4.5 shows the terminal propagation performed for this cut. The node p_1 represents the terminal propagated from o, k, and g. These nodes are the external connections of n and j and are located outside the mid-third window. Nodes i and j are connected to nodes e, f, and a outside,

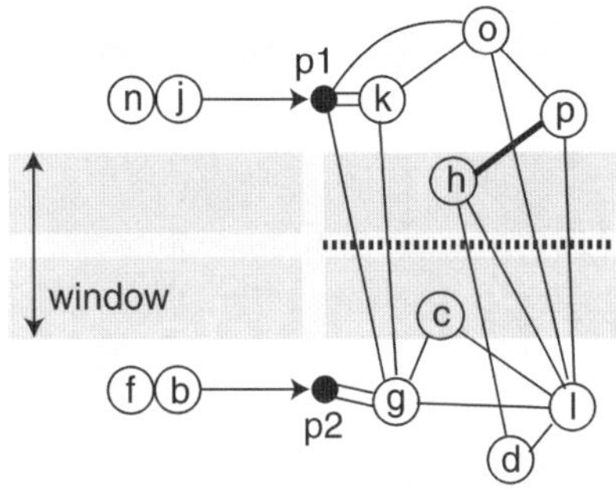

Figure 4.4. Terminal propagation for the partitioning shown in Figure 4.9(c). Terminal p_1 is propagated from nodes n and j and is pulling nodes k, o, and g to the top partition. Terminal p_2 is propagated from nodes f and b and is pulling node g to the bottom partition.

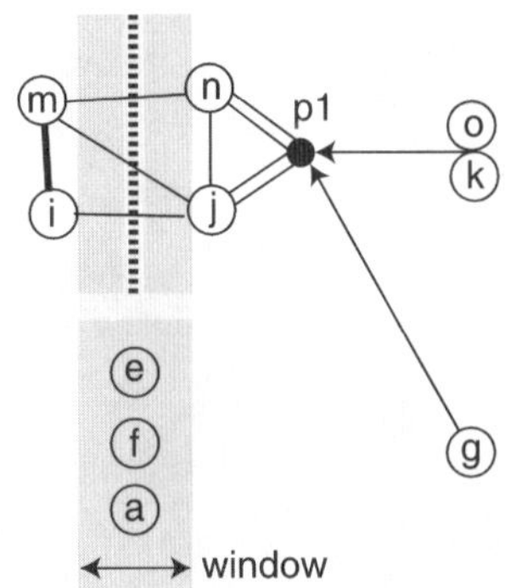

Figure 4.5. Terminal propagation for the partitioning shown in Figure 4.9(d). Terminal p_1 is propagated from nodes o, k, and g and is pulling nodes n and j to the right partition. Nodes i and j are connected to nodes e, f, and a, but no terminal is propagated because e, f, and a are located within the mid-third window.

but no terminal is propagated because they are located within the mid-third window. The nodes n and j are partitioned to the right due to the propagated terminal p_1.

- Cut 5: Figure 4.10(a) shows the placement after the fifth cut. Figure 4.6 shows the terminal propagation performed for this cut. Node i is propagated to terminal p_1, and nodes e and a are connected to p_1. Node j is propagated to terminal p_2, and nodes e and f are connected to p_2. Lastly, node g is propagated to terminal p_3, and nodes f and b are connected to p_3. All of these propagated terminals are located outside the

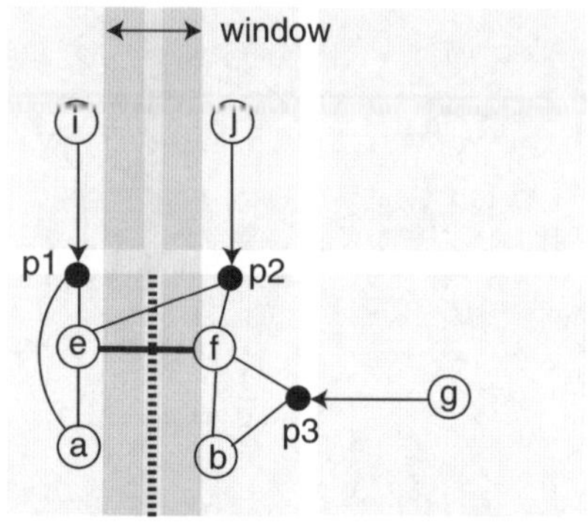

Figure 4.6. Terminal propagation for the partitioning shown in Figure 4.10(a). Three terminals p_1, p_2, and p_3 are propagated. p_1 is pulling nodes a and e to the left partition, and p_2 and p_3 are pulling nodes e, f, and b to the right partition.

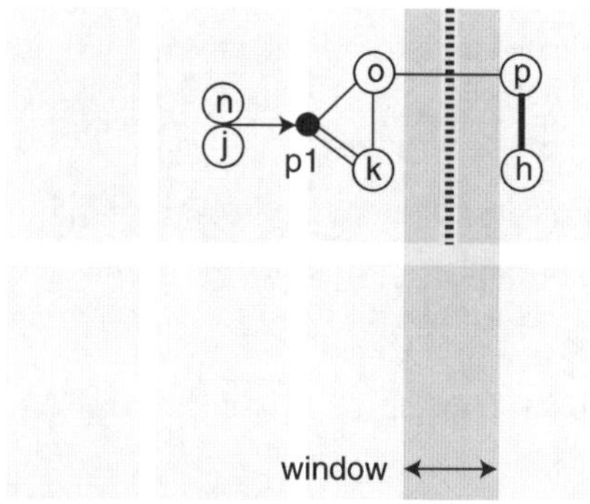

Figure 4.7. Terminal propagation for the partitioning shown in Figure 4.10(b). Terminal p_1 is propagated from nodes n and j and is pulling nodes o and k to the left partition.

mid-third window. Due to these propagated terminals, nodes e and a are partitioned to the left, and nodes f and b to the right.

- Cut 6: Figure 4.10(b) shows the placement after the sixth cut. Figure 4.7 shows the terminal propagation performed for this cut. Nodes n and j are propagated to terminal p_1, and nodes o and k are connected to p_1. p_1 is located outside the mid-third window. Nodes o and k are partitioned to the left due to their connection to p_1.

- Cut 7: Figure 4.10(c) shows the placement after the seventh cut. Figure 4.8 shows the terminal propagation performed for this cut. Nodes j, f, and b are propagated to terminal p_1, and node g has three connections to p_1. Nodes o and k are propagated to terminal p_2, and nodes g and l are connected to p_2. Lastly, nodes h and p are propagated to terminal

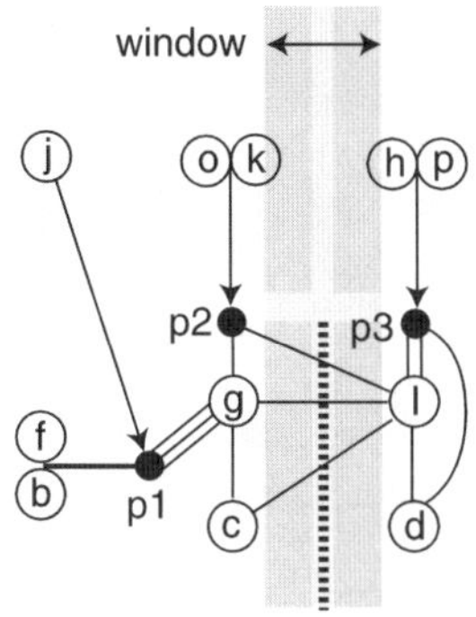

Figure 4.8. Terminal propagation for the partitioning shown in Figure 4.10(c). Three terminals p_1, p_2, and p_3 are propagated. p_1 is pulling g to the left, p_2 is pulling g and l to the left, and p_3 is pulling l and d to the right.

p_3, and nodes l and d are connected to p_3. All of these propagated terminals are located outside the mid-third window. Due to these propagated terminals, nodes g and c are partitioned to the left, and nodes l and d to the right.

- Cuts 8–15: Figure 4.10(d) shows the placement after all the remaining cuts inserted. The terminal propagation results for these cuts are straightforward and omitted.

From Figure 4.10(d), we compute the half-perimeter of the bounding box of the nets as follows: $w(n_1) = 1$, $w(n_2) = 2$, $w(n_3) = 2$, $w(n_4) = 2$, $w(n_5) = 2$, $w(n_6) = 2$, $w(n_7) = 1$, $w(n_8) = 2$, $w(n_9) = 3$, $w(n_{10}) = 1$, $w(n_{11}) = 1$, $w(n_{12}) = 2$, $w(n_{13}) = 2$. Thus, the total wirelength cost is 23. This is lower than the wirelength of the quadrature mincut shown in Figure 4.3(f), which is 27.

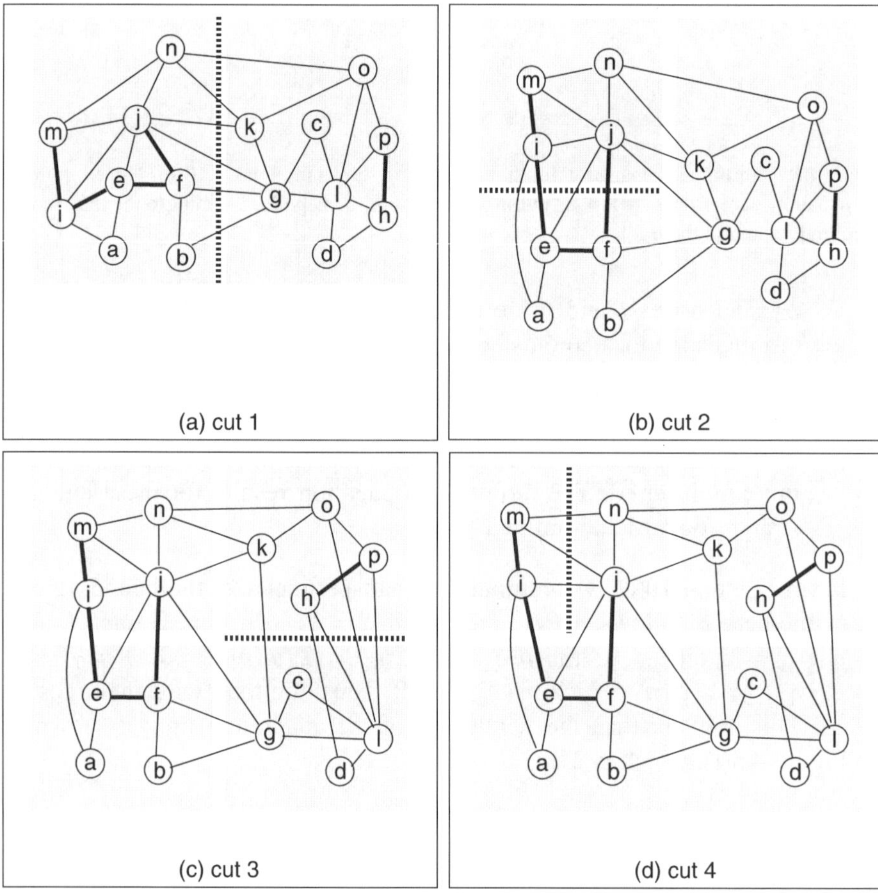

Figure 4.9. Recursive bipartitioning mincut placement. The thick edges have a weight of 1, and the thin edges have 0.5. The dotted lines show the current partitioning.

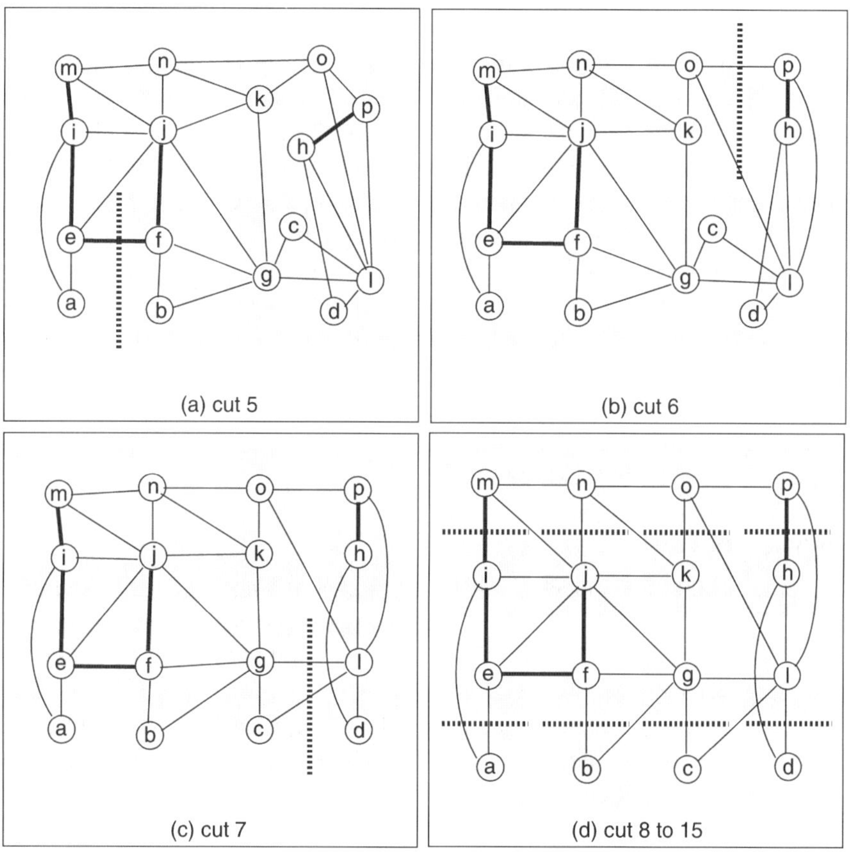

Figure 4.10. Recursive bipartitioning mincut placement (continued from Figure 4.9).

2. GORDIAN Algorithm

GORDIAN placement [Kleinhans et al., 1991] is one of the most successful works that adopt quadratic programming (QP) into circuit placement problem. In GORDIAN the placement problem is formulated as a sequence of QP derived from the connectivity information of the circuit. A set of constraints restricting the freedom of movement of the gates is imposed at every iteration, causing the placement solution to reduce the amount of overlap among the cells gradually. In this case, a top-down partitioning is performed so that the cells grouped into the same partition satisfy so called the "center of gravity" constraint, which requires that the area-weighted center of the cells coincide with the center location of the partition. This alternating sequence between QP and top-down partitioning repeats until the sizes of the partitions is small enough. Note that there may still exist overlaps among the cells once GORDIAN terminates. Thus, a post-process such as DOMINO [Doll et al., 1994] is used to remove the overlap and derive a standard cell placement.

The goal of GORDIAN placement is to minimize the *squared* wirelength among the cells, which necessitates a quadratic objective function. Later a pseudo-linear objective is introduced in GORDIAN-L [Sigl et al., 1991a] to minimize the linear wirelength while still using a quadratic solver. An important point to note is that GORDIAN uses QP as a *global* optimizer unlike other existing QP/partitioning-based placers. In other words, QP is applied to the entire circuit instead of local regions.

Quick Overview

Given a circuit to be placed, GORDIAN first constructs a clique-based undirected graph model, where each edge in a k-clique gets a weight of $2/k$. GORDIAN requires that some of the IO cells are fixed along the boundary of the placement region; otherwise, GORDIAN places all cells at one location. Once the fixed and the movable cells are identified, GORDIAN computes one matrix $(= C)$ and two vectors $(= d_x, d_y)$, where C is so called the Laplacian matrix that shows the connectivity among the cells. This is the same as the Q matrix used in the eigenvector-based partitioner [Hagen and Kahng, 1992] presented in Chapter 2, Section 3. d_x and d_y show the connectivity between the movable and fixed cells. Each entry in these vectors basically denotes the weighted sum of the edges that connect each cell to the fixed cells. We then perform the iterations as follows:

1. At the optimization level $l = 0$, we solve the following QP to obtain x that specifies the x-coordinates of the cells: Minimize $1/2 \cdot x^T Cx + {d_x}^T x$. The y-coordinates of the cells are computed similarly using d_y and y.

2. At the optimization level $l = 1$, we first partition the placement region into two using either vertical or horizontal cutline. We compute u_x and

u_y vectors that contain the center location of the sub-partitions. Next, we compute the matrix A that specifies the center-of-gravity constraint. Each entry in A basically shows (i) which partition each cell belongs to, and (ii) the ratio of the cell area to the partition area. Lastly, we solve the following Linearly constrained QP (LQP) to obtain x: Minimize $1/2 \cdot x^T C x + {d_x}^T x$ subject to $Ax = u_x$. The y-coordinates of the cells are computed similarly using d_y, y, and u_y.

3. At the optimization level $l = 2$, we add two cutlines that are perpendicular to the cutline inserted at $l = 1$. The rest of the steps is the same as those in $l = 1$.

GORDIAN terminates after an enough number of cutlines are added so that the size of the partitions is small enough, i.e., it is used for global placement.

Practice Problem

Consider the circuit shown in Figure 4.11. Assume the following: (1) cells have unit area, (2) pins are located at the center of the cells, (3) the weight of all nets is 1, (4) IO pins are fixed at the boundary of the placement region as shown in Figure 4.12, (5) when performing bipartitioning, use the area balance factor $\alpha = 0.5$ whenever possible.

1. Set up the clique-based graph model of the netlist.

   Each net of size k in the netlist forms a k-clique, and all the edges in the clique receives a weight of $2/k$. Figure 4.13 shows the graph.

2. Build the system matrix C and fixed node vectors d_x and d_y. Show all the intermediate matrices and vectors involved.

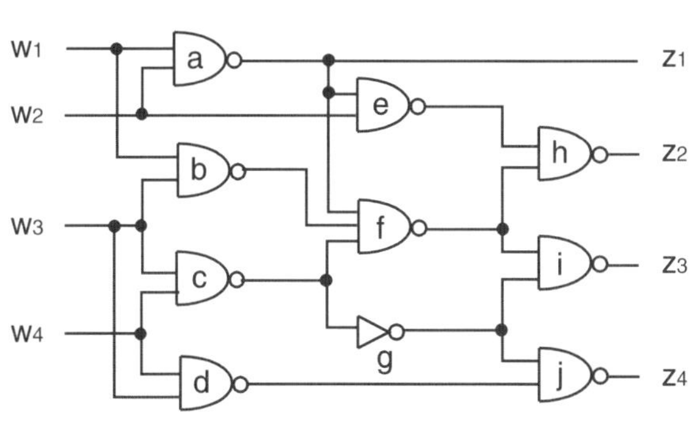

Figure 4.11. Gate-level circuit used for GORDIAN algorithm.

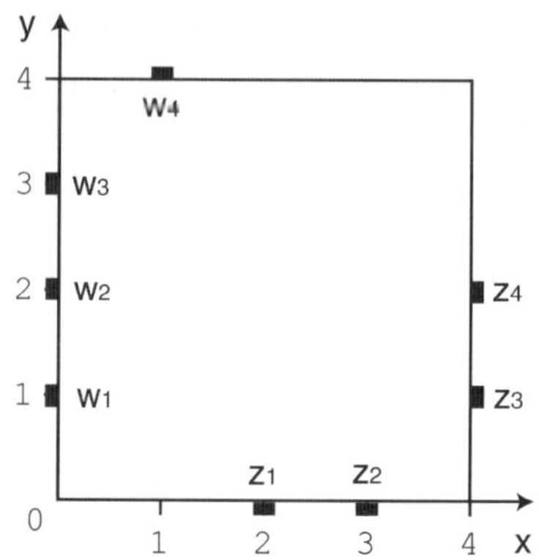

Figure 4.12. Fixed IO pin location.

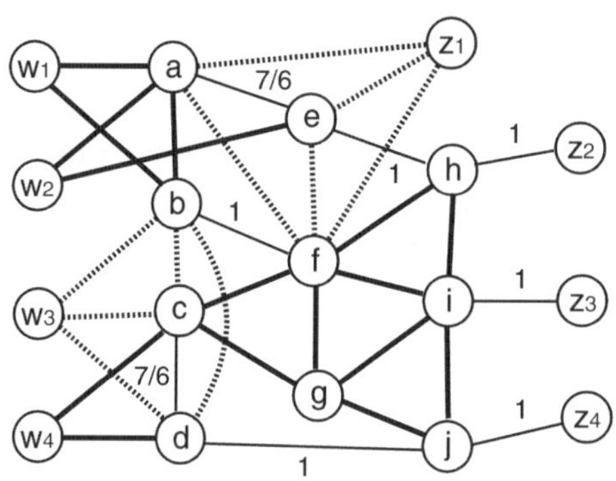

Figure 4.13. Undirected graph model of the circuit in Figure 4.11. The thick edges have a weight of 2/3, and the dotted edges have a weight of 0.5.

We need the following matrices and vectors:

(a) Adjacency matrix: the connections among the movable nodes (= a to j) are specified as follows:

$$\begin{pmatrix} 0 & \frac{2}{3} & 0 & 0 & \frac{7}{6} & \frac{1}{2} & 0 & 0 & 0 & 0 \\ \frac{2}{3} & 0 & \frac{1}{2} & \frac{1}{2} & 0 & 1 & 0 & 0 & 0 & 0 \\ 0 & \frac{1}{2} & 0 & \frac{7}{6} & 0 & \frac{2}{3} & \frac{2}{3} & 0 & 0 & 0 \\ 0 & \frac{1}{2} & \frac{7}{6} & 0 & 0 & 0 & 0 & 0 & 0 & 1 \\ \frac{7}{6} & 0 & 0 & 0 & 0 & \frac{1}{2} & 0 & 1 & 0 & 0 \\ \frac{1}{2} & 1 & \frac{2}{3} & 0 & \frac{1}{2} & 0 & \frac{2}{3} & \frac{2}{3} & \frac{2}{3} & 0 \\ 0 & 0 & \frac{2}{3} & 0 & 0 & \frac{2}{3} & 0 & 0 & \frac{2}{3} & \frac{2}{3} \\ 0 & 0 & 0 & 0 & 1 & \frac{2}{3} & 0 & 0 & \frac{2}{3} & 0 \\ 0 & 0 & 0 & 0 & 0 & \frac{2}{3} & \frac{2}{3} & \frac{2}{3} & 0 & \frac{2}{3} \\ 0 & 0 & 0 & 1 & 0 & 0 & \frac{2}{3} & 0 & \frac{2}{3} & 0 \end{pmatrix}$$

(b) Pin connection matrix: the connections between the movable nodes (a to j) and the fixed nodes (= w_i and z_i, $1 \leq i \leq 4$) are specified as follows:

$$\begin{pmatrix}
\frac{2}{3} & \frac{2}{3} & 0 & 0 & \frac{1}{2} & 0 & 0 & 0 \\
\frac{2}{3} & 0 & \frac{1}{2} & 0 & 0 & 0 & 0 & 0 \\
0 & 0 & \frac{1}{2} & \frac{2}{3} & 0 & 0 & 0 & 0 \\
0 & 0 & \frac{1}{2} & \frac{2}{3} & 0 & 0 & 0 & 0 \\
0 & \frac{2}{3} & 0 & 0 & \frac{1}{2} & 0 & 0 & 0 \\
0 & 0 & 0 & 0 & \frac{1}{2} & 0 & 0 & 0 \\
0 & 0 & 0 & 0 & 0 & 0 & 0 & 0 \\
0 & 0 & 0 & 0 & 0 & 1 & 0 & 0 \\
0 & 0 & 0 & 0 & 0 & 0 & 1 & 0 \\
0 & 0 & 0 & 0 & 0 & 0 & 0 & 1
\end{pmatrix}$$

The rows correspond to the movable nodes, and the columns correspond to the fixed nodes. For example, the entry in row 1, column 1 is the weight of the edge that connects cell a and IO pin w_1, which is 2/3 from Figure 4.13.

(c) Degree matrix: this matrix is built based on both the adjacency matrix and the pin connection matrix. Each entry in this diagonal matrix is the summation of the entries in the corresponding rows in both the adjacency and pin connection matrices.

$$\begin{pmatrix}
\frac{25}{6} & 0 & 0 & 0 & 0 & 0 & 0 & 0 & 0 & 0 \\
0 & \frac{23}{6} & 0 & 0 & 0 & 0 & 0 & 0 & 0 & 0 \\
0 & 0 & \frac{25}{6} & 0 & 0 & 0 & 0 & 0 & 0 & 0 \\
0 & 0 & 0 & \frac{23}{6} & 0 & 0 & 0 & 0 & 0 & 0 \\
0 & 0 & 0 & 0 & \frac{23}{6} & 0 & 0 & 0 & 0 & 0 \\
0 & 0 & 0 & 0 & 0 & \frac{31}{6} & 0 & 0 & 0 & 0 \\
0 & 0 & 0 & 0 & 0 & 0 & \frac{8}{3} & 0 & 0 & 0 \\
0 & 0 & 0 & 0 & 0 & 0 & 0 & \frac{10}{3} & 0 & 0 \\
0 & 0 & 0 & 0 & 0 & 0 & 0 & 0 & \frac{11}{3} & 0 \\
0 & 0 & 0 & 0 & 0 & 0 & 0 & 0 & 0 & \frac{10}{3}
\end{pmatrix}$$

(d) System (= Laplacian) matrix: this matrix is simply the degree matrix minus the adjacency matrix:

$$C = \begin{pmatrix} \frac{25}{6} & \frac{2}{3} & 0 & 0 & \frac{7}{6} & \frac{1}{2} & 0 & 0 & 0 & 0 \\ -\frac{2}{3} & \frac{23}{6} & -\frac{1}{2} & -\frac{1}{2} & 0 & -1 & 0 & 0 & 0 & 0 \\ 0 & -\frac{1}{2} & \frac{25}{6} & -\frac{7}{6} & 0 & -\frac{2}{3} & -\frac{2}{3} & 0 & 0 & 0 \\ 0 & -\frac{1}{2} & -\frac{7}{6} & \frac{23}{6} & 0 & 0 & 0 & 0 & 0 & -1 \\ -\frac{7}{6} & 0 & 0 & 0 & \frac{23}{6} & -\frac{1}{2} & 0 & -1 & 0 & 0 \\ -\frac{1}{2} & -1 & -\frac{2}{3} & 0 & -\frac{1}{2} & \frac{31}{6} & -\frac{2}{3} & -\frac{2}{3} & -\frac{2}{3} & 0 \\ 0 & 0 & -\frac{2}{3} & 0 & 0 & -\frac{2}{3} & \frac{8}{3} & 0 & -\frac{2}{3} & -\frac{2}{3} \\ 0 & 0 & 0 & 0 & -1 & -\frac{2}{3} & 0 & \frac{10}{3} & -\frac{2}{3} & 0 \\ 0 & 0 & 0 & 0 & 0 & -\frac{2}{3} & -\frac{2}{3} & -\frac{2}{3} & \frac{11}{3} & -\frac{2}{3} \\ 0 & 0 & 0 & -1 & 0 & 0 & -\frac{2}{3} & 0 & -\frac{2}{3} & \frac{10}{3} \end{pmatrix}$$

(e) Fixed pin vectors: we compute d_x and d_y based on the pin connection matrix and the IO pin location (= Figure 4.12). Each row i in d_x, denoted $d_{x,i}$, is computed as follows:

$$d_{x,i} = -\sum_j p_{ij} \cdot x(p_j)$$

where p_{ij} denotes the entry of the pin connection matrix at row i, column j. $x(p_j)$ denotes the x-coordinate of the corresponding IO pin j. From Figure 4.12, we see that the x-coordinates of the pins w_i and z_i $(1 \leq i \leq 4)$ are 0, 0, 0, 1, 2, 3, 4, and 4. Thus,

$$d_{x,1} = -(\frac{2}{3} \cdot 0 + \frac{2}{3} \cdot 0 + 0 \cdot 0 + 0 \cdot 1 + \frac{1}{2} \cdot 2 + 0 \cdot 3 + 0 \cdot 4 + 0 \cdot 4) = -1$$

By examining the remaining 9 movable cells, we get

$$d_x^T = \begin{pmatrix} -1 & 0 & -\frac{2}{3} & -\frac{2}{3} & -1 & -1 & 0 & -3 & -4 & -4 \end{pmatrix}$$

Each row i in d_y, denoted $d_{y,i}$, is computed as follows:

$$d_{y,i} = -\sum_j p_{ij} \cdot y(p_j)$$

where $y(p_j)$ denotes the y-coordinate of the corresponding IO pin j. From Figure 4.12, we see that the y-coordinates of the pins w_i and z_i $(1 \leq i \leq 4)$ are 1, 2, 3, 4, 0, 0, 1, and 2. Thus,

$$d_{y,1} = -(\frac{2}{3} \cdot 1 + \frac{2}{3} \cdot 2 + 0 \cdot 3 + 0 \cdot 4 + \frac{1}{2} \cdot 0 + 0 \cdot 0 + 0 \cdot 1 + 0 \cdot 2) = -2$$

By examining the remaining 9 movable cells, we get

$$d_y^T = \begin{pmatrix} -2 & -\frac{13}{6} & -\frac{25}{6} & -\frac{25}{6} & -\frac{4}{3} & 0 & 0 & 0 & -1 & -2 \end{pmatrix}$$

3. Perform placement at the top optimization level ($l = 0$).

   We first compute the x-coordinate of the movable cells by minimizing the following objective function:

$$\phi(x) = \frac{1}{2}x^T Cx + {d_x}^T x$$

   Since this is the top-level placement, there is no "center-of-gravity" constraint applied. Similarly, the y-coordinates of the movable cells are found by minimizing the following objective function:

$$\phi(y) = \frac{1}{2}y^T Cy + {d_y}^T y$$

   Using, MOSEK, a popular quadratic programming solver, we obtain the following solution[11]:

$$x^T = \begin{pmatrix} 0.95 & 0.92 & 1.21 & 1.32 & 1.32 & 1.61 & 1.98 & 2.13 & 2.59 & 2.51 \end{pmatrix}$$
$$y^T = \begin{pmatrix} 1.27 & 1.83 & 2.48 & 2.61 & 1.16 & 1.45 & 1.84 & 0.92 & 1.41 & 2.03 \end{pmatrix}$$

   Figure 4.14 shows the corresponding placement result.

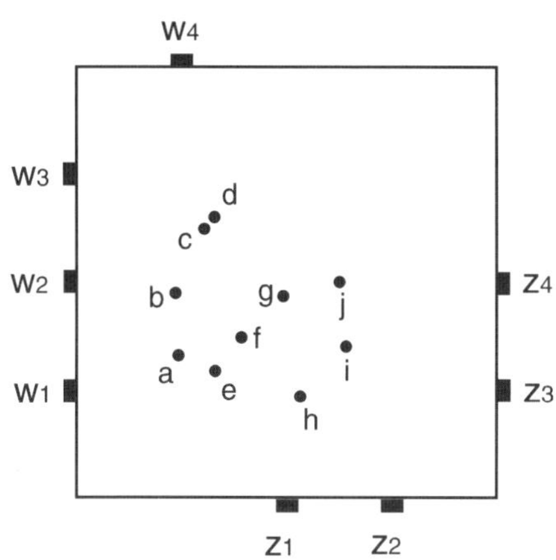

Figure 4.14. GORDIAN placement at level $l = 0$.

[11]The source files for all of the quadratic programming formulations presented in this section are available for download at: `http://users.ece.gatech.edu/limsk/book`

4. Perform placement at the optimization level ($l = 1$). Perform vertical partitioning with area balance factor $\alpha = 0.5$.

   We first sort the nodes based on their x-coordinates and obtain the following order:

$$\{b, a, c, e, d, f, g, h, j, i\}$$

   Since the area of the nodes is uniform, we place the first five nodes to left partition and the remaining to the right partition under $\alpha = 0.5$:

$$S_{\rho'} = \{b, a, c, e, d\}$$
$$S_{\rho''} = \{f, g, h, j, i\}$$

   Since we divide the 4×4 placement region into equal parts, the center location of the partitions are $(1, 2)$ and $(3, 2)$. Thus, the center location vectors are:

$$u_x^{(1)} = \begin{pmatrix} 1 \\ 3 \end{pmatrix}, \quad u_y^{(1)} = \begin{pmatrix} 2 \\ 2 \end{pmatrix}$$

   Next we build the matrix $A^{(1)}$ for the center-of-gravity constraint at level $l = 1$:

$$A^{(1)} = \begin{pmatrix} \frac{1}{5} & \frac{1}{5} & \frac{1}{5} & \frac{1}{5} & \frac{1}{5} & 0 & 0 & 0 & 0 & 0 \\ 0 & 0 & 0 & 0 & 0 & \frac{1}{5} & \frac{1}{5} & \frac{1}{5} & \frac{1}{5} & \frac{1}{5} \end{pmatrix}$$

   where the first (second) row corresponds to the left (right) partition. The columns corresponds to the cells a through j. For example, the entry in row 1, column 1 is the ratio between the area of cell a to the area of left partition, which is 1/5.

   We now solve the following Linearly constrained QP (LQP) to obtain the new x-coordinates of the movable nodes:

$$\text{Minimize } \phi(x) = \frac{1}{2} x^T C x + {d_x}^T x, \text{ subject to } A^{(1)} \cdot x = u_x^{(1)}$$

   Similarly, we solve the following LQP to obtain the new y-coordinates of the movable nodes:

$$\text{Minimize } \phi(y) = \frac{1}{2} y^T C y + {d_y}^T y, \text{ subject to } A^{(1)} \cdot y = u_y^{(1)}$$

   The solutions are as follows:

$$x^T = \begin{pmatrix} 0.70 & 0.71 & 1.17 & 1.21 & 1.22 & 2.17 & 3.10 & 2.84 & 3.56 & 3.33 \end{pmatrix}$$
$$y^T = \begin{pmatrix} 1.34 & 1.94 & 2.66 & 2.76 & 1.30 & 1.83 & 2.45 & 1.32 & 1.91 & 2.49 \end{pmatrix}$$

   Figure 4.15 shows the corresponding placement results.

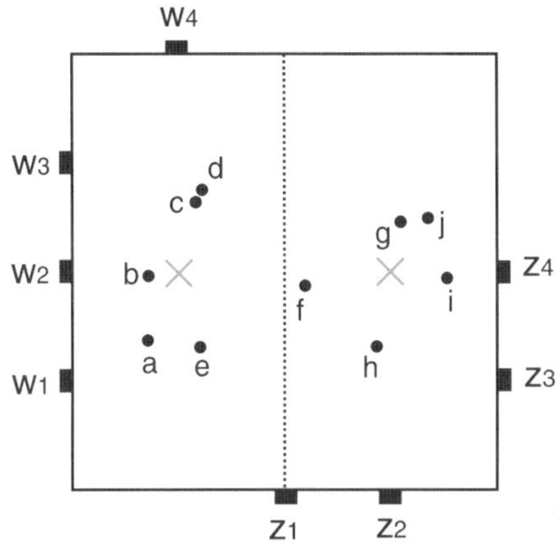

Figure 4.15. GORDIAN placement at level $l = 1$ with a vertical cut. X denotes the center location of the partitions.

5. Verify that the center of gravity constraint is satisfied in the left partition from level 1 placement.

   The following cells are partitioned to the left: $a(0.70, 1.34)$, $b(0.71, 1.94)$, $c(1.17, 2.66)$, $d(1.21, 2.76)$, and $e(1.22, 1.30)$. Thus, the center of gravity is located at:

$$\frac{0.70 + 0.71 + 1.17 + 1.21 + 1.22}{5} = 1.00$$
$$\frac{1.34 + 1.94 + 2.66 + 2.76 + 1.30}{5} = 2.00$$

   This agrees with the center location $(1, 2)$.

6. Perform placement at the optimization level $(l = 2)$.

   We add two horizontal cutlines as shown in Figure 4.16. Note that area-balanced partitioning is not possible. The gates in the left partition in Figure 4.15 are partitioned into top = (d, c) and bottom = (a, b, e) based on their y-coordinates. Similarly, we split the nodes in the right partition into top = (g, j) and bottom = (f, h, i) based on their y-coordinates. These level 2 cutlines result in the following center locations of the partitions (see Figure 4.16):

$$(x_{p1}, y_{p1}) = (1, 3.2)$$
$$(x_{p2}, y_{p2}) = (1, 1.2)$$
$$(x_{p3}, y_{p3}) = (3, 3.2)$$
$$(x_{p4}, y_{p4}) = (3, 1.2)$$

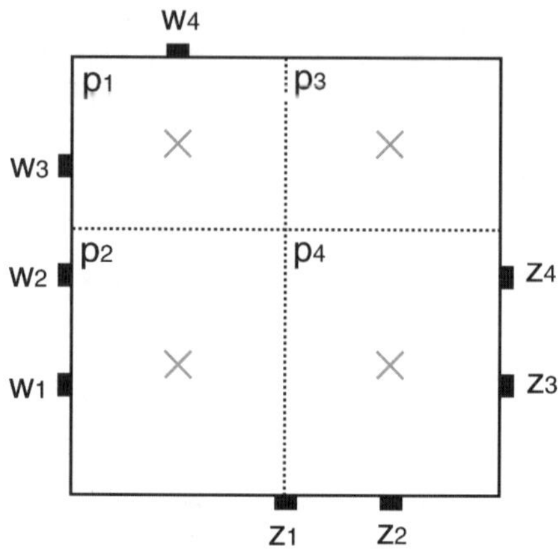

Figure 4.16. The 4 partitions (p_1 to p_4) and their center locations at level $l = 2$.

Thus,

$$u_x^{(2)} = \begin{pmatrix} 1 \\ 1 \\ 3 \\ 3 \end{pmatrix}, \; u_y^{(2)} = \begin{pmatrix} 3.2 \\ 1.2 \\ 3.2 \\ 1.2 \end{pmatrix}$$

Next, we build the matrix $A^{(2)}$ for the center-of-gravity constraint at level $l = 2$. Recall that $p_1 = \{c, d\}$, $p_2 = \{a, b, e\}$, $p_3 = \{g, j\}$, $p_4 = \{f, h, i\}$. Thus,

$$A^{(2)} = \begin{pmatrix} 0 & 0 & \frac{1}{2} & \frac{1}{2} & 0 & 0 & 0 & 0 & 0 & 0 \\ \frac{1}{3} & \frac{1}{3} & 0 & 0 & \frac{1}{3} & 0 & 0 & 0 & 0 & 0 \\ 0 & 0 & 0 & 0 & 0 & 0 & \frac{1}{2} & 0 & 0 & \frac{1}{2} \\ 0 & 0 & 0 & 0 & 0 & \frac{1}{3} & 0 & \frac{1}{3} & \frac{1}{3} & 0 \end{pmatrix}$$

where the rows denote the partitions p_1 through p_4, and the columns denote the cells a through j.

We now solve the following LQP to obtain the new x-coordinates of the movable nodes:

$$\text{Minimize } \phi(x) = \frac{1}{2} x^T C x + {d_x}^T x, \text{ subject to } A^{(2)} \cdot x = u_x^{(2)}$$

Similarly, we solve the following LQP to obtain the new y-coordinates of the movable nodes:

$$\text{Minimize } \phi(y) = \frac{1}{2} y^T C y + {d_y}^T y, \text{ subject to } A^{(2)} \cdot y = u_y^{(2)}$$

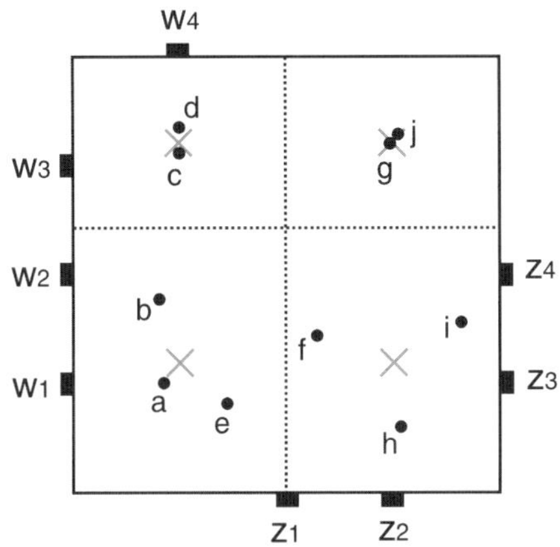

Figure 4.17. GORDIAN placement at level $l = 2$ with a vertical cut. X denotes the center location of the partitions.

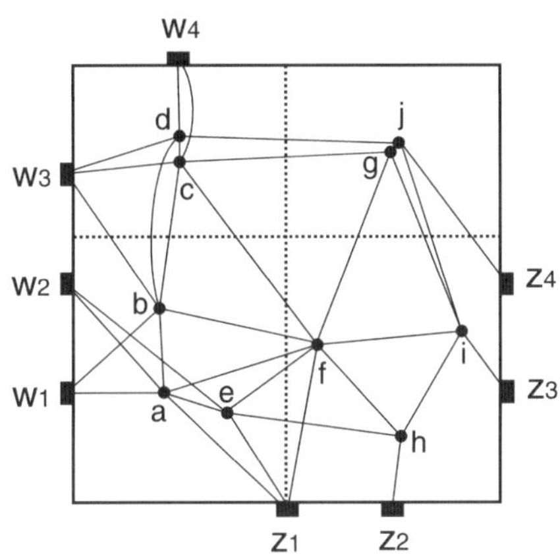

Figure 4.18. GORDIAN placement with wires shown.

The solutions are as follows:

$$x^T = \begin{pmatrix} 0.83 & 0.78 & 1.00 & 1.00 & 1.39 & 2.28 & 2.89 & 3.06 & 3.66 & 3.11 \end{pmatrix}$$
$$y^T = \begin{pmatrix} 1.01 & 1.78 & 3.08 & 3.32 & 0.82 & 1.44 & 3.18 & 0.59 & 1.57 & 3.22 \end{pmatrix}$$

Figure 4.17 shows the corresponding placement results.

7. Show the connection among the cells and IO pads using cliques.

   Figure 4.18 shows the placement with wiring.

3. TimberWolf Algorithm

The TimberWolf package [Sechen and Sangiovanni-Vincentelli, 1985] is a widely used placement and routing tool based on Simulated Annealing [Kirkpatrick et al., 1983]. The latest version TimberWolf 7.0 [Sun and Sechen, 1995] includes two important enhancements added to its standard cell placement engine. First, a hierarchical placement approach is used, where the netlist is clustered twice in a recursive fashion. The top-level (= second-level) clusters are first placed during the high annealing temperature region. These clusters are then decomposed to reveal the first-level clusters. The placement among the first-level clusters is refined during the mid annealing temperature region. Lastly, the first-level clusters are decomposed to reveal the original gate-level netlist. TimberWolf 7.0 refines this gate-level placement during the low annealing temperature region. This so called "multi-level" approach is also used in hMetis partitioning algorithm [Karypis et al., 1997] presented in Chapter 1, Section 3.

The second enhancement is the introduction of overlap-free placement. In general, cell moves/swaps introduce overlaps among the cells due to the non-uniform cell width. Since the removal of overlap is a time-consuming process, TimberWolf 6.0 [Swartz and Sechen, 1990] allows cell overlap and tries to minimize it by utilizing a penalty function. In case the overlap is not completely removed, cells are shifted at the end of annealing process to obtain a legal placement. It is during this post-process step that the solution quality degrades significantly in many cases. Thus, the authors of TimberWolf 7.0 [Sun and Sechen, 1995] proposed a way to maintain overlap-free placement during annealing while performing cell shifting efficiently. This section focuses on this overlap-free cell placement scheme.

Quick Overview

Given a pair of cells (x, y) to swap, we compute the change in the wirelength cost as follows:

$$\Delta C = \Delta W + \Delta W_S$$

where ΔW denotes the wirelength change due to swapping x and y, and ΔW_S is the wirelength change due to shifting the cells to remove overlap. After the swap, we shift the cells in the row that receives the cell with larger width between x and y. In this case, we choose the shift direction, either to the right or left, so that the number of cells shifted is minimized, i.e., shift the cells on the *shorter* side of the row. Note that the row that receives a narrower cell from the swap will have a small gap, but we do not attempt to shift the cells to remove the gap. The reason is that this gap is likely to be filled by the subsequent swaps. Cell swap is chosen carefully so that the difference among the row lengths is minimal. This has an effect of keeping the overall amount of gap to be small.

After we choose the set of cells to be shifted, we first compute ΔW accurately by examining the nets incident on the swapped cells. Next, we *estimate* ΔW_S instead of computing the exact value. The reason is that the runtime involved with the exact computation could be significant if the number of cells shifted is large. Given a cell z to be shifted, the authors of TimberWolf 7.0 proposed two different ways for this estimation, namely, Model A and Model B.

- Model A: for each net incident to z, we find two "break points" a and b, which defines a range $[a, b]$. The wirelength of the net (1) does not change if z is located within the range, (2) increases if z is located to the right of b, and (3) decreases if z is located to the left of a. The superposition of these break points for all nets incident to z efficiently estimates the change in the wirelength from shifting z.

- Model B: we first compute the "gradient" of z as follows:

$$gradient(z) = \sum_{i \in N_z} D_i(0)$$

  where N_z denotes the nets incident to z, and $D_i(0)$ denotes the "rate of wirelength change of net i measured at the origin." If z is at the left boundary of the bounding box of net i, $D_i(0) = -1$. If at the right boundary, $D_i(0) = 1$. Otherwise, $D_i(0) = 0$. Once the gradient of all cells to be shifted is computed, we *estimate* ΔW_S as follows:

$$\Delta W_S = \sum_{j \in shifted\_cell} gradient(j) \cdot shift\_amount(j)$$

  where $shift\_amount(j)$ denotes the distance cell j is shifted by (right is positive, left is negative).

It is shown in [Sun and Sechen, 1995] that Model B is as accurate as Model A but runs faster.

Practice Problem

Consider the gate-level netlist shown in Table 4.2 and its standard cell placement shown in Figure 4.19(a). Use the "Model B" of TimberWolf 7.0 [Sun and Sechen, 1995] when estimating the wirelength change from cell shifting. Use the lower left corner of each cell to represent the cell location.

1. Compute the half-perimeter wirelength of each net.

Table 4.2. Gate-level netlist used in TimberWolf algorithm.

$$
\begin{aligned}
n_1 &= \{a, e, g\} \\
n_2 &= \{f, o\} \\
n_3 &= \{b, c, k, n\} \\
n_4 &= \{d, h, i\} \\
n_5 &= \{j, l, m\} \\
n_6 &= \{d, k, j\} \\
n_7 &= \{c, e, f, h, n\} \\
n_8 &= \{d, l\} \\
n_9 &= \{b, g, i, m\} \\
n_{10} &= \{a, k, o\}
\end{aligned}
$$

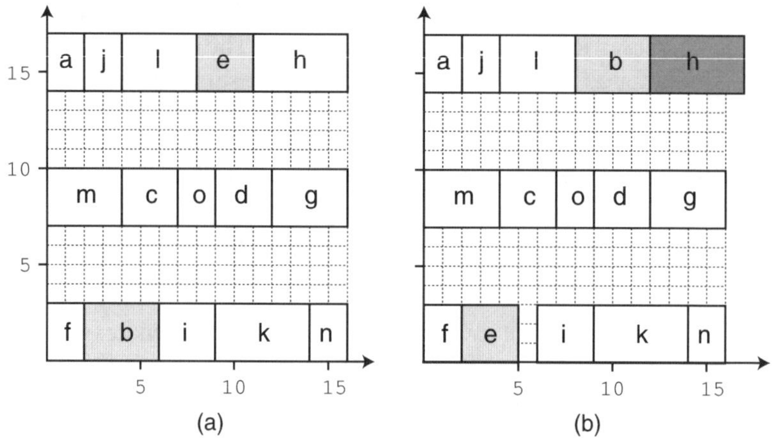

Figure 4.19. (**a**) Before swapping (b, e), (**b**) after the swap. Cell h is shifted to the right.

We compute the half-perimeter of the bounding box as follows:

$n_1 = 12+7 = 19$, $n_2 = 7+7 = 14$, $n_3 = 12+7 = 19$, $n_4 = 5+14 = 19$, $n_5 = 4+7 = 11$, $n_6 = 7+14 = 21$, $n_7 = 14+14 = 28$, $n_8 = 5+7 = 12$, $n_9 = 12+7 = 19$, $n_{10} = 9+14 = 23$.

2. Perform swap (b, e). What is the change on the wirelength cost?

   Figure 4.19(b) shows the placement after the swap. We shift the cells on "the shorter side of the row that receives the wider cell", which explains why cell h is shifted. The change in the cost function is calculated as

$$\Delta C = \Delta W + \Delta W_S$$

   (a) Computation of ΔW: In this case, we examine the swapped cells and compute their wirelength change. The set of nets containing b is $\{n_3, n_9\}$, and e is $\{n_1, n_7\}$. Let $w(x)$ and $w'(x)$ respectively denote the wirelength before and after the swap. Then,

$$\Delta(n_3) = w'(n_3) - w(n_3) = 24 - 19 = 5$$
$$\Delta(n_9) = w'(n_9) - w(n_9) = 26 - 19 = 7$$
$$\Delta(n_1) = w'(n_1) - w(n_1) = 26 - 19 = 7$$
$$\Delta(n_7) = w'(n_7) - w(n_7) = 28 - 28 = 0$$

Thus,

$$\Delta W = \Delta(n_3) + \Delta(n_9) + \Delta(n_1) + \Delta(n_7) = 19$$

(b) Estimation of ΔW_S: In this case, we examine the shifted cells and *estimate* the wirelength change as follows:

$$\Delta W_S = \sum_{i \in shifted} gradient(i) \cdot shift\_amount(i)$$

We see that h is the only shifted cell. The nets that contain h are $n_4 = \{d, h, i\}$ and $n_7 = \{c, e, f, h, n\}$. Figure 4.20 shows the bounding boxes of these nets. We start with $gradient(h) = 0$:

- n_4: h is located on the right boundary, so we increment $gradient(h)$.
- n_7: h is located on neither the left nor the right boundary, so we do not change $gradient(h)$.

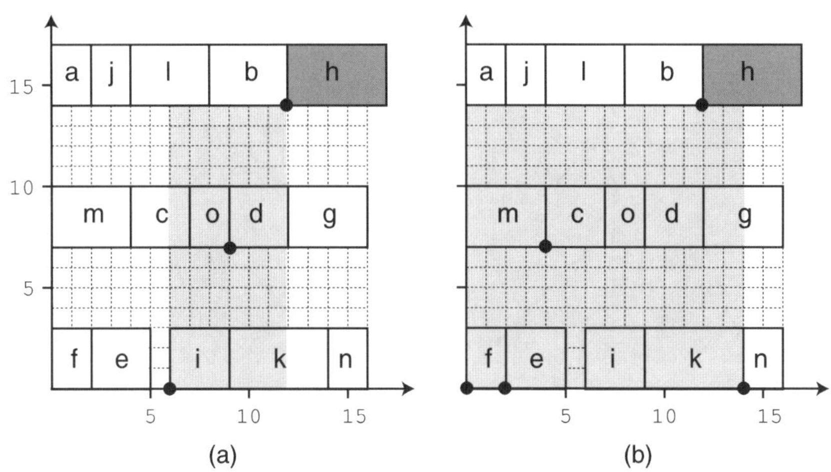

Figure 4.20. (**a**) Bounding box of $n_4 = \{d, h, i\}$ with h on the right boundary, (**b**) bounding box of $n_7 = \{c, e, f, h, n\}$ with h not on any boundary.

Thus, $gradient(h) = 1$. Since h is shifted to the right by 1

$$shift\_amount(h) = 1$$

Thus,

$$\Delta W_S = gradient(h) \cdot shift\_amount(h) = 1 \cdot 1 = 1 \qquad (4.1)$$

Based on the calculation of ΔW and ΔW_S, we get

$$\Delta C = \Delta W + \Delta W_S = 19 + 1 = 20$$

3. How accurate is the "Model B" estimation of ΔW_S in Equation (4.1)?

   From Figure 4.19 we note that $w(n_4) = 19$, $w'(n_4) = 20$, $w(n_7) = 28$, and $w'(n_7) = 28$. Thus, the actual change on the wirelength from shifting h is

   $$w'(n_4) - w(n_4) + w'(n_7) - w(n_7) = 20 - 19 + 28 - 28 = 1$$

   Thus, we see that the estimation results in a correct calculation.

4. Use the "Model A" instead to estimate the wirelength change from shifting cell h.

   Figure 4.21 shows the W_n graph of n_4 and n_7. The "break points" for n_4 are cell i and d, and for n_7 are cell f and n. The wirelength of n_4 and n_7

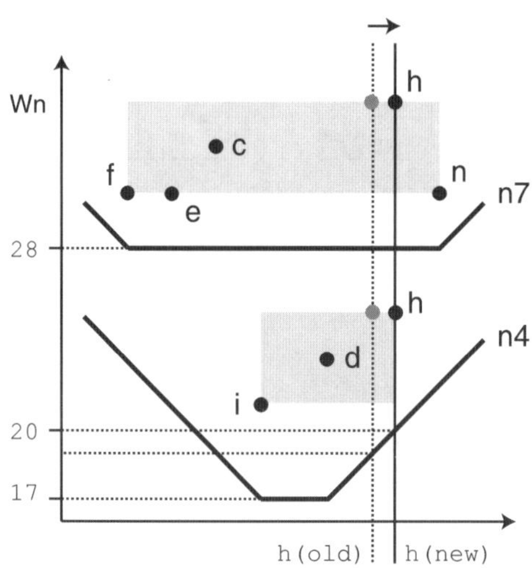

Figure 4.21. Piecewise-linear W_n graph for n_4 and n_7. Cell h is shifted to the right by 1, causing the wirelength of n_4 to increase by one and no change on n_7.

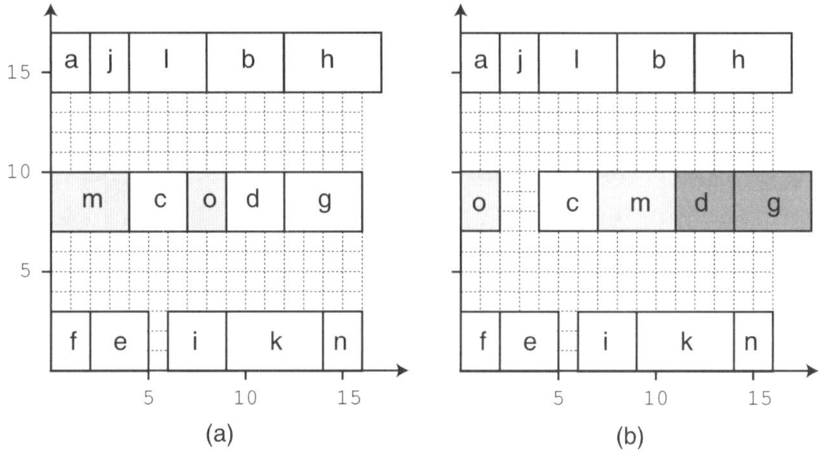

Figure 4.22. (**a**) Before swapping (m, o), (**b**) after the swap. Cell d and g are shifted to the right.

increase outside the range $[i, d]$ and $[f, n]$. We note that cell h is shifted to the right by 1, causing the wirelength of n_4 to increase by one (from 19 to 20) and no change on n_7 (staying at 28).

5. Perform swap (m, o). What is the change on the wirelength cost?

   Figure 4.22 shows the placement before and after the swap. We shift cell d and g.

   (a) Computation of ΔW: The set of nets containing m is $\{n_5, n_9\}$, and o is $\{n_2, n_{10}\}$. Then,

$$\Delta(n_5) = w'(n_5) - w(n_5) = 12 - 11 = 1$$
$$\Delta(n_9) = w'(n_9) - w(n_9) = 22 - 26 = -4$$
$$\Delta(n_2) = w'(n_2) - w(n_2) = 7 - 14 = -7$$
$$\Delta(n_{10}) = w'(n_{10}) - w(n_{10}) = 23 - 23 = 0$$

   Thus,

$$\Delta W = \Delta(n_5) + \Delta(n_9) + \Delta(n_2) + \Delta(n_{10}) = -10$$

   (b) Estimation of ΔW_S: Recall that cell d and g are shifted. First, the nets that contain d are $n_4 = \{d, h, i\}$, $n_6 = \{d, k, j\}$, and $n_8 = \{d, l\}$. Figure 4.23 shows the bounding boxes of these nets. We start with $gradient(d) = 0$:

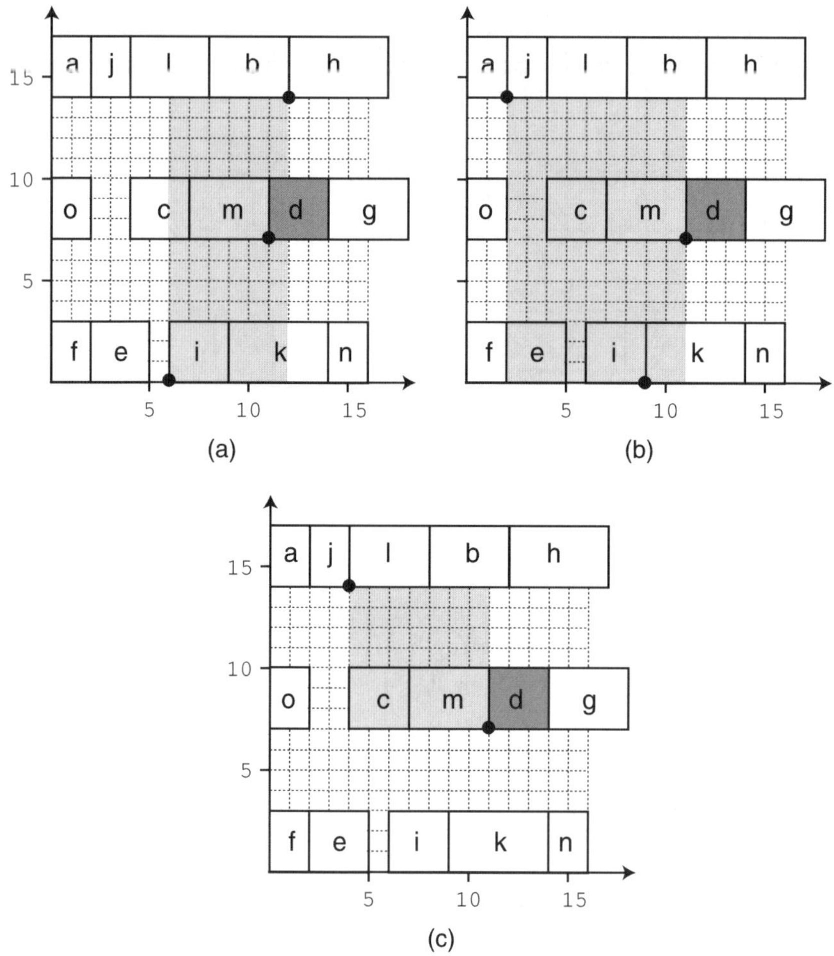

Figure 4.23. (**a**) Bounding box of $n_4 = \{d, h, i\}$, where d is not on any boundary, (**b**) bounding box of $n_6 = \{d, k, j\}$, where d is on the right boundary, (**c**) bounding box of $n_8 = \{d, l\}$, where d is on the right boundary.

- n_4: d is not on any boundary, so we do not change $gradient(d)$.
- n_6: d is on the right boundary, so we increment $gradient(d)$.
- n_8: d is on the right boundary, so we increment $gradient(d)$.

Thus, $gradient(d) = 2$. Second, the nets that contain g are $n_1 = \{a, e, g\}$ and $n_9 = \{b, g, i, m\}$. Figure 4.24 shows the bounding boxes of these nets. We start with $gradient(g) = 0$:

- n_1: g is on the right boundary, so we increment $gradient(g)$.
- n_9: g is on the right boundary, so we increment $gradient(g)$.

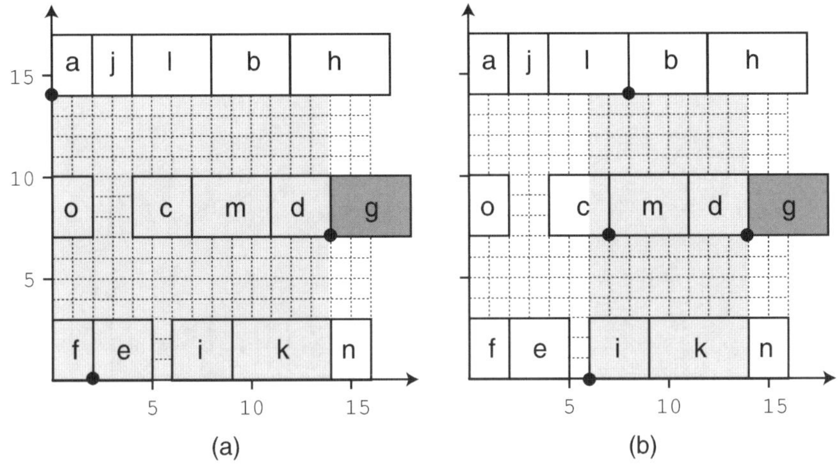

Figure 4.24. (**a**) Bounding box of $n_1 = \{a, e, g\}$, where g is on the right boundary, (**b**) bounding box of $n_9 = \{b, g, i, m\}$, where g is on the right boundary.

We have $gradient(g) = 2$. Both cell d and g are shifted to the right by 2. Thus,

$$\begin{aligned}\Delta W_S &= gradient(d) \cdot shift\_amount(d) + \\ &\quad gradient(g) \cdot shift\_amount(g) = 2 \cdot 2 + 2 \cdot 2 = 8\end{aligned}$$

Based on the calculation of ΔW and ΔW_S, we get

$$\Delta C = \Delta W + \Delta W_S = -10 + 8 = -2$$

6. Perform swap (k, m). What is the change on the wirelength cost?

   Figure 4.25 shows the placement before and after the swap. We shift cell c to the left because the middle row receives the wider cell (= k), and cell o and c are on the shorter side of the row.

   (a) Computation of ΔW: The set of nets containing k is $\{n_3, n_6, n_{10}\}$, and m is $\{n_5, n_9\}$. Then,

$$\Delta(n_3) = w'(n_3) - w(n_3) = 25 - 24 = 1$$
$$\Delta(n_6) = w'(n_6) - w(n_6) = 16 - 23 = -7$$
$$\Delta(n_{10}) = w'(n_{10}) - w(n_{10}) = 13 - 23 = -10$$
$$\Delta(n_5) = w'(n_5) - w(n_5) = 21 - 12 = 9$$
$$\Delta(n_9) = w'(n_9) - w(n_9) = 22 - 22 = 0$$

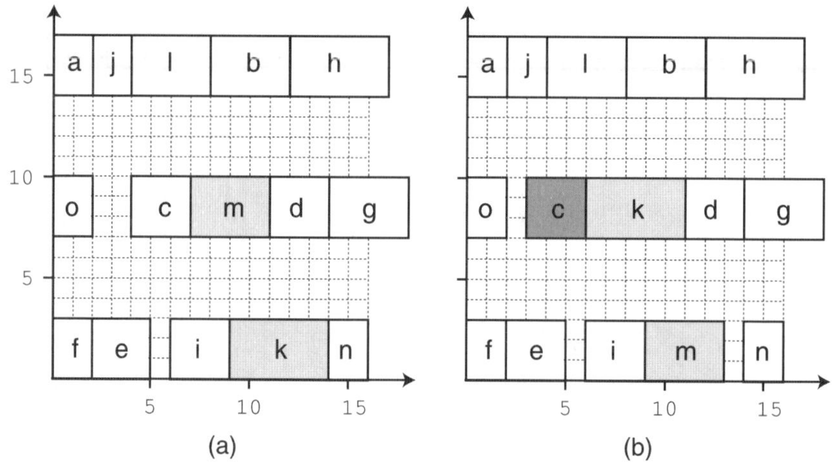

Figure 4.25. (**a**) Before swapping (k, m), (**b**) after the swap. Cell c is shifted to the left.

Thus,

$$\Delta W = \Delta(n_3) + \Delta(n_6) + \Delta(n_{10}) + \Delta(n_5) + \Delta(n_9) = -7$$

(b) Estimation of ΔW_S: The nets that contain c are $n_3 = \{b, c, k, n\}$, and $n_7 = \{c, e, f, h, n\}$. Figure 4.26 shows the bounding boxes of these nets. We start with $gradient(c) = 0$:

- n_3: c is on the left boundary, so we decrement $gradient(c)$.
- n_7: c is not on any boundary, so we do not change $gradient(c)$.

Thus, $gradient(c) = -1$. Since c is shifted to the left by 1,

$$shift\_amount(c) = -1$$

Lastly,

$$\Delta W_S = gradient(c) \cdot shift\_amount(c) = -1 \cdot -1 = 1$$

Based on the calculation of ΔW and ΔW_S, we get

$$\Delta C = \Delta W + \Delta W_S = -7 + 1 = -6$$

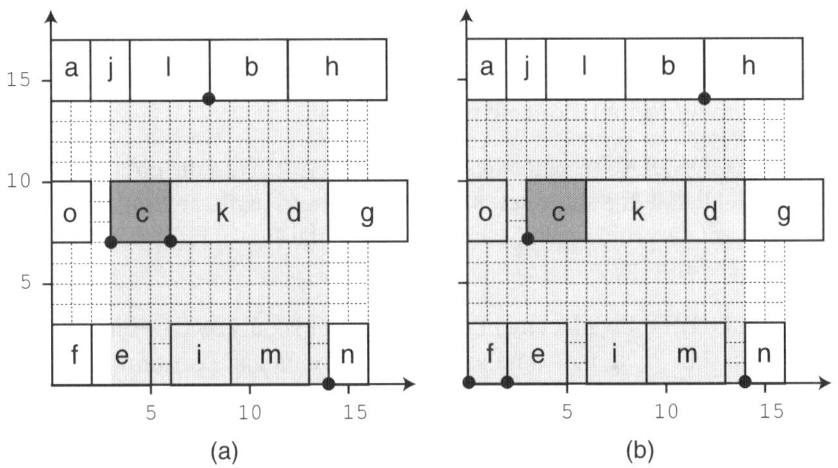

Figure 4.26. (**a**) Bounding box of $n_3 = \{b, c, k, n\}$, where c is on the left boundary, (**b**) bounding box of $n_7 = \{c, e, f, h, n\}$, where c is not on any boundary.

4. More Practice Problems

1. Perform quadrature mincut placement on the circuit shown in Figure 4.27 starting with a vertical cut first, and place the gates into 2×4 grid. Show the placement after each cut. The area skew is set to zero.

2. Perform the recursive bipartitioning mincut on the circuit shown in Figure 4.27 starting with a vertical cut first, and place the gates into 2×4 grid. Use terminal propagation, where the terminals located within the mid-third window should be ignored. Show the placement after each cut. The area skew is set to zero.

3. Perform GORDIAN placement at the optimization level 3 by adding four vertical cutlines into the partitioning result shown in Figure 4.16 and Figure 4.17. Partitions p_1[12] and p_3 are to be divided evenly, p_2 into 2-to-1 (= left partition contains 2 cells), and p_4 into 1-to-2. Show the quadratic programming formulation and placement figure.

4. Perform GORDIAN-L placement [Sigl et al., 1991a] on the circuit shown in Figure 4.11. Use the same set of assumptions and partitioning patterns used in the practice problem in Section 2. Show the first three-level placement results ($l = 0, 1, 2$).

5. Perform GORDIAN placement (first three levels only) on the circuit shown in Figure 4.27. Use the same set of assumptions and partitioning patterns used in the practice problem in Section 2. Assume that the IO pads are fixed at the following locations: $w_1(0, 2)$, $w_2(0, 3)$, $w_3(1, 4)$, $z_1(4, 1)$, $z_2(4, 2)$.

6. Consider the gate-level netlist shown in Table 4.2 and its standard cell placement shown in Figure 4.19(a). Perform the following swaps in the given order and compute the change in the wirelength:

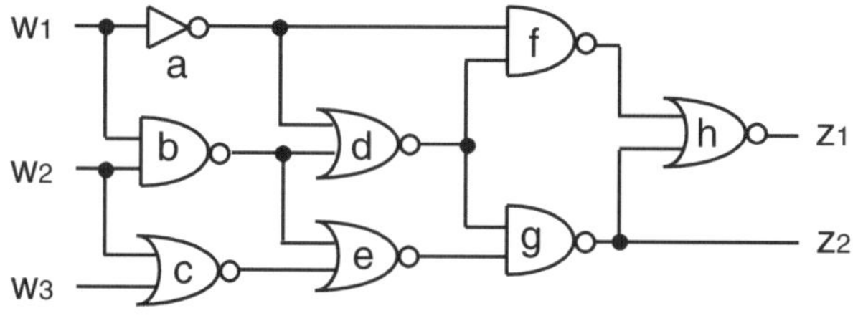

Figure 4.27. A gate-level circuit.

---
[12] Break the tie by partitioning c to the left.

(a) Swap (a, k)

(b) Swap (h, j)

(c) Swap (m, f)

Use the "Model B" of TimberWolf 7.0 [Sun and Sechen, 1995] when estimating the wirelength change from cell shifting. Use the lower left corner of each cell to represent the cell location. When selecting the cells to be shifted, try to minimize the number of cells being shifted. The x-coordinates of the cells should be non-negative.

5. Probing Further

Disclaimer: The list here is meant to be representative, not comprehensive. A comprehensive survey on circuit placement algorithms is provided in [Shahookar and Mazumder, 1991; Cong et al., 2005; Nam and Cong, 2007].

Mincut Placement

The authors of [Suaris and Kedem, 1989] developed a standard cell placement procedure based on recursively partitioning the netlist into four parts (instead of two parts as in the traditional approaches), while minimizing the partitioning cost. Their cost function prefers vertical wiring compared to horizontal connections, which is applicable in standard cell placement.

The authors of [Zhong and Dutt, 2000] improved the traditional mincut placement in the following ways: (1) they partition the entire circuit globally in every level of the partitioning tree, across the current cutlines, (2) cell gain computation is based on a global view of entire nets, thereby obviating terminal propagation [Dunlop and Kernighan, 1985], (3) cell gain is minimizing the half-perimeter cost instead of cutsize.

Capo [Caldwell et al., 2000a] is a congestion-driven recursive bisection placer. It incorporates a multilevel min-cut partitioner [Alpert et al., 1998], techniques for partitioning with small tolerance [Caldwell et al., 2000b], and end-case min-wirelength placers [Caldwell et al., 2000c]. The authors show that despite a potential mismatch of objectives, improved mincut bisection can still lead to improved wirelength and congestion.

The authors of [Ou and Pedram, 2000] presented a timing-driven mincut placer that is based on solving a quadratic programming problem iteratively. In its timing-driven partitioning, the authors control the number of times that a path in the circuit can be cut. In addition, a pre-locking mechanism and timing-aware terminal propagation are integrated into the flow. The detailed placement step is formulated as a constrained quadratic program and solved efficiently.

Dragon [Wang et al., 2000] performs placement in two steps: global and detailed placement. During the global placement step, a hMetis-based [Karypis et al., 1997] top-down quadrisection and terminal propagation is performed first, followed by simulated annealing based refinement. During the detailed placement step, cell overlaps are removed first, and a greedy cell exchange algorithm is used to further reduce wirelength.

Fengshui [Yildiz and Madden, 2001] improves the traditional mincut placement by adopting a dynamic programming approach to cut sequence generation. The authors developed a mathematical foundation for wirelength estimation in mincut placement, allowing the determination of an optimal cut sequence under a simplified model. Based on this study of the optimal sequences, they proposed a simple method to construct sequences that are near optimal.

The authors of [Alpert et al., 2003b] proposed an enhancement to mincut placement called analytic constraint generation (ACG). ACG utilizes an analytic engine to distribute available free space appropriately by determining balance constraints for each partitioning step. This is useful in handling IP blocks and large macro cells since they tend to cause an increase in the available free space.

The authors of [Kahng and Reda, 2006] improved the accuracy of terminal propagation by a "placement feedback" mechanism. At each level of mincut placement, all blocks are partitioned first. Then they "undo" all the partitioning but keep the node locations as assigned by the partitioning. Lastly, they use these node locations to "redo" terminal propagation and block partitioning. This yields substantial reductions in wirelength and congestion.

GORDIAN Algorithm

The quadratic wirelength objective in Gordian [Kleinhans et al., 1991] is replaced with a linear wirelength objective in Gordian-L [Sigl et al., 1991b]. It is generally believed that the quadratic objective tends to make long nets shorter than the linear objective at the expense of short nets. Thus, the total wirelength tends to be shorter with the linear objective. The authors presented an iterative method to minimize the linear wirelength using quadratic programming.

The author of [Vygen, 1997] combined quadratic placement with quadrisection approach. The quadrisection partitions the circuit into 4 regions based on the positions obtained by the quadratic optimization so that the capacity constraint is satisfied and the total displacement is minimized. The goal is to preserve the placement quality obtained with the quadratic placement while removing the overlap among the cells.

Kraftwerk [Eisenmann and Johannes, 1998] is a force-directed global placer. Besides the traditional wirelength dependent forces, which is formulated as a quadratic objective, the authors use additional forces to reduce cell overlaps and to consider the placement area. This iterative approach is shown to be flexible to handle various objectives such as area, timing, congestion and heat distribution.

Mongrel [Hur and Lillis, 2000] is a two-step standard cell placer. Their global placement is based on the Relaxation-Based Local Search (RBLS) framework, in which a combinatorial search mechanism is driven by an analytical engine. This enables a more global view of the problem and results in complex modifications of the placement in a single search move. When a global placement has converged, a detailed placement is formed and further optimized by an optimal interleaving technique.

mPL [Chan et al., 2003] is a multi-level placer, where the original circuit is coarsened in a bottom-up hierarchical fashion and placed in a top-down recursive manner. The enhancements added to mPL include (1) unconstrained quadratic relaxation at every level of the hierarchy; (2) improved interpolation (de-clustering) method, and (3) better quadratic placement refinement using additional geometric information for aggregation in subsequent refinement.

Grid-warping [Xiu et al., 2004] is a unique placement algorithm based on a simple idea: rather than moving the gates to optimize their location, they elastically deform the 2D grid surface on which the gates have been roughly placed. They stretch the grid until the gates arrange themselves to meet the goals. Deforming the elastic grid is a simple, low-dimensional nonlinear optimization, and augments a traditional quadratic formulation.

APlace [Kahng and Wang, 2005] is a multi-level analytic placer for mixed-size placement. The authors presented an effective objective function for spreading cells over the placement area while reducing wirelength and congestion. Timing objective is also considered during placement. They extended the placer to perform I/O-core co-placement and handle various geometric constraints for mixed-signal designs.

mFAR [Hu and Marek-Sadowska, 2005] is a multi-level analytic placer, where the placement problem is formulated as a quadratic program with non-linear constraints. Since these non-linear constraints are difficult to handle, the authors solve the unconstrained version that causes overlaps among the cells. The authors introduce fixed points into the non-constrained quadratic-programming formulation. These fixed points act as pseudo-cells to pull cells away from the dense regions to reduce overlapping.

FastPlace [Viswanathan and Chu, 2005] is a quadratic placer, where the wirelength minimization objective is formulated as a convex quadratic program. The authors presented an efficient cell shifting technique to remove cell overlap from the quadratic program solution and also accelerate the convergence of the solver. They also proposed a hybrid net model that is a combination of the traditional clique and star models. This net model results in a significant speedup of the solver.

DPlace [Ren et al., 2007] is a diffusion-based placement refinement method based on a discrete approximation to the closed-form solution of the continuous diffusion equation. It has the advantage of smooth spreading, which helps preserve neighborhood characteristics of the original placement. Applying this technique to placement legalization demonstrates significant improvements in wirelength and timing compared with other commonly used techniques.

TimberWolf Algorithm

The authors of [Tsay and Lin, 1995] presented a standard cell placer that is based on cone-based clustering. They first extract cones based on signal direction and group them to form fragments. They then perform a macro-cell placement using TimberWolf-MC [Swartz and Sechen, 1990], treating each cone as a soft macro. Next, they map the resulting macro-cell placement into a row-based placement. Finally, they apply TimberWolf-SC [Swartz and Sechen, 1990] to refine the row-based placement.

TimberWolf 7.0 [Sun and Sechen, 1995] was extended to perform timing optimization by the authors of [W. Swartz, 1995]. They used a timing graph to obtain path delay equations and added to the annealing cost function to address wirelength and timing optimization simultaneously. During the annealing process, they primarily target a set of critical paths for delay improvement, which is regularly updated at each iteration.

The authors of [Sun and Sechen, 1997] presented a parallel version of TimberWolf 7.0 [Sun and Sechen, 1995]. Their algorithm is targeted toward networks of Unix workstations connected with a local area network. This is the first reported parallel algorithm for standard cell placement which yields as good or better placement results than its serial version. The processor utilization is high, up to 98% for two processors and 90% for six processors.

The authors of [Chandy et al., 1997] presented another parallel version of TimberWolf-SC [Sechen and Sangiovanni-Vincentelli, 1985]. They proposed a parallel move strategy that is based on dynamic message sizing, message prioritization, and error control. They also investigated two approaches to parallel cell placement: multiple Markov chains and speculative computation. They show that parallel moves and multiple Markov chains are effective approaches to parallel simulated annealing, yet speculative computation is inadequate.

MGP [Chang et al., 2003] is a multi-level global placer for congestion management. MGP integrates a fast incremental global routing for accurately updating and optimizing congestion cost during physical hierarchy generation. A hierarchical area density control is developed for placing objects with significant size variations. TimberWolf 7.0 [Sun and Sechen, 1995] is used to place the clusters at the top-level hierarchy.

Chapter 5

STEINER ROUTING

Given a set of points P in a 2D plane, the Steiner tree problem seeks a set of additional points S so that the wirelength of minimum spanning tree (MST) of $P \cup S$ is minimum. Additional objectives include performance-related metrics such as radius and delay. The rectilinear version of this problem—rectilinear Steiner tree (RST)—has an important application in VLSI routing and has seen a huge volume of works. This chapter presents sample problems related to the following works:

- L-shaped Steiner routing algorithm [Ho et al., 1990]
- 1-Steiner algorithms by Kahng and Robins [Kahng and Robins, 1992] and by Borah, Owens, and Irwin [Borah et al., 1994]
- BPRIM (Bounded Prim) and BRBC (Bounded Radius Bounded Cost) algorithms [Cong et al., 1992]
- A-tree algorithm [Cong et al., 1993]
- ERT (Elmore Routing Tree) and SERT (Steiner Elmore Routing Tree) algorithms [Boese et al., 1995]

The first two works focus on wirelength minimization, while the last three are delay-oriented works. The first work constructs an MST first and rectilinearize it by transforming the edges in the MST into L-shapes. The second work transforms a given MST into a Steiner tree by adding Steiner points one-by-one. The third work addresses the wirelength and delay (= measured by so called radius) trade-off during rectilinear MST construction. The fourth work builds the minimum-cost rectilinear Steiner arborescence, where the source-sink path length is the shortest for all sinks. The last work constructs Steiner trees that directly minimize Elmore delay objective.

S.K. Lim, *Practical Problems in VLSI Physical Design Automation*

1. L-Shaped Steiner Routing Algorithm

One of the most popular ways to solve the RST (Rectilinear Steiner Tree) problem is to construct a R-MST (Rectilinear Minimum Spanning Tree) on the original set of points and refine this tree in some way. The rationale behind this approach is the well-known fact that the quality of any R-MST is not worse than 3/2 of that of the *optimal* RST [Hwang, 1976]. In addition, the construction of R-MST can be done efficiently by the classical Prim [Prim, 1957] or Kruskal [Kruskal, 1956] algorithm.

The L-RST (L-shaped Rectilinear Steiner Tree) algorithm [Ho et al., 1989; Ho et al., 1990] refines an initial R-MST by replacing the Euclidean edges in the initial R-MST with either an upper or a lower rectilinear L-shape. The goal is to maximize the overlap between the L-shapes and the existing tree during the replacement. If done naively, the complexity of optimally replacing the edges for wirelength minimization is $O(2^n)$, where n is the number of edges in the initial R-MST. However, L-RST algorithm solves this problem in *linear time* by exploiting the "separability" property, which states that it is possible to build an R-MST so that the L-shapes among non-adjacent edges do not overlap with each other. This property, together with the well-known fact that the maximum degree among all nodes in any R-MST is 6, enables an efficient (= linear time) recursive method for an optimal L-shape replacement.

Quick Overview

The first step of L-RST algorithm is to build a separable MST. Given a set of points P in a 2D plane, we first construct the $|P|$-clique using the points in P. Each edge (i, j) in this complete graph is associated with the following 3-tuple as its weight[13]:

$$weight(i, j) = (D(i, j), -|y(i) - y(j)|, -\max\{x(i), x(j)\}) \qquad (5.1)$$

where $D(i, j)$ is the rectilinear distance between i and j, and $x(i)/y(i)$ is the x/y coordinate of i. This 3-tuple prefers the edges that are shorter, taller, and located further right during the MST construction. We apply Prim [Prim, 1957] or Kruskal [Kruskal, 1956] algorithm using the 3-tuples to construct a separable MST.

Next, we choose any node in the separable MST as the root and build a rooted tree. The process of converting the edges in an initial separable MST into L-shape consists two phases, namely, bottom-up and top-down tree traversal. Given a node v during the bottom-up traversal of a rooted tree, let T_v

[13]The authors first suggested 4-tuple in their conference version [Ho et al., 1989] and later changed it to 3-tuple in their journal version [Ho et al., 1990].

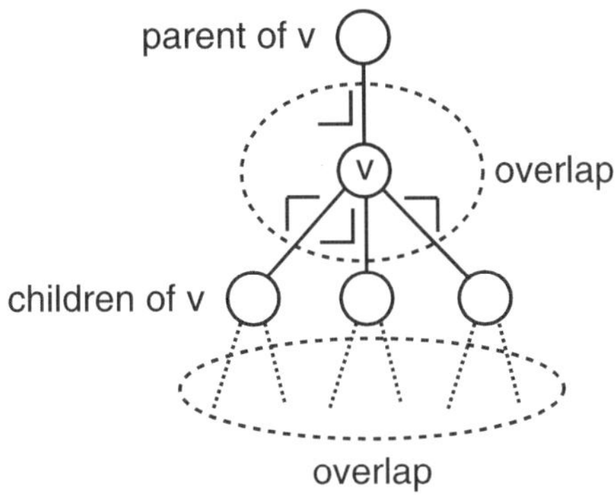

Figure 5.1. Two sources of overlap in L-RST: (1) among the edges incident on v, (2) overlaps in the sub-trees rooted at the children of v.

denote the sub-tree rooted at v, e_v denote the edge between v and its parent, and $T_v^+ = T_v \cup e_v$.[14] Our goal is to compute $\Phi_l(v)$ and $\Phi_u(v)$, where $\Phi_l(v)$ denotes the L-RST of T_v^+ with the minimum wirelength (= maximum overlap) under the constraint that e_v is fixed to lower-L shape. $\Phi_u(v)$ is defined similarly except that we use upper-L for e_v. The total amount of overlap in T_v^+ is the sum of the following two factors:

- Z_1: overlap in T_v^S, where T_v^S denotes the set of edges incident to v.
- Z_2: overlap in the $\Phi(w_i)$, where w_i denotes the children of v.

Figure 5.1 provides an illustration. Given a certain set of L-shape choices for the edges in T_v^S, we first compute Z_1. Then, for each edge (v, w_i), we compute Z_2 by utilizing $\Phi_l(w_i)$ or $\Phi_u(w_i)$ that we have already computed during the bottom-up traversal. Lastly, we compute $\Phi_l(v)$ by fixing e_v to lower-L and examining all 2^d L-shape choices for the children, where d is the total number of the children. We compute $\Phi_u(v)$ similarly using upper-L for e_v. When we reach the root node, we first choose the L-shape orientations for the edges incident to the root node that result in the maximum overlap. We then visit the rest of the nodes in the tree in top-down fashion to retrieve their L-shape orientations (see the sample problem for details).

Practice Problem

Consider the routing problem instance shown in Figure 5.2.

[14]We use the terminologies used in the conference version [Ho et al., 1989] instead of the journal version [Ho et al., 1990].

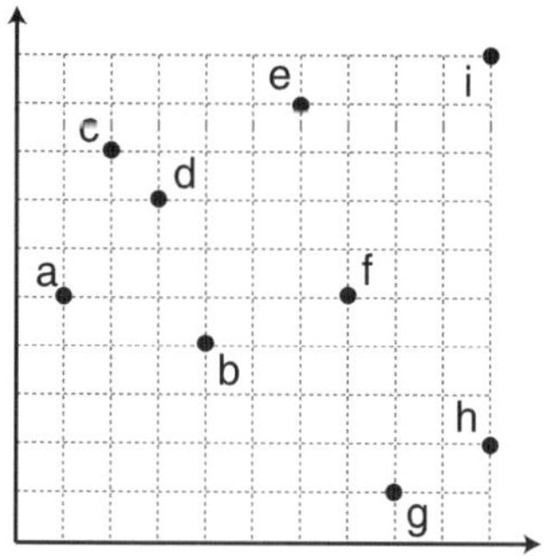

Figure 5.2. Routing problem instance for L-RST algorithm. Node b is the root node for the separable MST and L-RST computation.

1. Construct a separable MST using node b as the root.

   We perform Prim's MST algorithm [Prim, 1957] using the 3-tuple weights defined in Equation (5.1). Edges that are shorter, taller, and located further right have the priority. We initially set our separable MST $T = \{b\}$.

   (a) Iteration 1: T contains three nearest neighbors: a, d, and f. These nodes can connect to T via the following edges (sorted based on their weights):

   - $(b, d) = (4, -3, -4)$
   - $(b, f) = (4, -1, -7)$
   - $(b, a) = (4, -1, -4)$

   Thus, we add (b, d) to T based on this lexicographical order. The resulting tree is shown in Figure 5.3(a).

   (b) Iteration 2: Node c is the nearest neighbor of the tree T shown in Figure 5.3(a). Thus, we add (c, d) to T. The resulting tree is shown in Figure 5.3(b).

   (c) Iteration 3: T shown in Figure 5.3(b) contains two nearest neighbors: a and f. These nodes can connect to T via the following edges (sorted based on their weights):

   - $(a, c) = (4, -3, -2)$
   - $(a, d) = (4, -2, -3)$
   - $(b, f) = (4, -1, -7)$
   - $(b, a) = (4, -1, -4)$

   Thus, we add (a, c) to T based on this lexicographical order. The resulting tree is shown in Figure 5.3(c).

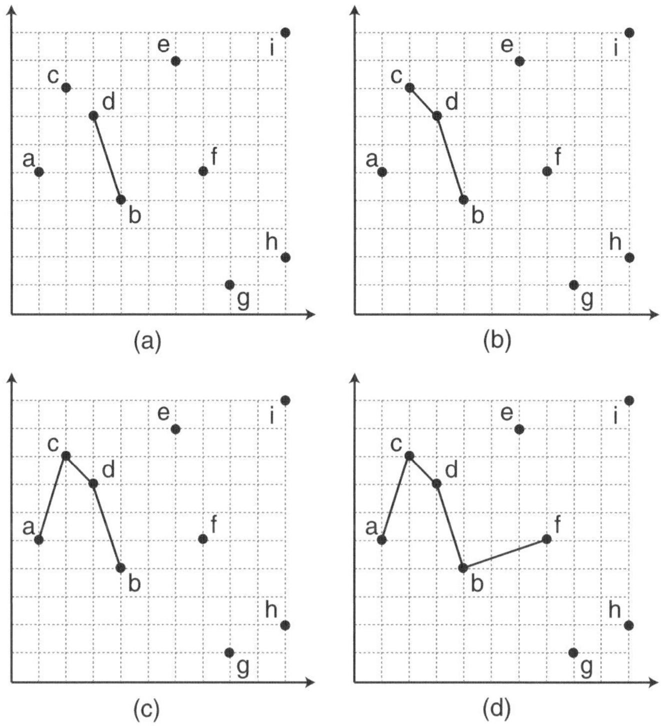

Figure 5.3. Adding the first four edges to the separable MST.

(d) Iteration 4: Node f is the nearest neighbor of the tree T shown in Figure 5.3(c). Thus, we add (b, f) to T. The resulting tree is shown in Figure 5.3(d).

(e) Iteration 5: T shown in Figure 5.3(d) contains two nearest neighbors: e and g. These nodes can connect to T via the following edges (sorted based on their weights):

- $(f, g) = (5, -4, -8)$
- $(f, e) = (5, -4, -7)$
- $(d, e) = (5, -2, -6)$

Thus, we add (f, g) to T based on this lexicographical order. The resulting tree is shown in Figure 5.4(e).

(f) Iteration 6: Node h is the nearest neighbor of the tree T shown in Figure 5.4(e). Thus, we add (g, h) to T. The resulting tree is shown in Figure 5.4(f).

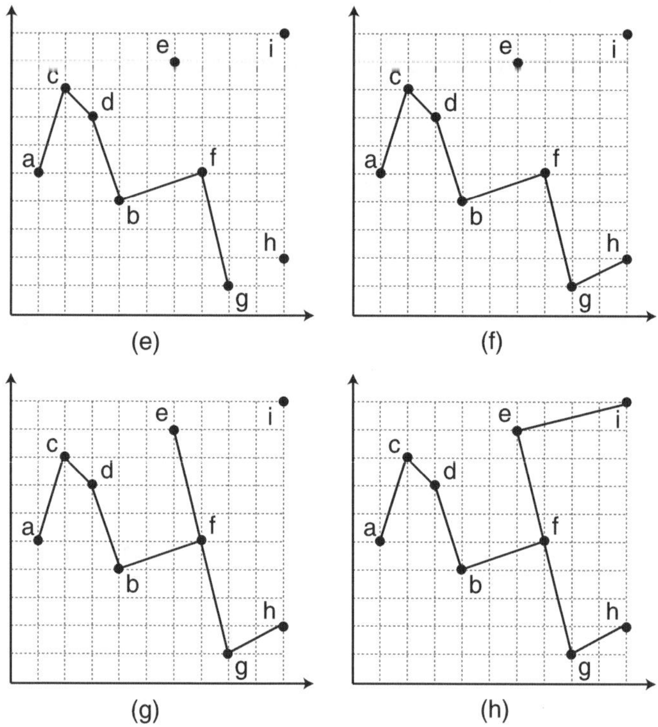

Figure 5.4. Adding the last four edges to the separable MST (continue from Figure 5.3).

(g) Iteration 7: Node e is the nearest neighbor of the tree T shown in Figure 5.4(f). This node can connect to T via the following edges (sorted based on their weights):

- $(f, e) = (5, -4, -7)$
- $(d, e) = (5, -2, -6)$

Thus, we add (f, e) to T based on this lexicographical order. The resulting tree is shown in Figure 5.4(g).

(h) Iteration 8: we add (e, i) to T. The resulting tree is shown in Figure 5.4(h).

Figure 5.4(h) shows the final separable MST with the wirelength of 32.

2. Perform Ho, Vijayan, and Wong algorithm and obtain an optimal L-RST using b as the root node.

   Figure 5.5 shows T_b, the tree rooted at node b. Let $Z(T)$ denote the amount of overlap in tree T. We visit each non-leaf node $v \in T_b$ in bottom-up fashion and compute $\Phi_l(v)$ and $\Phi_u(v)$ as follows:

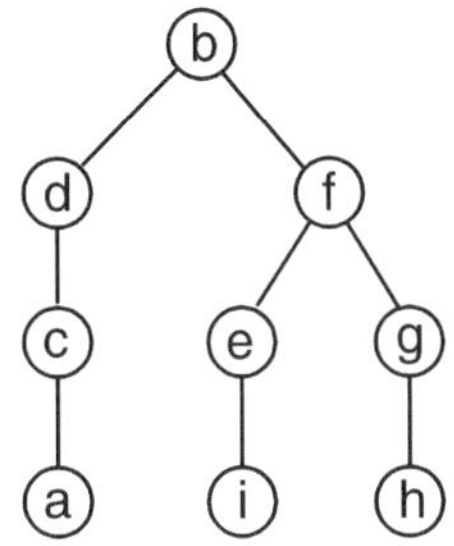

Figure 5.5. Rooted tree T_b derived from the separable MST.

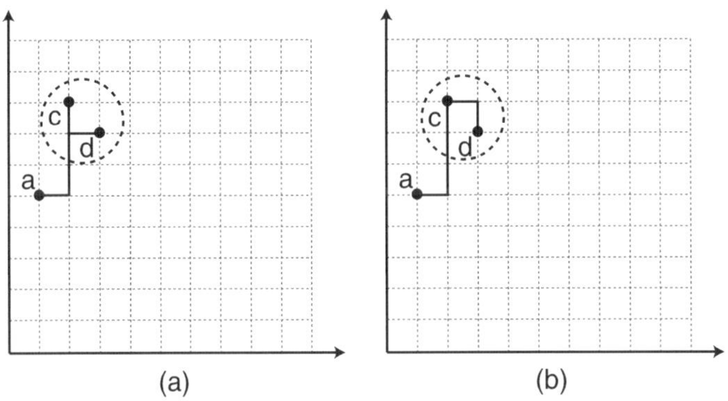

Figure 5.6. Partial L-RSTs for node c, where $e_c = (c, d)$. (**a**) $\Phi_l(c)$, (**b**) $\Phi_u(c)$.

(a) Node c: First, we obtain the edge $e_c = (c, d)$.

- $\Phi_l(c)$: (c, d) is fixed to lower-L. Figure 5.6(a) shows $\Phi_l(c)$. We assign lower-L to (c, a) in order to maximize the overlap in T_c^S. Thus, $Z(\Phi_l(c)) = 1$.
- $\Phi_u(c)$: (c, d) is fixed to upper-L. Figure 5.6(b) shows $\Phi_u(c)$. The orientation of (c, a) is irrelevant because no overlap occurs in T_c^S. Thus, $Z(\Phi_u(c)) = 0$.

(b) Node e: First, we obtain the edge $e_e = (e, f)$.

- $\Phi_l(e)$: (e, f) is fixed to lower-L. Figure 5.7(a) shows $\Phi_l(e)$. The orientation of (e, i) is irrelevant because no overlap occurs in T_e^S. Thus, $Z(\Phi_l(e)) = 0$.
- $\Phi_u(e)$: (e, f) is fixed to upper-L. Figure 5.7(b) shows $\Phi_u(e)$. We assign lower-L to (e, i) in order to maximize the overlap in T_e^S. Thus, $Z(\Phi_u(e)) = 1$.

(c) Node g: First, we obtain the edge $e_g = (g, f)$.

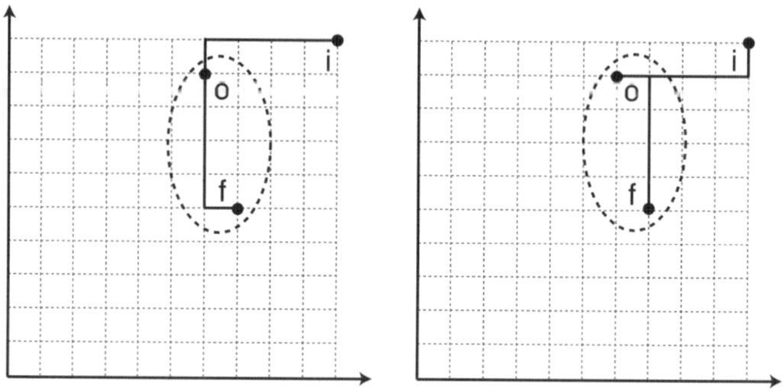

Figure 5.7. Partial L-RSTs for node e, where $e_e = (e, f)$. **(a)** $\Phi_l(e)$, **(b)** $\Phi_u(e)$.

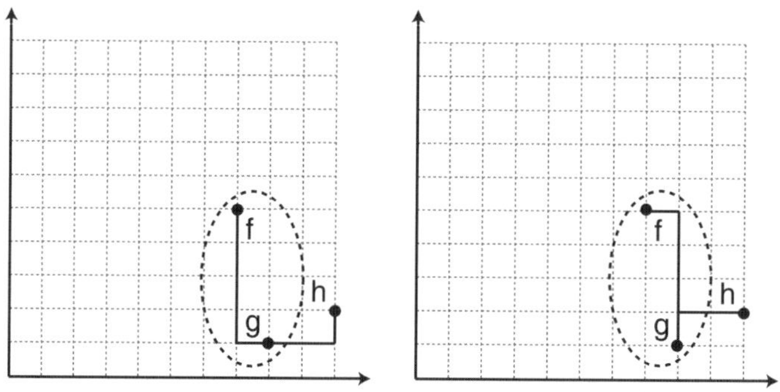

Figure 5.8. Partial L-RSTs for node g, where $e_g = (g, f)$. **(a)** $\Phi_l(g)$, **(b)** $\Phi_u(g)$.

- $\Phi_l(g)$: (g, f) is fixed to lower-L. Figure 5.8(a) shows $\Phi_l(g)$. The orientation of (g, h) is irrelevant because no overlap occurs in T_g^S. Thus, $Z(\Phi_l(g)) = 0$.
- $\Phi_u(g)$: (g, f) is fixed to upper-L. Figure 5.8(b) shows $\Phi_u(g)$. We assign upper-L to (g, h) in order to maximize the overlap in T_g^S. Thus, $Z(\Phi_u(g)) = 1$.

(d) Node d: First, we obtain the edge $e_d = (d, b)$.

- $\Phi_l(d)$: (d, b) is fixed to lower-L. Figure 5.9(a) shows $\Phi_l(d)$. There is no overlap in T_d^S, but we choose lower-L for (d, c) because $Z(\Phi_l(c)) > Z(\Phi_u(c))$ from Part (a). Thus,

$$
\begin{aligned}
Z(\Phi_l(d)) &= Z(T_d^S) + \max\{Z(\Phi_l(c)), Z(\Phi_u(c))\} \\
&= 0 + \max\{1, 0\} = 1
\end{aligned}
$$

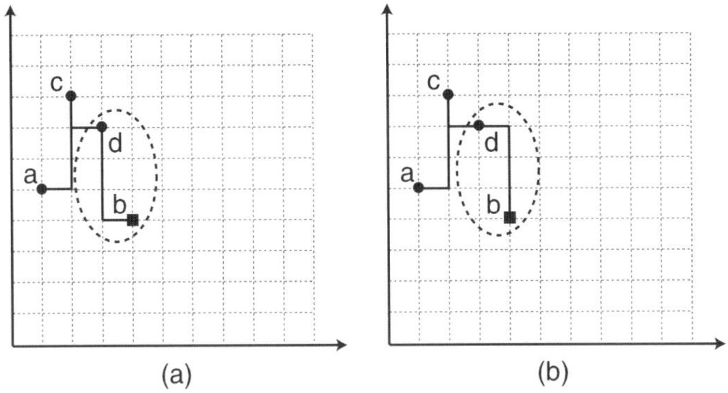

Figure 5.9. Partial L-RSTs for node d, where $e_d = (d, b)$. **(a)** $\Phi_l(d)$, **(b)** $\Phi_u(d)$.

- $\Phi_u(d)$: (d, b) is fixed to upper-L. Figure 5.9(b) shows $\Phi_u(d)$. There is no overlap in T_d^S, but we choose lower-L for (d, c) because $Z(\Phi_l(c)) > Z(\Phi_u(c))$ from Part (a). Thus,

$$\begin{aligned} Z(\Phi_u(d)) &= Z(T_d^S) + \max\{Z(\Phi_l(c)), Z(\Phi_u(c))\} \\ &= 0 + \max\{1, 0\} = 1 \end{aligned}$$

(e) Node f: First, we obtain the edge $e_f = (f, b)$.

- $\Phi_l(f)$: (f, b) is fixed to lower-L. Since f has two children e and g, we consider the following four cases to compute the total amount of overlap in $\Phi_l(f)$. We use the results from Part (b) and Part (c):

  (i) lower-L for both (f, e) and (f, g): we have

$$\begin{aligned} Z(\Phi_l(f)) &= Z(T_f^S) + Z(\Phi_l(e)) + Z(\Phi_l(g)) \\ &= 1 + 0 + 0 = 1 \end{aligned}$$

  (ii) lower-L for (f, e) and upper-L for (f, g): we have

$$\begin{aligned} Z(\Phi_l(f)) &= Z(T_f^S) + Z(\Phi_l(e)) + Z(\Phi_u(g)) \\ &= 0 + 0 + 1 = 1 \end{aligned}$$

  (iii) upper-L for (f, e) and lower-L for (f, g): we have

$$\begin{aligned} Z(\Phi_l(f)) &= Z(T_f^S) + Z(\Phi_u(e)) + Z(\Phi_l(g)) \\ &= 1 + 1 + 0 = 2 \end{aligned}$$

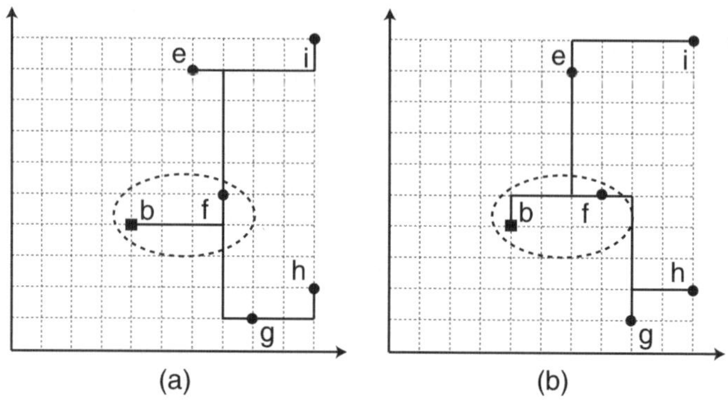

Figure 5.10. Partial L-RSTs for node f, where $e_f = (f, b)$. **(a)** $\Phi_l(f)$, **(b)** $\Phi_u(f)$. Node f has two children e and g.

(iv) upper-L for both (f, e) and (f, g): we have

$$Z(\Phi_l(f)) = Z(T_f^S) + Z(\Phi_u(e)) + Z(\Phi_u(g))$$
$$= 0 + 1 + 1 = 2$$

Thus, the maximum overlap occurs in two cases. We break the tie randomly and choose the first case, where we use upper-L for (f, e) and lower-L for (f, g). This $\Phi_l(f)$ is shown in Figure 5.10(a).

- $\Phi_u(f)$: (f, b) is fixed to upper-L. We again consider the following four cases to compute the total amount of overlap in $\Phi_u(f)$:

(i) Lower-L for both (f, e) and (f, g): we have

$$Z(\Phi_u(f)) = Z(T_f^S) + Z(\Phi_l(e)) + Z(\Phi_l(g))$$
$$= 1 + 0 + 0 = 1$$

(ii) Lower-L for (f, e) and upper-L for (f, g): we have

$$Z(\Phi_u(f)) = Z(T_f^S) + Z(\Phi_l(e)) + Z(\Phi_u(g))$$
$$= 1 + 0 + 1 = 2$$

(iii) Upper-L for (f, e) and lower-L for (f, g): we have

$$Z(\Phi_u(f)) = Z(T_f^S) + Z(\Phi_u(e)) + Z(\Phi_l(g))$$
$$= 0 + 1 + 0 = 1$$

(iv) Upper-L for both (f, e) and (f, g): we have

$$Z(\Phi_u(f)) = Z(T_f^S) + Z(\Phi_u(e)) + Z(\Phi_u(g))$$
$$= 0 + 1 + 1 = 2$$

Thus, the maximum overlap occurs in two cases. We break the tie randomly and choose the first case, where we use lower-L for (f, e) and upper-L for (f, g). This $\Phi_u(f)$ is shown in Figure 5.10(b).

(f) Node b: b has two children d and f. So, we examine the following four cases and pick the best solution. We use the results from Part (d) and Part (e):

(i) Lower-L for both (b, d) and (b, f): we have

$$\begin{aligned} Z(\Phi(b)) &= Z(T_b^S) + Z(\Phi_l(d)) + Z(\Phi_l(f)) \\ &= 0 + 1 + 2 = 3 \end{aligned}$$

(ii) Lower-L for (b, d) and upper-L for (b, f): we have

$$\begin{aligned} Z(\Phi(b)) &= Z(T_b^S) + Z(\Phi_l(d)) + Z(\Phi_u(f)) \\ &= 0 + 1 + 2 = 3 \end{aligned}$$

(iii) Upper-L for (b, d) and lower-L for (b, f): we have

$$\begin{aligned} Z(\Phi(b)) &= Z(T_b^S) + Z(\Phi_u(d)) + Z(\Phi_l(f)) \\ &= 0 + 1 + 2 = 3 \end{aligned}$$

(iv) Upper-L for both (b, d) and (b, f): we have

$$\begin{aligned} Z(\Phi(b)) &= Z(T_b^S) + Z(\Phi_u(d)) + Z(\Phi_u(f)) \\ &= 1 + 1 + 2 = 4 \end{aligned}$$

Thus, the maximum overlap of 4 occurs when we assign upper-L for both (b, d) and (b, f).

3. Construct the final L-RST.

We visit each node $v \in T_b$ shown in Figure 5.5 in top-down fashion and obtain the L-shape orientations of the edges in T_b.

(a) Node b: we choose $\Phi_u(d)$ and $\Phi_u(f)$ to maximize the overlap. Thus, we assign upper-L to both (d, b) and (f, b).

(b) Node d: $\Phi_u(d)$ requires $\Phi_l(c)$. Thus, we assign lower-L to (c, d).

(c) Node f: $\Phi_u(f)$ requires $\Phi_l(e)$ and $\Phi_u(g)$. Thus, we assign lower-L to (e, f) and upper-L to (g, f).

(d) Node c: $\Phi_l(c)$ requires $\Phi_l(a)$. Thus, we assign lower-L to (a, c).

(e) Node e: $\Phi_l(e)$ requires either $\Phi_l(i)$ or $\Phi_u(i)$. Thus, we can assign either lower-L or upper-L to (i, e).

(f) Node g: $\Phi_u(g)$ requires $\Phi_u(h)$. Thus, we assign upper-L to (h, g).

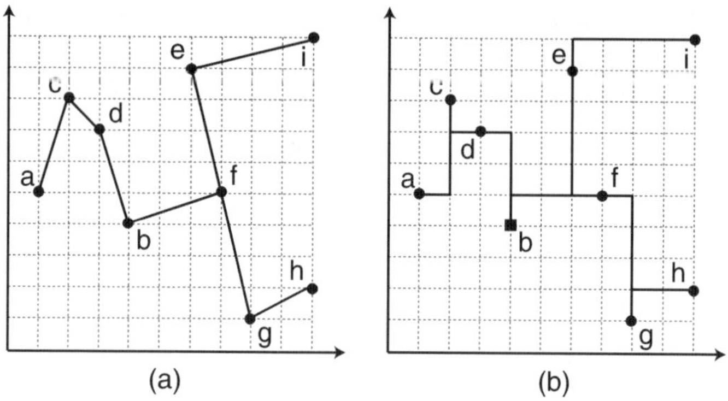

Figure 5.11. (**a**) Initial separable MST, (**b**) final L-RST.

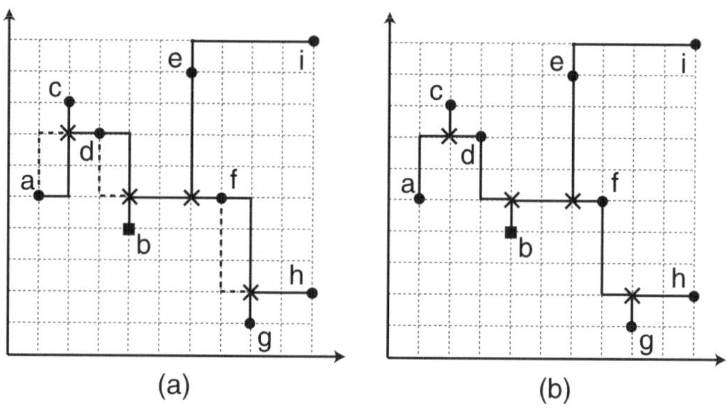

Figure 5.12. (**a**) L-MST with Steiner points shown in X and alternate staircase segments shown in dotted line, (**b**) staircase rerouting does not cause any additional overlap.

Figure 5.11 shows the final L-RST. The total wirelength is 28. The wirelength of the initial separable MST is 32, and the total overlap is 4. This agrees with the calculation $28 = 32 - 4$.

4. Show that the resulting L-RST is stable under re-routing.

   Figure 5.12(a) shows the L-RST we obtained. The Steiner points are marked with X, and alternate *staircase segments* are shown in dotted line. Figure 5.12(b) shows that using these alternate paths do not result in any additional overlaps (= wirelength reduction).

2. 1-Steiner Routing Algorithms

Given a set of points P along with its Minimum Spanning Tree (MST), denoted $MST(P)$, the 1-Steiner problem seeks one additional Steiner point s so that the wirelength of $MST(P \cup s)$ is shorter than that of $MST(P)$. The point s is called 1-Steiner point. Note that the solution to this problem can be used to iteratively insert as many 1-Steiner points until no more wirelength improvement is possible. This approach—adding one Steiner point at a time—is quite different from the L-RST algorithm [Ho et al., 1990], where the edges in the given initial MST is rectilinearized into L-shapes. The authors of [Georgakopoulos and Papadimitriou, 1987] solved the 1-Steiner problem by a computational geometry-based method for Euclidean space, and the authors of [Kahng and Robins, 1992] extended this method to rectilinear space for VLSI routing application.

Kahng and Robins presented two 1-Steiner algorithms [Kahng and Robins, 1992]: naive and sophisticated. The $O(n^5)$-time naive algorithm constructs MST for *all* possible 1-Steiner point locations and choose the best location. The sophisticated algorithm adopts the method in [Georgakopoulos and Papadimitriou, 1987] for more efficient computation of 1-Steiner computation and runs in $O(n^3)$.[15] It has been shown in [Kahng and Robins, 1992] that these 1-Steiner algorithms provide better quality solutions compared with L-RST algorithm [Ho et al., 1990] but at the cost of longer runtime. The 1-Steiner heuristic presented by Borah, Owens, and Irwin [Borah et al., 1994] produces results that are comparable to [Kahng and Robins, 1992] but with a complexity of only $O(n^2)$. This method improves an initial MST by finding the shortest edge between a node in P and any point along any MST edge. If the edge is inserted, a cycle is formed. Then, the removal of the longest edge on this cycle may result in wirelength reduction of the MST.

Quick Overview

The "naive" 1-Steiner algorithm presented in [Kahng and Robins, 1992] works in an iterative fashion: given an initial MST we first identify all possible locations of 1-Steiner points, usually based on Hanan grid [Hanan, 1966]. We then insert a 1-Steiner point at *each* candidate location and build its MST. The last step is to choose the MST with the minimum wirelength, which serves as the initial MST for the next 1-Steiner point insertion. This entire exhaustive search repeats until there is no more wirelength improvement.

The 1-Steiner heuristic presented by Borah, Owens, and Irwin [Borah et al., 1994] starts by computing the gain of all (node, edge) pairs in the MST. For a

[15] This book provides a sample problem on the naive algorithm only. The sophisticated algorithm is outside the scope of this book.

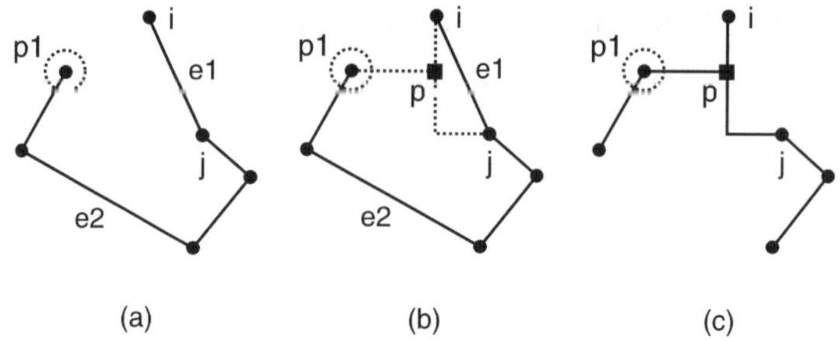

Figure 5.13. (**a**) Node p_1 and edge (i, j) are paired, (**b**) connecting p_1 and $e_1 = (i, j)$ creates a cycle. e_2 is the longest edge in the cycle. p is the point along the rectilinear layout of (i, j) that is closest to p_1, (**c**) insertion of p causes e_1 and e_2 to be removed and $(p, p1)$, (p, i), and (p, j) to be added.

given pair $\{p_1, e_1 = (i, j)\}$, the gain of 1-Steiner point insertion is computed as follows (see Figure 5.13):

$$gain\{p_1, (i, j)\} = length(e_2) - length(p, p_1) \tag{5.2}$$

where e_2 is the longest edge along the cycle formed by connecting p_1 and (i, j), and p, the 1-Steiner point, is the point along the "rectilinear" layout of (i, j) that is the closest to p_1. All length and distance values are based on rectilinear space. Upon the insertion of p as the 1-Steiner point, we remove (i, j) and e_2 and connect p to p_1, i, and j. Note that e_2 is removed and (p, p_1) is added after inserting p, which explains the gain computation in Equation (5.2).

The algorithm consists of multiple passes. In each pass we first compute the gain for all (node, edge) pairs. Then we select only the maximum gain pair for each edge and sort them in a descending order of their gain values. Note that some edges are removed during 1-Steiner point insertion, namely, e_1 and e_2. Thus, some pairs are *infeasible* if some of the required edges are removed during the earlier 1-Steiner point insertion. We process only the feasible edges in the sorted order in each pass. After all the feasible pairs are processed, we attempt the next pass using the final MST from the current pass as its initial MST. We terminate the algorithm if the current pass cannot reduce the wirelength any further.

Practice Problem

Consider the routing problem instance shown in Figure 5.14(a). Figure 5.14(b) shows an initial MST with rectilinear wirelength of 20.

1. Perform 1-Steiner point insertion using the "naive" method presented in [Kahng and Robins, 1992].

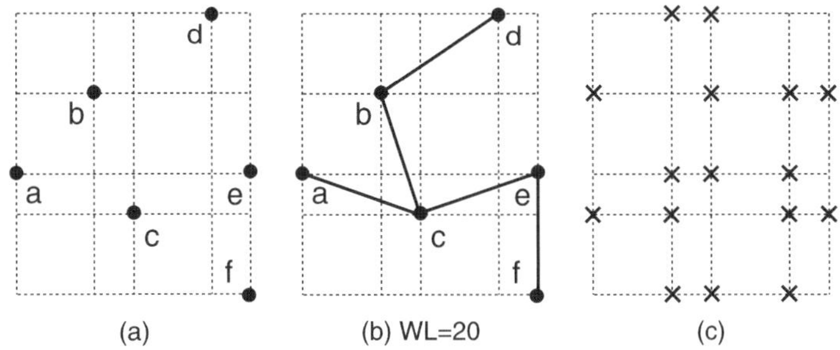

Figure 5.14. (**a**) Routing problem instance for the 1-Steiner algorithm shown in Hanan grid, (**b**) initial MST with rectilinear wirelength of 20, (**c**) candidate locations (shown in X) for Steiner point insertion.

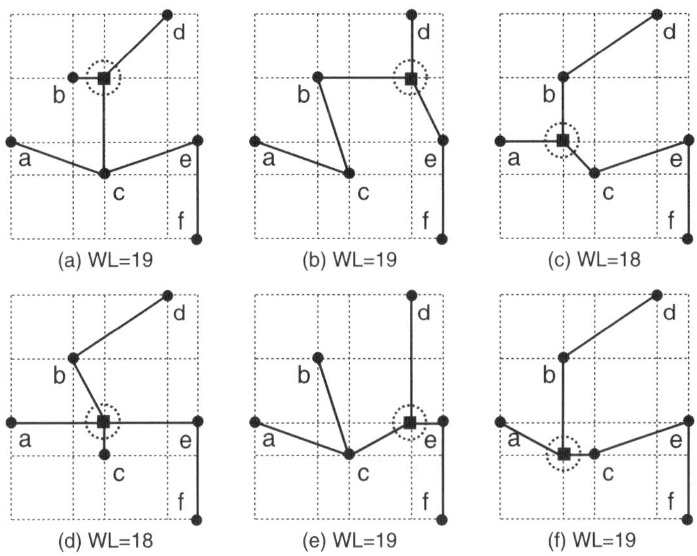

Figure 5.15. Insertion of the first 1-Steiner point. (**a–f**) 1-Steiner points (shown in dotted circles) that reduce the wirelength of the initial MST.

(a) First 1-Steiner point: The initial MST shown in Figure 5.14(b) contains 16 candidate locations for 1-Steiner point insertion as shown in Figure 5.14(c). Figure 5.15 shows 6 1-Steiner points that reduce the wirelength of this initial MST. The MSTs shown in Figure 5.15(c) and Figure 5.15(d) have the same minimum wirelength of 18.

(b) Second 1-Steiner point: We start with the MST shown in Figure 5.15(c).[16] Figure 5.16 shows three 1-Steiner points that reduce the

[16]See the related practice problem #6 on page 190.

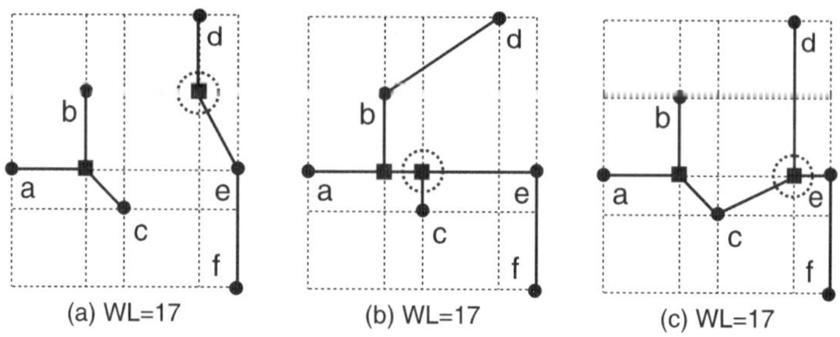

Figure 5.16. Insertion of the second 1-Steiner point. (**a–c**) 1-Steiner points (shown in dotted circles) that reduce the wirelength of the initial MST.

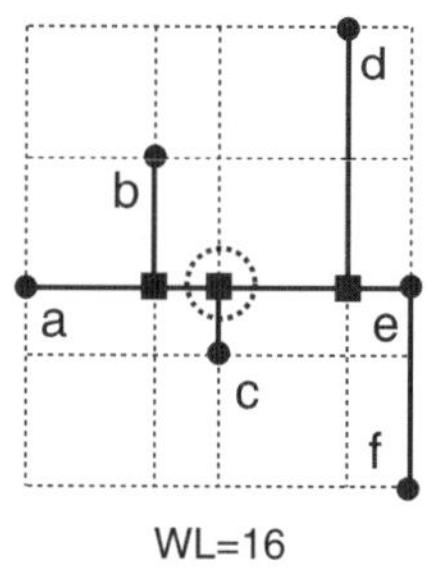

Figure 5.17. Insertion of the third 1-Steiner point.

wirelength of this initial MST. All three MSTs have the same minimum wirelength of 17.

(c) Third 1-Steiner point: The first two MST shown in Figure 5.16(a) and Figure 5.16(b) do not contain a 1-Steiner point that reduces the wirelength further. However, the MST shown in Figure 5.16(c) contains a 1-Steiner point. Figure 5.17 shows the MST after inserting the third 1-Steiner point. This final MST has wirelength of 16, and all of the MST edges are rectilinearized.

2. Perform a single pass of 1-Steiner point insertion using the edge-based heuristic presented in [Borah et al., 1994].

   First, we compute the maximum gain (node, edge) pair for each edge in the initial MST shown in Figure 5.14(b) as follows:

   (a) Edge (a, c): Out of the four nodes (= b, d, e, f), b is the only node that can pair up with (a, c) because d and e are "blocked" by edge (b, c). In case of f, the location of Steiner point coincides with c, making it impossible for 1-Steiner point insertion.

   Figure 5.18 shows how to compute the gain for the $\{b, (a, c)\}$ pair. We first let $p_1 = b$ and $e_1 = (a, c)$. Next, we compute the shortest

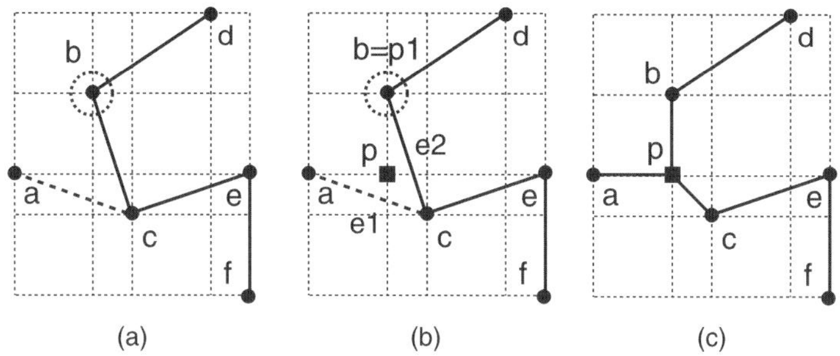

Figure 5.18. Computing the gain for the $\{b, (a, c)\}$ pair for 1-Steiner point insertion. (**a**) Initial MST with wirelength 20, (**b**) Steiner point p, the nearest point between b and (a, c), is identified. Also, e_2 is the longest edge on the b-to-a path. (**c**) Tree after inserting p and deleting e_1 and e_2. The wirelength is now reduced to 18.

Manhattan distance between p_1 and e_1, which is 2 in this case. The node p shown in Figure 5.18(b) corresponds to the nearest point on a "rectilinear layout" of e_1 to p_1. Next, we look for e_2, the longest edge on p_1-to-a path, which is $e_2 = (b, c)$. Thus,

$$gain\{b, (a, c)\} = length(e_2) - length(p, p_1) = 4 - 2 = 2$$

We connect p to p_1 and the two end points of e_1 (= a and c) as shown in Figure 5.18(c). The total rectilinear wirelength of this final tree is 18. This verifies our gain computation $18 = 20 - 2$, where the 20 is the rectilinear wirelength of the initial MST tree shown in Figure 5.18(a), and the 2 is the gain value we computed. Thus, $\{b, (a, c)\}$ is the only pair with positive gain.

(b) Edge (b, c): There are three nodes that can pair up with (b, c) for 1-Steiner point insertion, namely, a, d, and e. We see from Figure 5.19 that all three nodes have positive gain as follows:

$$gain\{a, (b, c)\} = length(a, c) - length(p, a) = 4 - 2 = 2$$
$$gain\{d, (b, c)\} = length(b, d) - length(p, d) = 5 - 4 = 1$$
$$gain\{e, (b, c)\} = length(c, e) - length(p, e) = 4 - 3 = 1$$

The $\{a, (b, c)\}$ pair has the maximum gain of 2.

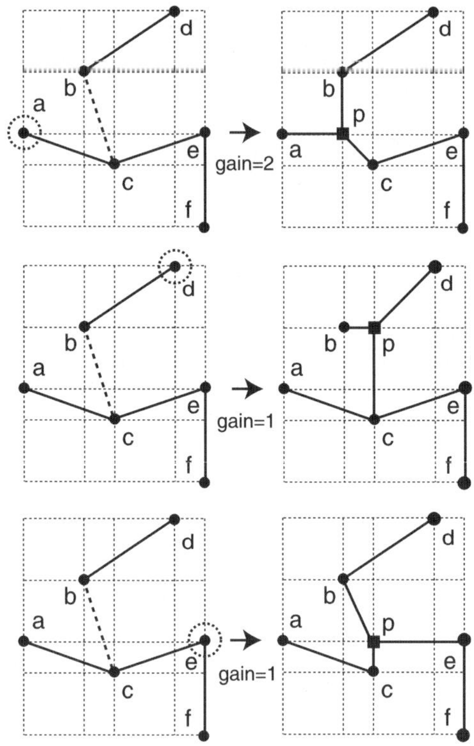

Figure 5.19. 1-Steiner point insertion for edge (b, c). The $\{a, (b, c)\}$ pair has the maximum gain of 2.

(c) Edge (b, d): There are two nodes that can pair up with (b, d) for 1-Steiner point insertion, namely, c and e. We see from Figure 5.20 that both c and e have positive gain as follows:

$$gain\{c, (b, d)\} = length(b, c) - length(p, c) = 4 - 3 = 1$$
$$gain\{e, (b, d)\} = length(b, c) - length(p, e) = 4 - 3 = 1$$

Both pairs have the maximum gain of 1.[17]

(d) Edge (c, e): There are three nodes that can pair up with (c, e) for 1-Steiner point insertion, namely, b, d, and f. We see from Figure 5.21 that all three of them have positive gain as follows:

$$gain\{b, (c, e)\} = length(b, c) - length(p, b) = 4 - 3 = 1$$
$$gain\{d, (c, e)\} = length(b, d) - length(p, d) = 5 - 4 = 1$$
$$gain\{f, (c, e)\} = length(e, f) - length(p, f) = 3 - 2 = 1$$

All three pairs have the maximum gain of 1.

[17] Note that edge (c, e) can be removed for the $\{e, (b, d)\}$ pair instead of (b, c).

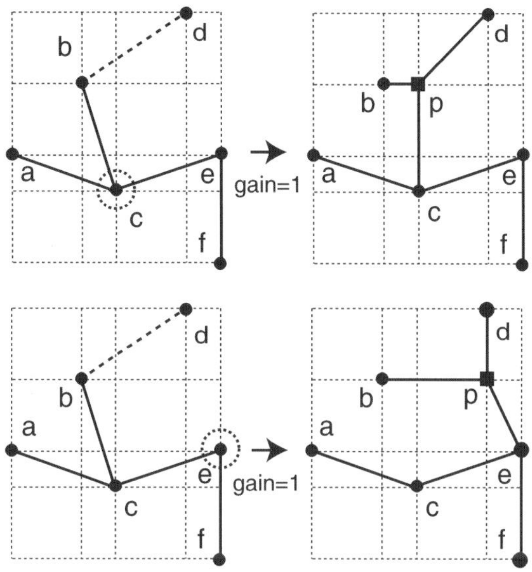

Figure 5.20. 1-Steiner point insertion for edge (b, d). Both pairs $\{c, (b, d)\}$ and $\{e, (b, d)\}$ have the maximum gain of 1.

(e) Edge (e, f): Node c is the only node that can pair up with (e, f) for 1-Steiner point insertion. We see from Figure 5.22 that the $\{c, (e, f)\}$ pair has positive gain as follows:

$$gain\{c, (e, f)\} = length(c, e) - length(p, c) = 4 - 3 = 1$$

Table 5.1 shows the maximum gain pair for each edge, where ties are broken randomly.[18] Our next step is to visit these pairs in a descending order of their gain values. We first choose $\{b, (a, c)\}$ and perform 1-Steiner insertion as shown in Figure 5.23(b). Since this insertion removes two edges, namely, $e_1 = (a, c)$ and $e_2 = (b, c)$, all other pairs in Table 5.1 that have these edges as their e_1 or e_2 cannot be used. Thus, we skip $\{a, (b, c)\}$, $\{c, (b, d)\}$, and $\{b, (c, e)\}$. Since the last pair $\{c, (e, f)\}$ is not involved with (a, c) nor (b, c), we choose this pair and perform 1-Steiner insertion as shown in Figure 5.23(c). This final MST has the wirelength of 17.

3. Compare the Steiner trees built by the two algorithms in terms of wirelength.

   Figure 5.24 shows the comparison. The tree built by Kahng and Robins algorithm [Kahng and Robins, 1992] has a shorter wirelength.

[18] See the related practice problem #6 on page 190.

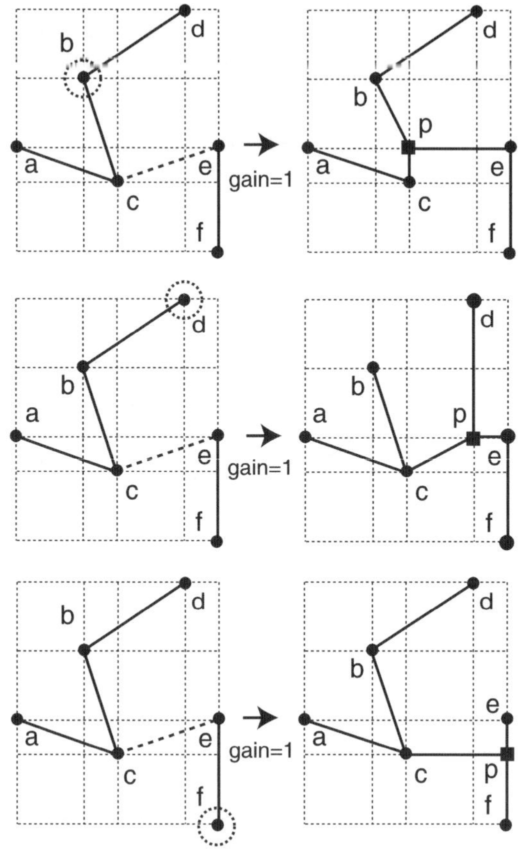

Figure 5.21. 1-Steiner point insertion for edge (c, e). All three pairs $\{b, (c, e)\}$, $\{d, (c, e)\}$, and $\{f, (c, e)\}$ have the maximum gain of 1.

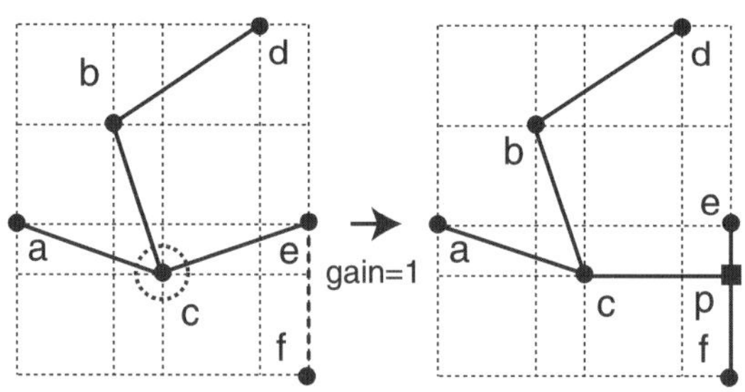

Figure 5.22. 1-Steiner point insertion for edge (e, f). This single pair $\{c, (e, f)\}$ has the maximum gain of 1.

Table 5.1. Maximum gain pair for each edge.

Pair	Gain	e_1	e_2
$\{b, (a, c)\}$	2	(a, c)	(b, c)
$\{a, (b, c)\}$	2	(b, c)	(a, c)
$\{c, (b, d)\}$	1	(b, d)	(b, c)
$\{b, (c, e)\}$	1	(c, e)	(b, c)
$\{c, (e, f)\}$	1	(e, f)	(c, e)

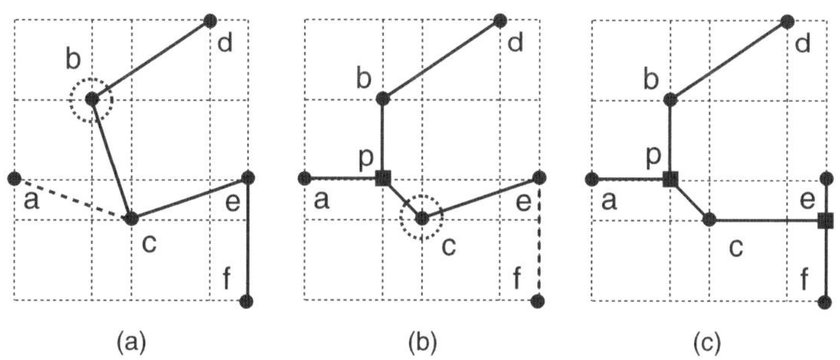

Figure 5.23. Single iteration of 1-Steiner point insertion. (**a**) Original MST with wirelength 20, (**b**) after utilizing $\{b, (a, c)\}$, (**c**) after utilizing $\{c, (e, f)\}$, where the final wirelength is 17.

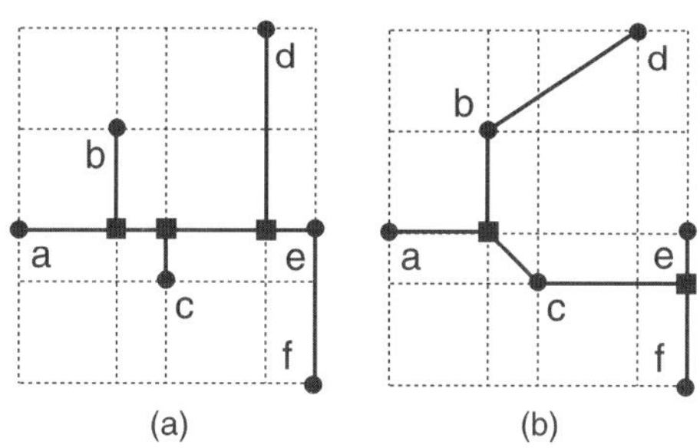

Figure 5.24. (**a**) Steiner tree built by the Kahng/Robins algorithm with wirelength of 16, (**b**) Steiner tree built by the Borah/Owens/Irwin algorithm with wirelength of 17.

3. Bounded Radius Routing Algorithms

Given a routing tree of a signal net, the *radius* is defined to be the longest source-sink path length among all sinks. The minimization of radius has been a focus of performance-oriented Steiner tree construction because the reduction of sink-source path length naturally translates to sink-source delay reduction and thus better performance. In addition to the radius reduction, the total wirelength is still important for performance optimization due to the capacitive effect of wires. The Bounded Radius Minimum Routing Tree (BR-MRT) problem seeks a rectilinear spanning tree with minimum wirelength under a given radius bound.

Cong, Kahng, Robins, Sarrafzadeh, and Wong presented two algorithms for the construction of BR-MST [Cong et al., 1992]: Bounded Prim (BPRIM), and Bounded Radius Bounded Cost (BRBC). BPRIM is a simple extension of the Prim's MST algorithm [Prim, 1957], where the growth of tree is controlled by the radius bound. Every time a node y is added to the growing tree via an edge (x, y), we make sure the radius bound is not violated. Otherwise, we seek another node x' in the tree, so called "appropriate" node, so that the radius bound is satisfied after inserting (x', y). BRBC algorithm extended the "shallow-light" tree construction algorithm by Awerbuch, Baratz, and Peleg [Awerbuch et al., 1990], which was originally designed for communication protocols. The goal is to construct spanning trees that have bounded wirelength and diameter values. Total wirelength is at most $(2 + 2/\epsilon)$ times that of a minimum spanning tree, while the diameter is at most $(1 + 2\epsilon)$ times that of the diameter of the node set. The ϵ parameter is used to trade off wirelength and diameter.

Quick Overview

Given a set of nodes P including a source s, BPRIM algorithm starts with an initial tree T that contains s only. The goal, as in Prim's MST algorithm, is to add edges one by one to grow T until T spans all nodes in P so that the radius is bounded and the wirelength is minimized. Every time we add an edge to grow T, we look for the closest neighbor to T among the nodes not in T. Assume we attempt to add an edge (x, y), where $x \in T$ and $y \notin T$ is the closest neighbor. The radius bound is violated if:

$$dist_T(s, x) + dist(x, y) > (1 + \epsilon) \cdot R$$

where $dist_T(s, x)$ denotes the path length between s and x measured on T, $dist(x, y)$ is the rectilinear distance between x and y, ϵ is the user-specified constant, and R is the radius of P. This equation is checking to see if the distance between the source and y measured on T is larger than the bound that is calculated based on R. If not, we add (x, y) into T. Otherwise, we seek another node x' in the tree, so called "appropriate" node, so that the radius bound is satisfied after inserting (x', y). In order to find x', we back-trace from x to s and find the *first* node that satisfy:

$$dist_T(s, x') + dist(x', y) \leq R$$

Note that we use a tighter condition than the original radius bound, i.e., R instead of $(1 + \epsilon) \cdot R$. The authors of [Cong et al., 1992] note that this helps reduce the number of appropriate edges used. Note that the more appropriate edges we use, the longer the wirelength tends to become.

We start BRBC algorithm by constructing an MST of P and setting it to $Q = MST(P)$. Next, we obtain a rooted tree version of $MST(P)$, denoted T_s. We then perform a depth-first traversal on T_s and obtain a node ordering L, where the nodes are listed in L based on the order they are visited during the traversal. Next, we traverse L while computing S, where S is simply the total cost of visited edges so far during the traversal. Given a node $L_i \in L$, we check to see if:

$$\epsilon \cdot dist(s, L_i) < S$$

where $dist(s, L_i)$ is the rectilinear distance between s and L_i. If so, we reset $S = 0$ and add the edge (s, L_i) into Q. After we traverse all nodes in L and finish adding edges to Q, we compute a shortest path tree (SPT) on Q. This SPT is the final BR-MST.

Practice Problem

Consider the routing problem instance shown in Figure 5.25. Break ties in alphabetical order.

1. What is the radius R of this signal net?

   The radius $R = 12$, which corresponds to the rectilinear distance between the source s and h, the farthest sink.

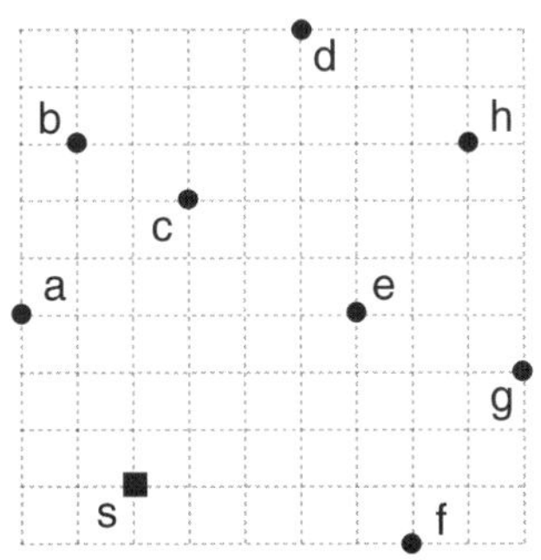

Figure 5.25. Problem instance for the bounded-radius routing algorithms. Node s is the source.

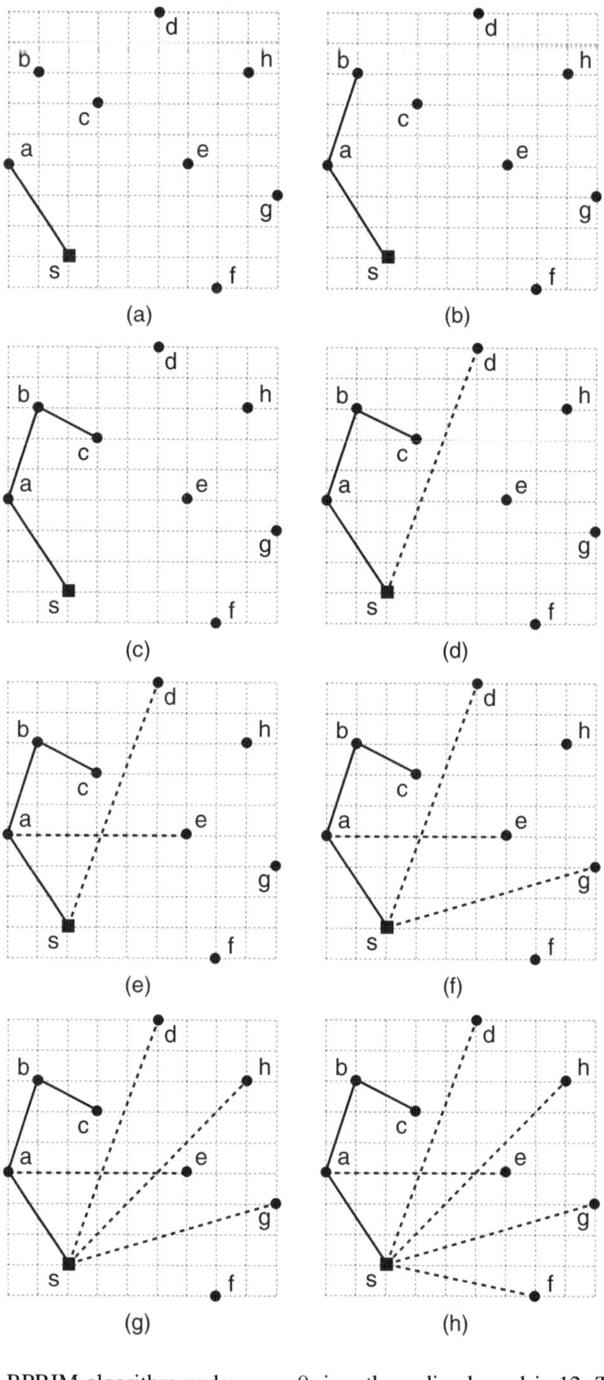

Figure 5.26. BPRIM algorithm under $\epsilon = 0$, i.e., the radius bound is 12. The dotted lines denote the "appropriate" edges [Cong et al., 1992].

2. Construct BR-MSTs using BPRIM algorithm under $\epsilon = 0$, $\epsilon = 0.5$, and $\epsilon = \infty$. Compare these trees in terms of radius and wirelength.

   (a) $\epsilon = 0$: Figure 5.26 shows the illustration of the BPRIM algorithm progress. In addition, Table 5.2 shows the summary. The entries under the "min $dist(x, y)$" are the edges to closest neighbors of the current tree that is growing. In case of ties, we choose the edge based on alphabetical order. The entries under the "$dist_T(s, x) + dist(x, y)$" are the distance from the source s to the selected closest neighbor y. If this distance is larger than the bound 12, we start back-tracing from x back to s and search for the first "appropriate" node x', where the $s \rightarrow x' \rightarrow y$ path is shorter than the radius bound.

   (b) $\epsilon = 0.5$: Figure 5.27 shows the illustration of the BPRIM algorithm progress. Table 5.3 shows the summary.

   (c) $\epsilon = \infty$: This case corresponds to Prim's MST construction. There is no appropriate edge used. Figure 5.28 shows the illustration of the algorithm progress.

   Figure 5.29 shows the comparison among the three BR-MSTs built.[19] We observe that as ϵ increases (0, 0.5, ∞), the radius bound increases (12, 18, ∞), the actual radius increases (12, 18, 22), and the wirelength decreases (56, 49, 36).

3. Construct a BR-MST using BRBC algorithm under $\epsilon = 0.5$. Use s as the root node.

Table 5.2. BPRIM algorithm under $\epsilon = 0$, i.e., the radius of the tree should not exceed 12. In case of tie among the edges under "min $dist(x, y)$", we choose the first entry based on alphabetical order.

min $dist(x, y)$	Chosen	$dist_T(s, x) + dist(x, y)$ of chosen edge	Appropriate edge
(s, a)	(s, a)	$0 + 5$	–
(a, b)	(a, b)	$5 + 4$	–
(b, c)	(b, c)	$9 + 3$	–
(c, d), (c, e)	(c, d)	$12 + 5$	(s, d)
(c, e), (d, h)	(c, e)	$12 + 5$	(a, e)
(e, g)	(e, g)	$11 + 4$	(s, g)
(d, h), (e, h), (e, f), (g, f)	(d, h)	$11 + 5$	(s, h)
(e, f), (g, f)	(e, f)	$11 + 5$	(s, f)

[19] See the related practice problem #6 on page 190.

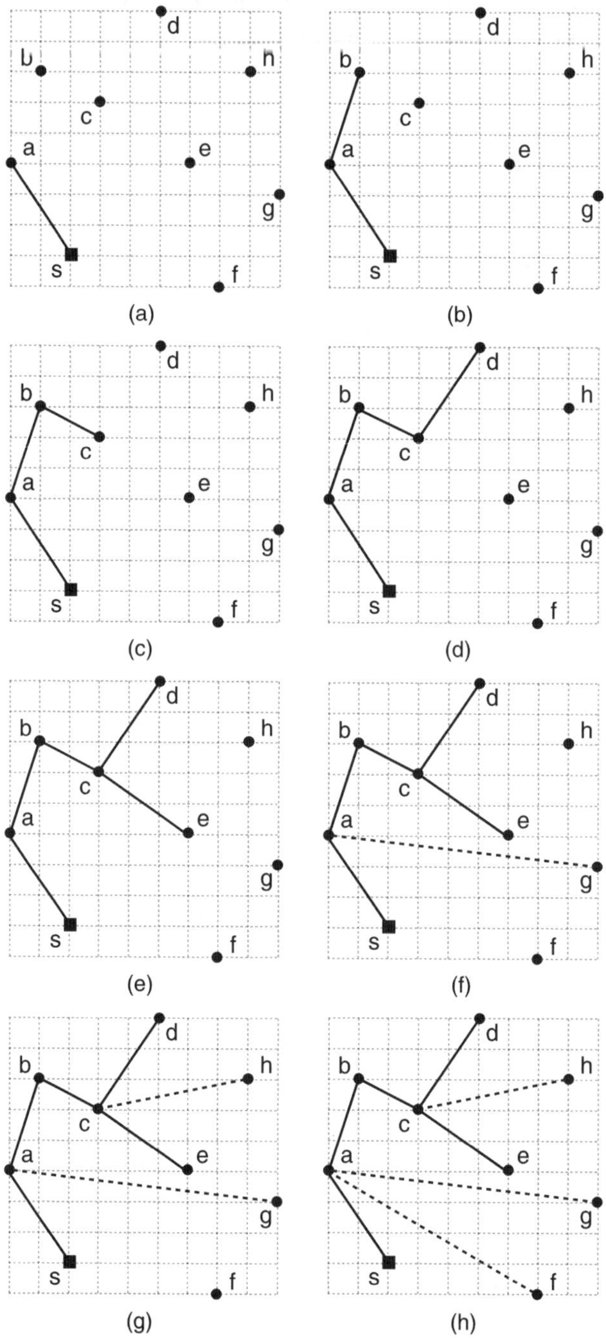

Figure 5.27. BPRIM algorithm under $\epsilon = 0.5$, i.e., the radius bound is 18. The dotted lines denote the "appropriate" edges.

Table 5.3. BPRIM algorithm under $\epsilon = 0.5$, i.e., the radius of the tree should not exceed 18. In case of tie among the edges under "min $dist(x, y)$", we choose the first entry based on alphabetical order.

min $dist(x, y)$	Chosen	$dist_T(s, x) + dist(x, y)$ of chosen edge	Appropriate edge
(s, a)	(s, a)	$0 + 5$	–
(a, b)	(a, b)	$5 + 4$	–
(b, c)	(b, c)	$9 + 3$	–
(c, d), (c, e)	(c, d)	$12 + 5$	–
(c, e), (d, h)	(c, e)	$12 + 5$	–
(e, g)	(e, g)	$17 + 4$	(a, g)
(d, h), (e, h), (g, h), (e, f), (g, f)	(d, h)	$17 + 5$	(c, h)
(e, f), (g, f)	(e, f)	$17 + 5$	(a, f)

The first step is to build an MST, which is shown in Figure 5.30(a). Next, we perform depth-first traversal on the rooted-MST shown in Figure 5.30(b) and obtain the following node ordering:

$$L = \{s, a, b, c, e, f, e, g, e, c, d, h, d, c, b, a, s\}$$

Next, we set the graph $Q = MST$ and augment it during the traversal of L. Table 5.4 shows the summary of this process. For example, if we visit node a via edge (s, a), the running total becomes $S = 5$. Note that $dist(s, a) = 5$ from Figure 5.25. Thus, $\epsilon \cdot dist(s, a) = 0.5 \cdot 5 = 2.5$, which is smaller than S. Thus, we reset $S = 0$ and add (s, a) to Q. Note that (s, a) already exists in Q.

Figure 5.31(a) shows the graph Q after the augmentation (added edges are shown in dotted lines) during the traversal of L. The last step is to construct a shortest path tree on Q to obtain the final BR-MST. Figure 5.31(b) shows the result, where the radius is 12, and the wirelength is 52.

4. Compare the BR-MSTs built by BPRIM and BRBC algorithms under $\epsilon = 0.5$ in terms of radius and wirelength.

   Figure 5.32 shows the two BR-MSTs. The tree generated by BPRIM algorithm has radius $= 18$ and wirelength $= 49$. The tree by BRBC algorithm has radius $= 12$ and wirelength $= 52$. We observe that BRBC has significantly shorter radius at the cost of slightly longer wirelength compared to BPRIM. This observation agrees with that made by the original authors of [Cong et al., 1992]:

   "For any given value of ϵ, the BPRIM approach, being inherently greedy, will yield a routing solution with radius approaching $(1 + \epsilon) \cdot R$, but with small tree weight. On the other hand, the BRBC approach, being more

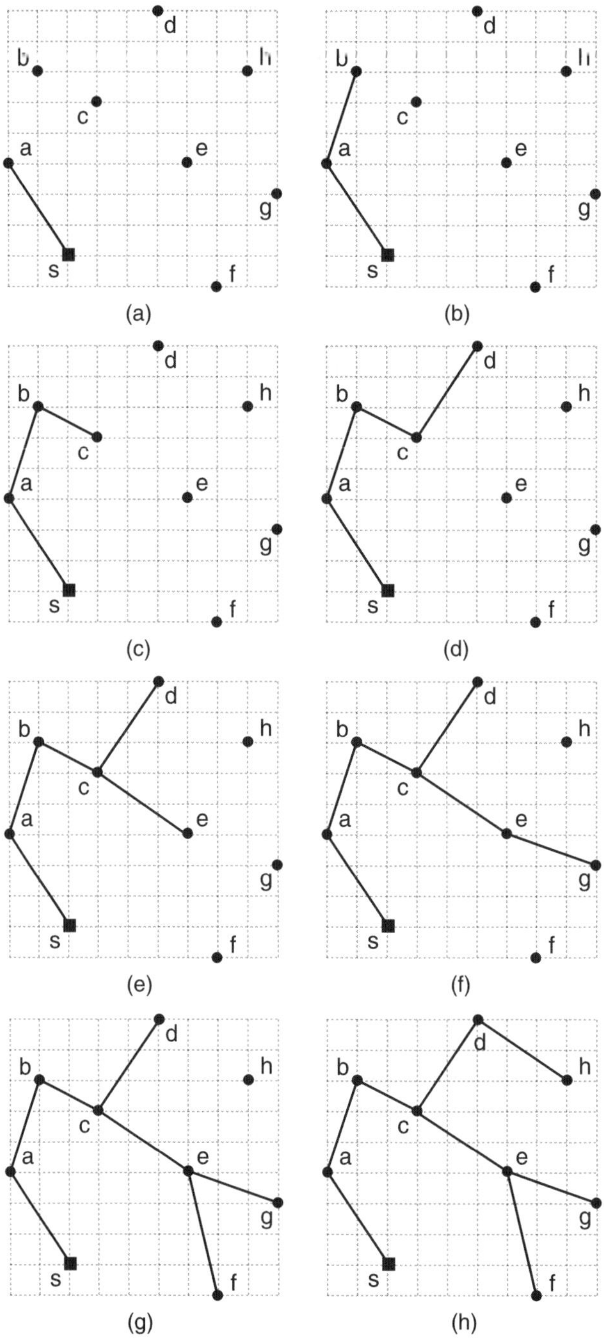

Figure 5.28. BPRIM algorithm under $\epsilon = \infty$. This case corresponds to Prim's MST construction. There is no "appropriate" edge used.

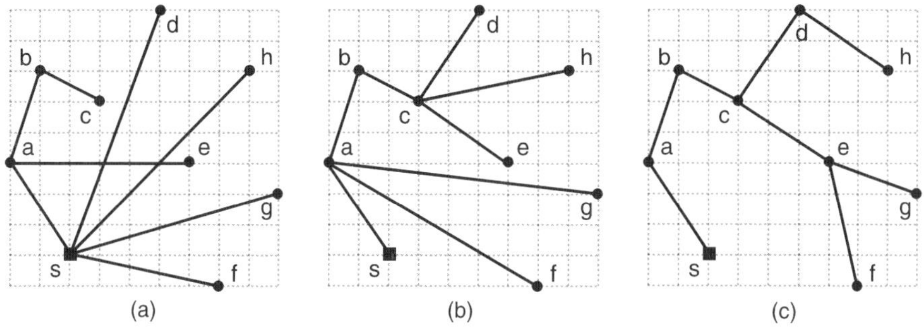

Figure 5.29. Comparison among the BR-MSTs built under various radius bounds. (**a**) $\epsilon = 0$, bound $= 12$, radius $= 12$, wirelength $= 56$, (**b**) $\epsilon = 0.5$, bound $= 18$, radius $= 18$, wirelength $= 49$, (**c**) $\epsilon = \infty$, bound $= \infty$, radius $= 22$, wirelength $= 36$.

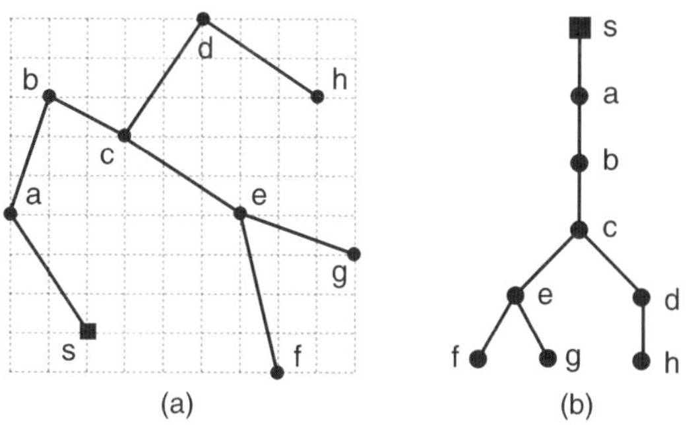

Figure 5.30. (**a**) Initial MST, (**b**) rooted tree of the initial MST for DFS traversal.

conservative, will yield a routing solution with radius noticeably smaller than $(1 + \epsilon) \cdot R$, but at the expense of slightly larger tree cost. In practice, the asymptotic efficiency of implementation and the provably good output provide compelling reasons to adopt the BRBC algorithm, rather than the BPRIM approach."

Table 5.4. DFS traversal of MST and augmentation of graph Q under $\epsilon = 0.5$.

Edge	L_i	$\epsilon \cdot dist(s, L_i)$	S	Reset S?
(s, a)	a	$0.5 \cdot 5$	5	yes
(a, b)	b	$0.5 \cdot 7$	4	yes
(b, c)	c	$0.5 \cdot 6$	3	no
(c, e)	e	$0.5 \cdot 7$	8	yes
(e, f)	f	$0.5 \cdot 6$	5	yes
(f, e)	e	$0.5 \cdot 7$	5	yes
(e, g)	g	$0.5 \cdot 9$	4	no
(g, e)	e	$0.5 \cdot 7$	8	yes
(e, c)	c	$0.5 \cdot 6$	5	yes
(c, d)	d	$0.5 \cdot 11$	5	no
(d, h)	h	$0.5 \cdot 12$	10	yes
(h, d)	d	$0.5 \cdot 11$	5	no
(d, c)	c	$0.5 \cdot 6$	10	yes
(c, b)	b	$0.5 \cdot 7$	3	no
(b, a)	a	$0.5 \cdot 5$	7	yes
(a, s)	s	$0.5 \cdot 0$	5	yes

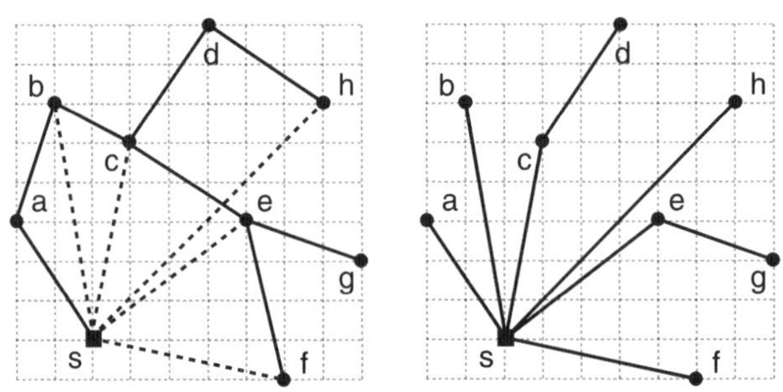

Figure 5.31. BRBC algorithm under $\epsilon = 0.5$. (**a**) Graph Q after adding additional edges (shown in dotted lines), (**b**) shortest path tree on Q, where the radius is 12, and the wirelength is 52.

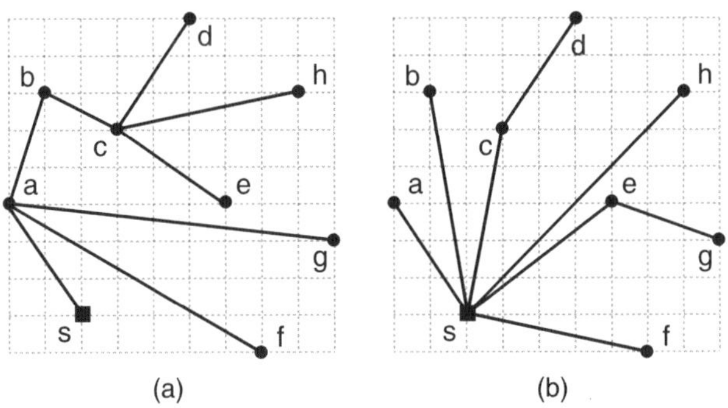

Figure 5.32. Bounded-radius MSTs under $\epsilon = 0.5$. (**a**) BPRIM algorithm, where the radius is 18 and the wirelength is 49, (**b**) BRBC algorithm, where the radius is 12, and the wirelength is 52.

4. A-tree Algorithm

Given an edge-weighted undirected graph $G(V, E)$, a set of nodes $N \subset V$, and a root node $r \in N$, the Minimum Shortest Path Steiner Arborescence (MSP-SA) of N is a Steiner tree rooted at r spanning all nodes in N such that every source-sink path is a shortest path in G, and the total tree weight is minimized. The rectilinear version of MSP-SA is called the Minimum Rectilinear Steiner Arborescence (MRSA) problem, where the underlying graph G is either the Hanan grid graph [Hanan, 1966] or uniform grid graph. The MSP-SA problem is shown to be NP-hard in [Hwang et al., 1992] and the MSRA problem in [Shi and Su, 2000]. The goal of MRSA is to simultaneously minimize the the source-sink path length for *all* sinks and the total wirelength of the routing tree. Note that a similar goal is achieved under the Bounded Radius Minimum Routing Tree (BR-MRT) problem presented in Section 3 of this chapter, where the radius (= maximum source-sink path length) is bounded and the wirelength is minimized.

Note that the delay of a path under the popular Elmode delay model presented in Section 5 of this chapter is quadratically proportional to its length. Under the deep sub-micron technology, however, various interconnect optimization techniques such as buffer insertion, wire/driver sizing, etc, are applied to reduce the interconnect delay. In this case, the path delay tends to become linearly proportional to its length. This phenomenon is one of the main reasons behind the recent popularity of the MRSA problem since it targets the minimization of the linear path length. The A-tree algorithm proposed by Cong, Leung, and Zhou [Cong et al., 1993] is one of the well-known algorithms that constructs an MRSA using a sequence of moves that minimizes the overall wirelength while maintaining the shortest linear path lengths for all sinks.

Quick Overview

The A-tree algorithm starts with an initial forest F_0, where each sink node of the given net becomes its own root node. The goal is to make a sequence of "moves" that either grow an existing rooted-tree or merge two rooted-trees until there is only one rooted-tree left. Let F_k denote the forest after the k-th move, and $R(F_k)$ denote the set of root nodes in F_k.[20] The A-tree algorithm utilizes three kinds of "safe" moves and two kinds of "heuristic" moves, where the safe moves are proven to keep the forest optimal in terms of wirelength. In case of the heuristic moves, the methods suggested in [Rao et al., 1992] are adopted, where the wirelength optimality is not guaranteed. We discuss the

[20] We use the following shorthand notations to improve readability: $R(F_k)$ denotes $ROOR(F_k)$ originally used in [Cong et al., 1993], $D(p, F_k)$ denotes $DOM(p, F_k)$, mf_w denotes mf_{west}, and mf_s denotes mf_{south}.

computation of dx/dy/df values that are used to determine the safe moves in the following section.

Node Blockage and dx/dy/df Value Computation

Given a node $p \in F_k$, the A-tree algorithm distinguishes the neighboring nodes of p into the following three sets: (1) northwest, $NW(p)$: set of nodes with x-coordinate strictly smaller than that of p and y-coordinate strictly larger than that of p. (2) southeast, $SE(p)$: set of nodes with x-coordinate strictly larger than that of p and y-coordinate strictly smaller than that of p. (3) dominated, $D(p, F_k)$: set of nodes with both x and y-coordinate smaller or equal to that of p. In this case, a node $q \in D(p, F_k)$ is said to be "dominated" by p.

Given a pair of nodes p and q, there are two ways q can be "blocked from" p. In case q is located in the northwest of p, we draw a vertical line from q towards p until it reaches the y-coordinate of p. If there is any other node intersecting with this vertical line, q is blocked from p. An illustration is shown in Figure 5.33(a). Similarly, in case q is located in the southeast of p, we draw a horizontal line from q towards p until it reaches the x-coordinate of p. If there is any other node intersecting with this horizontal line, q is again blocked from p. An illustration is shown in Figure 5.33(b).

Given a pair of nodes p and q, let $d_H(p, q) = |x_p - x_q|$ denote the horizonal distance between nodes p and q. We define $d_V(p, q)$, the vertical distance, using y-coordinates similarly. Given a root node $p \in R(F_k)$, the A-tree algorithm utilizes the following three values:

- $dx(p, F)$: we first compute the set of root nodes located in the northwest of p that are not blocked from p. From this set, we choose $q = mx(p, F_k)$

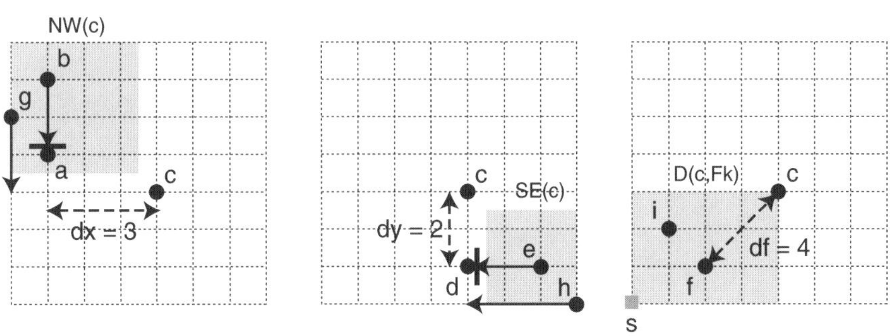

Figure 5.33. (**a**) Node b is blocked from c (= by a), while a and g are not. We have $mx(c, F_k) = a$, and $dx(c, F_0) = 3$. (**b**) Node e is blocked from c (= by d), while h is not. We have $my(c, F_k) = d$, and $dy(c, F_0) = 2$. (**c**) We have $MF(c, F_0) = \{f, i\}$, $df(c, F_0) = 4$, $mf_w = i$, and $wf_s = f$.

with the minimum horizontal distance $d_H(p, q)$. $dx(p, F_k)$ is this minimum $d_H(p, q)$ value. An illustration is shown in Figure 5.33(a).

- $dy(p, F_k)$: we first compute the set of root nodes located in the southeast of p that are not blocked from p. From this set, we choose $q = my(p, F_k)$ with the minimum vertical distance $d_V(p, q)$. $dy(p, F_k)$ is this minimum $d_V(p, q)$ value. An illustration is shown in Figure 5.33(b).

- $df(p, F_k)$: we first compute $MF(p, F_k)$, the set of nodes (= not necessarily root nodes) that are dominated by p and are separated by p with the minimum rectilinear distance. $df(p, F_k)$ is this minimum rectilinear distance value. In addition, we compute mk_w, the node in $MF(p, F_k)$ with the minimum x-coordinate. Similarly, mk_s is the node in $MF(p, F_k)$ with the minimum y-coordinate. An illustration is shown in Figure 5.33(c).

Safe and Heuristic Moves

Based on the $dx/dy/df$ values computed, we define the three safe moves as follows:

- Type-1 (S1): if $dx(p, F_k) \geq df(p, F_k)$ and $dy(p, F_k) \geq df(p, F_k)$ for a $p \in R(F_k)$, we add a path that connects p to mf_w. We remove p from $R(F_k)$. This move merges two trees. Figure 5.34 shows an illustration.

- Type-2 (S2): if $dx(p, F_k) \geq df(p, F_k)$ and $dy(p, F_k) < df(p, F_k)$ for a $p \in R(F_k)$, we add a down-ward vertical path of length p' from p, where p' is the minimum between (1) the vertical distance between p and $mf_s(p, F_k)$, and (2) $dy(p, F_k)$. We remove p from $R(F_k)$ and add p'. This move grows the tree rooted at p. Figure 5.35 shows an illustration.

- Type-3 (S3): if $dx(p, F_k) < df(p, F_k)$ and $dy(p, F_k) \geq df(p, F_k)$ for a $p \in R(F_k)$, we add a left-ward horizontal path of length p' from p,

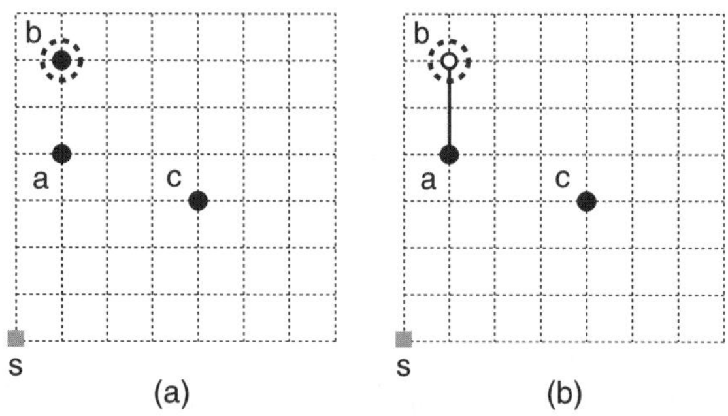

Figure 5.34. Type-1 safe move for node b. (**a**) Before the move, where $dx = \infty$, $dy = 3$, $df = 2$, and $mf_w = a$, (**b**) after the move, where b is no longer a root node.

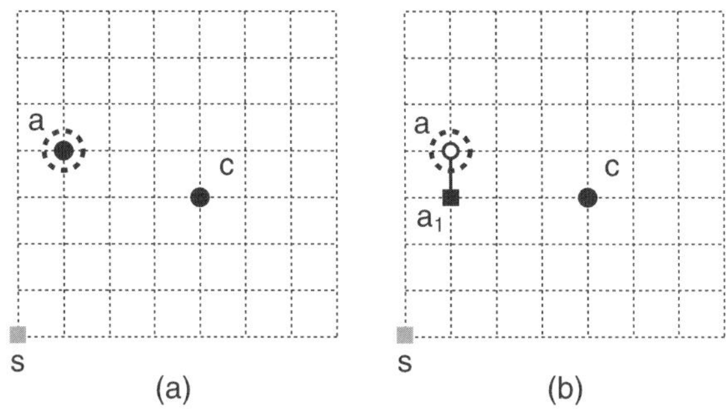

Figure 5.35. Type-2 safe move for node a. (**a**) Before the move, where $dx = \infty$, $dy = 1$, $df = 5$, and $mf_s = s$, (**b**) after the move, where the p-to-p' length is computed as $\min\{d_V(a, s), dy\} = 1$. a_1 is the new root node.

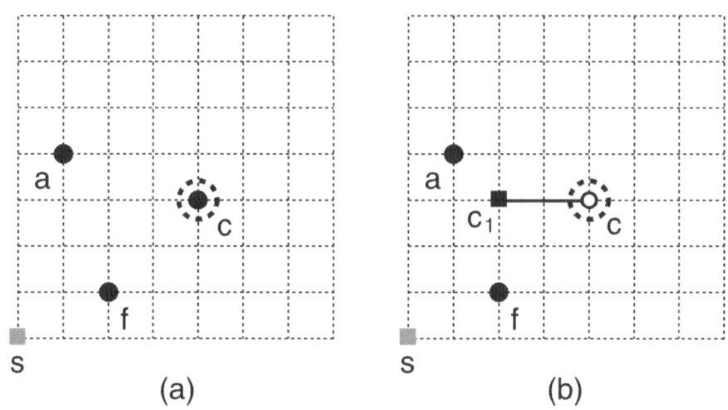

Figure 5.36. Type-3 safe move for node c. (**a**) Before the move, where $dx = 3$, $dy = \infty$, $df = 4$, and $mf_w = f$, (**b**) after the move, where the p-to-p' length is computed as $\min\{d_H(c, f), dx\} = 2$. c_1 is the new root node.

where p' is the minimum between (1) the horizontal distance between p and $mf_w(p, F_k)$, and (2) $dx(p, F_k)$. We remove p from $R(F_k)$ and add p'. This move grows the tree rooted at p. Figure 5.36 shows an illustration.

The definitions of two heuristic moves [Rao et al., 1992] are as follows:

- Type-1 (H1): select a node $p \in R(F_k)$ such that its $mf_w(p, F_k)$, called p', is the farthest away from the source. We add a path between p and p', and remove p from $R(F_k)$. This move merges two trees.

- Type-2 (H2): select two nodes $p, q \in R(F_k)$ such that p', the lower left corner of the lower L-shaped edge connecting p and q, is the farthest from the source. We connect p and p', and then q and p'. We remove p and q from $R(F_k)$ and add p'. This move merges two trees.

The algorithm starts with computing the $dx/dy/mf$ values for all root nodes in F_0. Then, a sequence of safe moves is applied to obtain the successive forests until we obtain the final rectilinear Steiner arborescence. Note that the order of the safe moves chosen is irrelevant since they maintain the wirelength optimality and shortest path length for all sinks. In case there is no more safe move available, heuristic moves H1 and H2 are performed. Preference is given to a move that results in the farthest p', the newly added root node after the tree merging. Note that each move, regardless of its type, requires an update of the $dx/dy/mf$ values of all root nodes in the current forest as well as the root node set itself.

Practice Problem

Consider the routing problem instance shown in Figure 5.37.

1. Compute $dx(c, F_0)$, $dy(c, F_0)$, and $df(c, F_0)$ for node c in the initial forest F_0.

   We begin with $R(F_0) = \{a, b, c, d, e, f\}$.[21] Figure 5.38 shows the illustration.

   - $dx(c, F_0)$: we see that $NW(c) \cap R(F_0) = \{a, b\}$ as shown in Figure 5.38(a). In this case, node b is blocked from node c (= by a) while a is not. Thus, we have $mx(c, F_0) = a$. Since $d_H(a, c) = 3$, we have $dx(c, F_0) = 3$.
   - $dy(c, F_0)$: we see that $SE(c) \cap R(F_0) = \emptyset$ as shown in Figure 5.38(b). Thus, we have $my(c, F_0) = \emptyset$, and $dy(c, F_0) = \infty$.
   - $df(c, F_0)$: we see that $D(c, F_0) = \{s, f\}$ as shown in Figure 5.38(c). Thus, we have $MF(c, F_0) = \{f\}$ and $df(c, F_0) = 4$. Since f is only node in $MF(c, F_0)$, we have $mf_w(c, F_0) = mf_s(c, F_0) = f$.

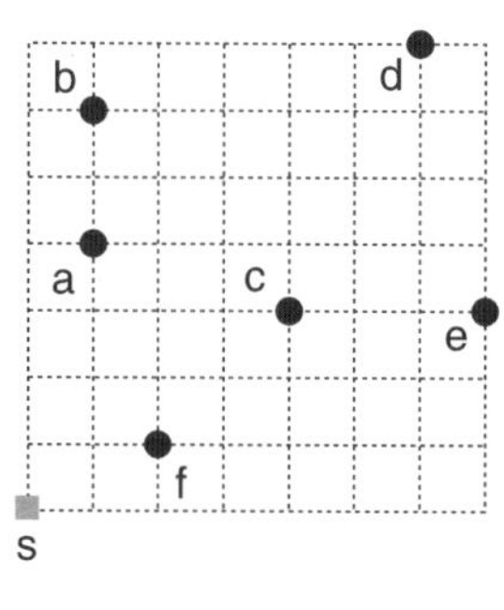

Figure 5.37. Routing problem instance for the A-tree algorithm with the source located at the origin. This is also the initial forest F_0, where the root set $R(F_0) = \{a, b, c, d, e, f\}$.

[21] The source node s is not included in $R(F_0)$ because it does not affect the routing result at all. Even if s is included, it causes a trivial safe move (of type-1) that creates a connection to itself with zero wirelength and then is removed from $R(F_0)$ afterwards.

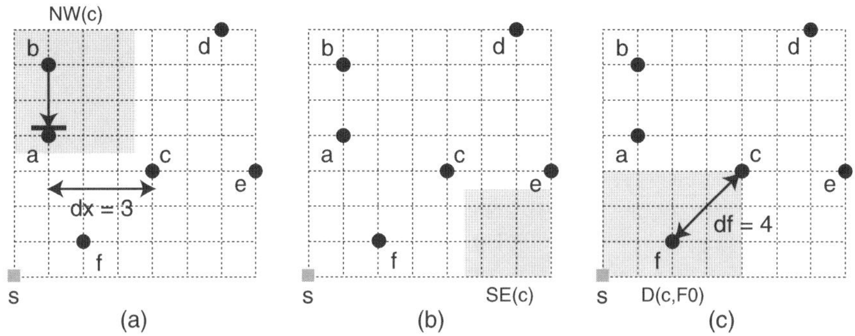

Figure 5.38. (**a**) Computing $dx(c, F_0)$, where the shaded region denotes $NW(c)$. Node b is blocked by a, so $mx(c, F_0) = a$ and $dx(c, F_0) = 3$. (**b**) Computing $dy(c, F_0)$, where the shaded region denotes $SE(c)$. We have $my(c, F_0) = \emptyset$, and $dy(c, F_0) = \infty$. (**c**) Computing $df(c, F_0)$, where the shaded region denotes $D(c, F_0)$. Thus, we have $MF(c, F_0) = \{f\}$ and $df(c, F_0) = 4$.

Table 5.5. $dx/dy/df$ values for $R(F_0)$ shown in Figure 5.37.

p	mx	dx	my	dy	MF	mf_w	mf_s	df
a	$\emptyset$	∞	c	1	$\{s\}$	s	s	5
b	$\emptyset$	∞	c	3	$\{a\}$	a	a	2
c	a	3	$\emptyset$	∞	$\{f\}$	f	f	4
d	$\emptyset$	∞	e	4	$\{b, c\}$	b	c	6
e	d	1	$\emptyset$	∞	$\{c\}$	c	c	3
f	a	1	$\emptyset$	∞	$\{s\}$	s	s	3

2. Compute $dx(p, F_0)$, $dy(p, F_0)$, and $df(p, F_0)$ for the remaining root nodes in the initial forest F_0.

   Table 5.5 shows the result.

3. What kind of safe moves does node a contain?

   From Table 5.5, we see that $dx(a, F_0) = \infty$, $dy(a, F_0) = 1$, and $df(a, F_0) = 5$.

   - Type-1: we check to see if $dx(a, F_0) \geq df(a, F_0)$ and $dy(a, F_0) \geq df(a, F_0)$. Since the second condition is not met, a does not contain type-1 safe move.
   - Type-2: we check to see if $dx(a, F_0) \geq df(a, F_0)$ and $dy(a, F_0) < df(a, F_0)$. Since both conditions are met, a contains type-2 safe move.

Table 5.6. Safe moves exist in F_0 shown in Figure 5.37.

Node	Type-1	Type-2	Type-3
a	No	Yes	No
b	Yes	No	No
c	No	No	Yes
d	No	Yes	No
e	No	No	Yes
f	No	No	Yes

- Type-3: we check to see if $dx(a, F_0) < df(a, F_0)$ and $dy(a, F_0) \geq df(a, F_0)$. Since neither condition is not met, a does not contain type-3 safe move.

Thus, node a contains type-2 safe move only.

4. Identify all safe moves exist in F_0.

   Table 5.6 shows the result.

5. Perform the first safe move using node a and update the related variables.

   Node a contains a type-2 safe move. First, we see that $d_V(mf_s(a, F_0), a) = d_V(s, a) = 4$, and $dy(a, F_0) = 1$ according to Table 5.5 and Figure 5.37. Thus, the length of vertical path to be added to node a is $\min\{4, 1\} = 1$. We connect a to a newly added root node a_1. We then update $R(F_1) = R(F_0) - \{a\} + \{a_1\} = \{a_1, b, c, d, e, f\}$. Figure 5.39(a) shows the resulting F_1.

   Table 5.7 shows the updated $dx/dy/df$ values for $R(F_1)$, the root nodes in F_1. We also show the safe moves for $R(F_1)$. We see that all nodes in $R(F_1)$ contain a safe move.

6. Perform the remaining moves on F_1 and obtain the final rectilinear Steiner arborescence. Choose the safe moves based on alphabetical order.[22]

   (a) Move 2: we choose the type-2 safe move for node a_1 in Table 5.7. First, we see that $d_V(mf_s(a_1, F_1), a_1) = d_V(s, a_1) = 3$, and $dy(a_1, F_1) = 2$ according to Table 5.7 and Figure 5.39(a). Thus, the length of vertical path to be added to node a_1 is $\min\{3, 2\} = 2$. We connect a_1 to a newly added root node a_2. We then update $R(F_2) = R(F_1) - \{a_1\} + \{a_2\} = \{a_2, b, c, d, e, f\}$. Figure 5.39(b) shows the resulting F_2. Table 5.8 shows the updated $dx/dy/df$ values and safe moves for $R(F_2)$.

   (b) Move 3: we choose the type-1 safe move for node a_2 in Table 5.8. In this case, we connect a_2 to $mf_w(a_2, F_2) = s$. We then update $R(F_3) =$

[22] See the related practice problem #6 on page 190.

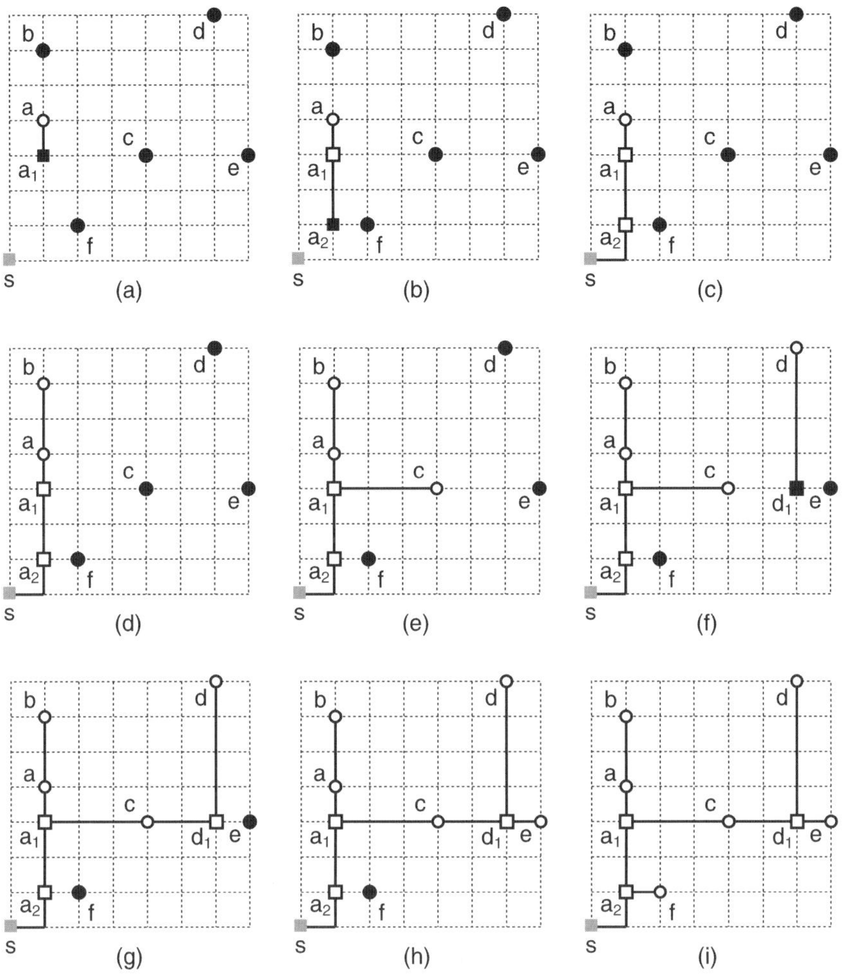

Figure 5.39. **(a–i)** Forests F_1 to F_9 obtained from a sequence of safe moves. F_9 in (i) is the final rectilinear Steiner arborescence, where all source-sink paths are shortest, and the overall wirelength is minimal. The black colored nodes correspond to the current root nodes.

Table 5.7. $dx/dy/df$ values and safe moves for $R(F_1)$ shown in Figure 5.39(a).

p	mx	dx	my	dy	MF	mf_w	mf_s	df	Type-1	Type-2	Type-3
a_1	$\emptyset$	∞	f	2	$\{s\}$	s	s	4	No	Yes	No
b	$\emptyset$	∞	c	3	$\{a\}$	a	a	2	Yes	No	No
c	$\emptyset$	∞	$\emptyset$	∞	$\{a_1\}$	a_1	a_1	3	Yes	No	No
d	$\emptyset$	∞	e	4	$\{b, c\}$	b	c	6	No	Yes	No
e	d	1	$\emptyset$	∞	$\{c\}$	c	c	3	No	No	Yes
f	a_1	1	$\emptyset$	∞	$\{s\}$	s	s	3	No	No	Yes

Table 5.8. $dx/dy/df$ values and safe moves for $R(F_2)$ shown in Figure 5.39(b).

p	mx	dx	my	dy	MF	mf_w	mf_s	df	Type-1	Type-2	Type-3
a_2	$\emptyset$	∞	$\emptyset$	∞	$\{s\}$	s	s	2	Yes	No	No
b	$\emptyset$	∞	c	3	$\{a\}$	a	a	2	Yes	No	No
c	$\emptyset$	∞	$\emptyset$	∞	$\{a_1\}$	a_1	a_1	3	Yes	No	No
d	$\emptyset$	∞	e	4	$\{b, c\}$	b	c	6	No	Yes	No
e	d	1	$\emptyset$	∞	$\{c\}$	c	c	3	No	No	Yes
f	$\emptyset$	∞	$\emptyset$	∞	$\{a_2\}$	a_2	a_2	1	Yes	No	No

Table 5.9. $dx/dy/df$ values and safe moves for $R(F_3)$ shown in Figure 5.39(c).

p	mx	dx	my	dy	MF	mf_w	mf_s	df	Type-1	Type-2	Type-3
b	$\emptyset$	∞	c	3	$\{a\}$	a	a	2	Yes	No	No
c	$\emptyset$	∞	$\emptyset$	∞	$\{a_1\}$	a_1	a_1	3	Yes	No	No
d	$\emptyset$	∞	e	4	$\{b, c\}$	b	c	6	No	Yes	No
e	d	1	$\emptyset$	∞	$\{c\}$	c	c	3	No	No	Yes
f	$\emptyset$	∞	$\emptyset$	∞	$\{a_2\}$	a_2	a_2	1	Yes	No	No

Table 5.10. $dx/dy/df$ values and safe moves for $R(F_4)$ shown in Figure 5.39(d).

p	mx	dx	my	dy	MF	mf_w	mf_s	df	Type-1	Type-2	Type-3
c	$\emptyset$	∞	$\emptyset$	∞	$\{a_1\}$	a_1	a_1	3	Yes	No	No
d	$\emptyset$	∞	e	4	$\{b, c\}$	b	c	6	No	Yes	No
e	d	1	$\emptyset$	∞	$\{c\}$	c	c	3	No	No	Yes
f	$\emptyset$	∞	$\emptyset$	∞	$\{a_2\}$	a_2	a_2	1	Yes	No	No

$R(F_2) - \{a_2\} = \{b, c, d, e, f\}$. Figure 5.39(c) shows the resulting F_3. Note that either L-shape is fine for the a_2-to-s connection. Table 5.9 shows the updated $dx/dy/df$ values and safe moves for $R(F_3)$.

(c) Move 4: we choose the type-1 safe move for node b in Table 5.9. In this case, we connect b to $mf_w(b, F_3) = a$. We then update $R(F_4) = R(F_3) - \{b\} = \{c, d, e, f\}$. Figure 5.39(d) shows the resulting F_4. Table 5.10 shows the updated $dx/dy/df$ values and safe moves for $R(F_4)$.

(d) Move 5: we choose the type-1 safe move for node c in Table 5.10. In this case, we connect c to $mf_w(c, F_4) = a_1$. We then update $R(F_5) = R(F_4) - \{c\} = \{d, e, f\}$. Figure 5.39(e) shows the resulting F_5. Table 5.11 shows the updated $dx/dy/df$ values and safe moves for $R(F_5)$.

(e) Move 6: we choose the type-2 safe move for node d in Table 5.11. First, we see that $d_V(mf_s(d, F_5), d) = d_V(c, d) = 4$, and $dy(d, F_5) = 4$

Table 5.11. $dx/dy/df$ values and safe moves for $R(F_5)$ shown in Figure 5.39(e).

p	mx	dx	my	dy	MF	mf_w	mf_s	df	Type-1	Type-2	Type-3
d	$\emptyset$	∞	e	4	$\{b, c\}$	b	c	6	No	Yes	No
e	d	1	$\emptyset$	∞	$\{c\}$	c	c	3	No	No	Yes
f	$\emptyset$	∞	$\emptyset$	∞	$\{a_2\}$	a_2	a_2	1	Yes	No	No

Table 5.12. $dx/dy/df$ values and safe moves for $R(F_6)$ shown in Figure 5.39(f).

p	mx	dx	my	dy	MF	mf_w	mf_s	df	Type-1	Type-2	Type-3
d_1	$\emptyset$	∞	$\emptyset$	∞	$\{c\}$	c	c	2	Yes	No	No
e	$\emptyset$	∞	$\emptyset$	∞	$\{d_1\}$	d_1	d_1	1	Yes	No	No
f	$\emptyset$	∞	$\emptyset$	∞	$\{a_2\}$	a_2	a_2	1	Yes	No	No

Table 5.13. $dx/dy/df$ values and safe moves for $R(F_7)$ shown in Figure 5.39(g).

p	mx	dx	my	dy	MF	mf_w	mf_s	df	Type-1	Type-2	Type-3
e	$\emptyset$	∞	$\emptyset$	∞	$\{d_1\}$	d_1	d_1	1	Yes	No	No
f	$\emptyset$	∞	$\emptyset$	∞	$\{a_2\}$	a_2	a_2	1	Yes	No	No

according to Table 5.11 and Figure 5.39(e). Thus, the length of vertical path to be added to node d is $\min\{4, 4\} = 4$. We connect d to a newly added root node d_1. We then update $R(F_6) = R(F_5) - \{d\} + \{d_1\} = \{d_1, e, f\}$. Figure 5.39(f) shows the resulting F_6. Table 5.12 shows the updated $dx/dy/df$ values and safe moves for $R(F_6)$.

(f) Move 7: we choose the type-1 safe move for node d_1 in Table 5.12. In this case, we connect d_1 to $mf_w(d_1, F_6) = c$. We then update $R(F_7) = R(F_6) - \{d_1\} = \{e, f\}$. Figure 5.39(g) shows the resulting F_7. Table 5.13 shows the updated $dx/dy/df$ values and safe moves for $R(F_7)$.

(g) Move 8: we choose the type-1 safe move for node e in Table 5.13. In this case, we connect e to $mf_w(e_1, F_7) = d_1$. We then update $R(F_8) = R(F_7) - \{e\} = \{f\}$. Figure 5.39(h) shows the resulting F_8. Table 5.14 shows the updated $dx/dy/df$ values and safe moves for $R(F_8)$.

(h) Move 9: we choose the last safe move remaining in Table 5.14. In this case, we connect f to $mf_w(f_1, F_8) = a_2$. Figure 5.39(i) shows the resulting F_9.

Table 5.14. $dx/dy/df$ values and safe moves for $R(F_8)$ shown in Figure 5.39(h).

p	mx	dx	my	dy	MF	mf_w	mf_s	df	Type-1	Type-2	Type-3
f	$\emptyset$	∞	$\emptyset$	∞	$\{a_2\}$	a_2	a_2	1	Yes	No	No

Figure 5.39(i) is the final rectilinear Steiner arborescence, where all source-sink path lengths are shortest. The total wirelength is 18.[23] The white square nodes correspond to the Steiner nodes used. Note that all moves performed are safe.

[23]See the related practice problem #6 on page 190.

5. Elmore Routing Tree Algorithms

Historically, wirelength and radius have been popular objectives for performance oriented Steiner routing. Reduction of radius naturally translates to lower source-sink resistance and thus better performance. In addition, wirelength reduction translates to lower capacitance to handle and thus better performance. The bounded-radius, bounded-wirelength algorithms [Cong et al., 1992] presented in Section 3 of this chapter is a state-of-the-art in this direction. Although these wirelength and radius metrics provide high-fidelity in performance optimization, they are still an indirect metric, i.e., their units are not in time domain.

Boese, Kahng, McCoy, and Robins presented a class of algorithms [Boese et al., 1995] that *directly* minimizes the popular Elmore delay metric [Elmore, 1948]. Their approach is iterative and resembles Prim's MST (Minimum Spanning Tree) algorithm in that the tree is grown by adding one node at a time. The major difference is on how to select the next node to be added. Given a candidate node v to be evaluated, we add v to the current tree and compute Elmore delay at all sinks. We also record the maximum Elmore delay among the sinks. After examining all possible candidates, we choose the node that results in the minimum maximum Elmore delay among the sinks. The ERT (Elmore Routing Tree) algorithm constructs an MST that minimizes Elmore delay, and the SERT (Steiner Elmore Routing Tree) algorithm builds a Steiner tree.

Quick Overview

The Elmore delay [Elmore, 1948] of a given node n_i in RC tree T is defined as follows:

$$t_{elmore}(n_i) = r_d \cdot C_{n_0} + \sum_{e_v \in path(n_0, n_i)} r_{e_v} \left(\frac{c_{e_v}}{2} + C_v \right)$$

where r_d is the driver output resistance, n_0 is the source (= driver) node, C_{n_i} is the total capacitance of a subtree of T rooted at n_i, e_v is an edge along the $n_0 \rightarrow n_i$ path, r_{e_v} is the resistance of e_v, and c_{e_v} is the capacitance of e_v. Note that the Elmore delay is almost quadratically proportional to the path length due to the multiple inclusions of the RC parasitics of some wires along the source-sin path into the calculation, especially the edges closer to the sink. The Elmore delay is computed by visiting each edge along the source-sink path, where the downstream capacitance at all intermediate nodes are pre-computed using a bottom-up tree traversal.

Given a net to be routed, we start ERT algorithm with an initial tree that contains the source node only. We grow this tree by adding a node at a time. Given a tree T and a set of nodes not included in the tree, our goal is to find the node pair (u, v), where $u \in T$ and $v \notin T$, so that the maximum Elmore delay among all sinks in $T \cup v$ is minimized. This is simply done by visiting each node $u \in T$ and finding its nearest neighbor (= shortest rectilinear distance) not included in T.[24] After examining all trees derived from the candidate nodes, we choose the one that minimizes the maximum Elmore delay. ERT terminates if the tree spans all nodes in the net.

Given a tree T and a set of nodes not included in the tree, our goal in SERT algorithm is to find the edge-node pair (e, v), where $e \in T$ and $v \notin T$, so that the maximum Elmore delay in $T \cup v$ is minimized. In addition, we allow v to directly connect to the source. In case of an edge-node pair (e, v) for evaluation, we connect v to e using the shortest distance. Note that this may require a Steiner node p along e so that (p, v) becomes the shortest. We examine $|E|+1$ connections for *each* node outside the tree and choose the edge-node pair (or source-node pair) that minimizes the maximum Elmore delay. SERT terminates if the tree spans all nodes in the net.

Practice Problem

Consider the routing problem instance shown in Figure 5.40. Assume that the minimum grid edge is $1mm$ long. Let C_x denote the total capacitance of the sub-tree rooted at x, z_x denote the input capacitance of sink x, and $r_{(x,y)}$ and $c_{(x,y)}$ denote the resistance and capacitance of edge (x, y).

1. Perform ERT algorithm using the following parameters (typical for global wires in $65nm$ technology): unit-length resistance is $r = 0.4\,\Omega/\mu m$, unit-length capacitance is $c = 0.2fF/\mu m$, driver output resistance is $r_d = 250\,\Omega$, and input capacitance of the sinks is $50fF$. Break ties based on alphabetical order.

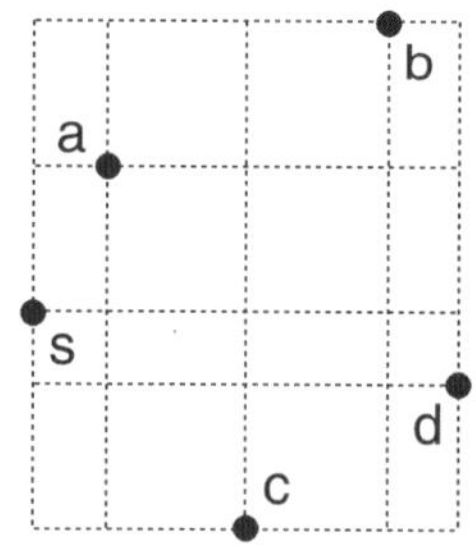

Figure 5.40. Routing problem instance for ERT/SERT algorithms in Hanan grid. Node s is the source.

[24] This approach works only when the sink capacitance is uniform among all sinks. Otherwise, this nearest neighbor may not minimize the Elmore delay. Thus, we have to examine all other nodes not included in T to find the pair for u. Note that this is a $O(n)$ operation as opposed to $O(1)$ in case of uniform sink capacitance.

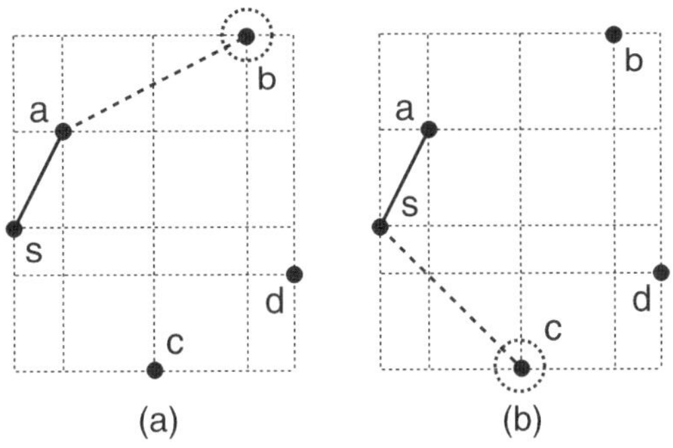

Figure 5.41. Second iteration of ERT algorithm. (**a**) Nearest neighbor of a, (**b**) nearest neighbor of s. (b) is the tree with minimum Elmore delay increase.

We add s to the initial tree and grow it by adding edges in the following sequence:

(a) First iteration: we simply add the nearest neighbor, node a.

(b) Second iteration: Figure 5.41 shows two candidate connections we consider: edge (a, b) because node b is the nearest neighbor of a, and edge (s, c) because node c is the nearest neighbor of s.

- Edge (a, b): From Figure 5.41(a) we compute $t(b)$ as follows (resistance is in $k\Omega$, and capacitance is in fF):

$$\begin{aligned} t(b) &= r_d \cdot C_s + r_{(s,a)}(0.5c_{(s,a)} + C_a) + r_{(a,b)}(0.5c_{(a,b)} + z_b) \\ &= r_d \cdot (c_{(s,a)} + z_a + c_{(a,b)} + z_b) + r_{(s,a)}(0.5c_{(s,a)} + z_a \\ &\quad + c_{(a,b)} + z_b) + r_{(a,b)}(0.5c_{(a,b)} + z_b) \\ &= 0.25(600 + 50 + 1200 + 50) + 1.2(300 + 50 + 1200 \\ &\quad + 50) + 2.4(600 + 50) \\ &= 3955ps \end{aligned}$$

- Edge (s, c): From Figure 5.41(b) we need both $t(a)$ and $t(c)$ to compute the maximum Elmore delay. Since it is easy to see that $t(c) > t(a)$, we only compute $t(c)$ and obtain $2035ps$.

Thus, adding (s, c) results in the minimum Elmore delay increase.

(c) Third iteration: Figure 5.42 shows three candidate connections we consider.

- Edge (a, b): From Figure 5.42(a) we note that node b has the maximum Elmore delay. We obtain $t(b) = 4267.5ps$.
- Edge (s, d): From Figure 5.42(b) we note that node d has the maximum Elmore delay. We obtain $t(d) = 2937.5ps$.

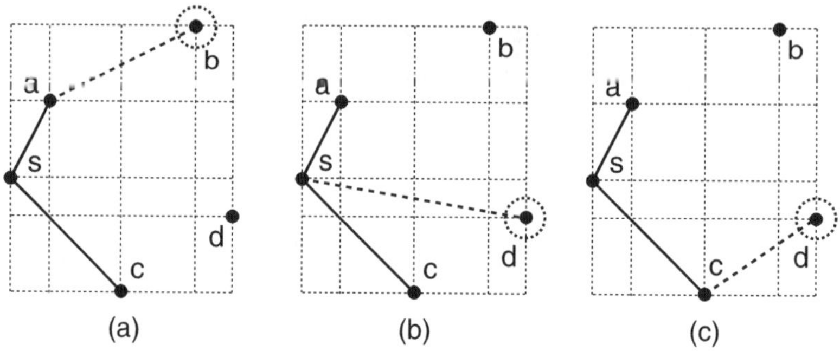

Figure 5.42. Third iteration of ERT algorithm. (**a**) Nearest neighbor of a, (**b**) nearest neighbor of s, (**c**) nearest neighbor of c. (**b**) is the tree with minimum Elmore delay increase.

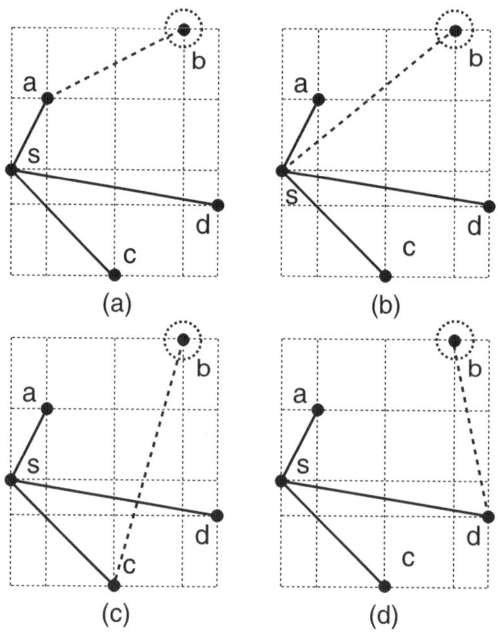

Figure 5.43. Fourth iteration of ERT algorithm. (**a**) Nearest neighbor of a, (**b**) nearest neighbor of s, (**c**) nearest neighbor of c, (**d**) nearest neighbor of d. (a) is the tree with minimum Elmore delay increase.

- Edge (c, d): From Figure 5.42(c) we note that node d has the maximum Elmore delay. We obtain $t(d) = 5917.5ps$.

Thus, adding (s, d) results in the minimum Elmore delay increase.

(d) Fourth iteration: Figure 5.43 shows four candidate connections we consider.

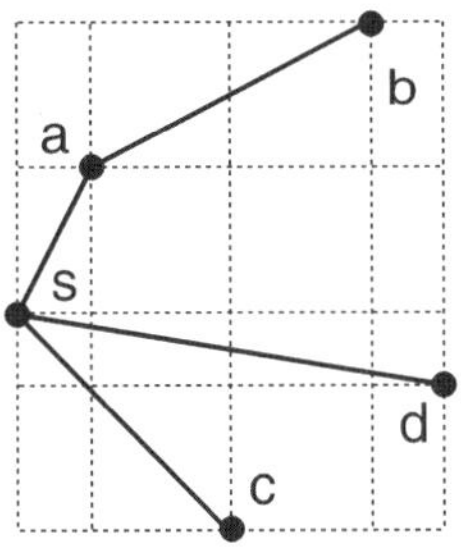

Figure 5.44. Final tree obtained by ERT algorithm with the maximum Elmore delay $t(b) = 4630ps$.

- Edge (a, b): From Figure 5.43(a) we note that node b has the maximum Elmore delay. We obtain $t(b) = 4630ps$.
- Edge (s, b): From Figure 5.43(b) we note that node b has the maximum Elmore delay. We obtain $t(b) = 4720ps$.
- Edge (c, b): From Figure 5.43(c) we note that node b has the maximum Elmore delay. We obtain $t(b) = 10720ps$.
- Edge (d, b): From Figure 5.43(d) we note that node b has the maximum Elmore delay. We obtain $t(b) = 8310ps$.

Thus, adding (a, b) results in the minimum Elmore delay increase.

Figure 5.44 shows the final tree built by ERT algorithm, where the maximum Elmore delay $t(b) = 4630ps$.

2. Perform SERT algorithm using the following parameters for $1.2\mu m$ technology (similar to the "IC2 technology" used in [Boese et al., 1995]): unit-length resistance is $r = 0.073\Omega/\mu m$, unit-length capacitance is $c = 0.083fF/\mu m$, driver output resistance is $r_d = 212\Omega$, and input capacitance of the sinks is $7.1fF$.[25]

   We add each sink and its Steiner point in the following sequence:

   (a) First iteration: we simply add the nearest neighbor of s, which is a. The L-shape orientation of edge (s, a) remain flexible.

   (b) Second iteration: Figure 5.45 shows the six candidate connections we consider. Each node outside the tree can connect to either the nearest point in the tree via a Steiner point or the source directly.

   - Edge (a, b): From Figure 5.45(a) we see that the nearest point between b and (s, a) along (s, a) is a. Thus, no additional Steiner

[25] See the related practice problem #6 on page 191.

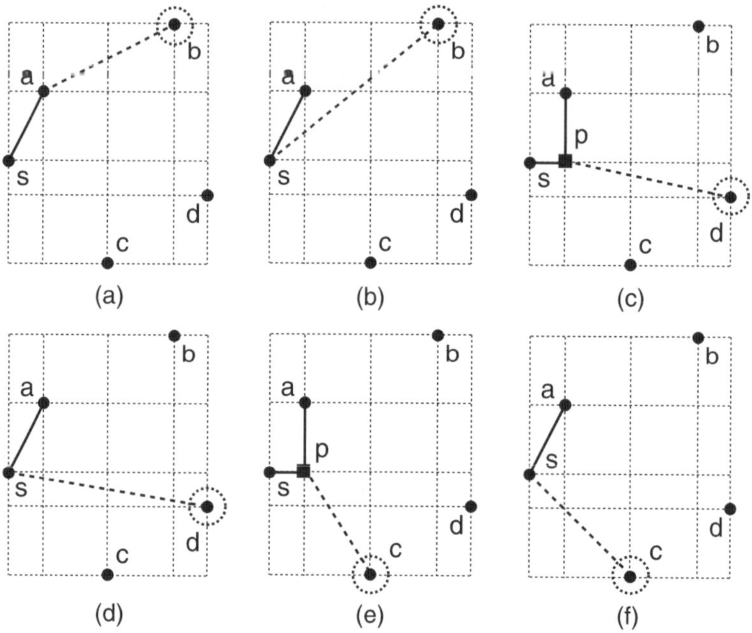

Figure 5.45. Second iteration of SERT algorithm. **(a–b)** Two ways node b can connect to the tree, **(c–d)** two ways node d can connect to the tree, **(e–f)** two ways node c can connect to the tree. (e) is the tree with the minimum Elmore delay increase.

point is needed. It is obvious that the maximum Elmore delay node is b, which is computed as follows:

$$\begin{aligned} t(b) &= r_d \cdot C_s + r_{(s,a)}(0.5c_{(s,a)} + C_a) + r_{(a,b)}(0.5c_{(a,b)} + z_b) \\ &= 0.212(3000 \cdot 0.083 + 7.1 + 6000 \cdot 0.083 + 7.1) \\ &\quad + 3 \cdot 0.073(3000 \cdot 0.083/2 + 7.1 + 6000 \cdot 0.083 + 7.1) \\ &\quad + 6 \cdot 0.073(6000 \cdot 0.083/2 + 7.1) \\ &= 413.0ps \end{aligned}$$

- Edge (s, b): Figure 5.45(b) shows the case, where node b connects to the source directly. It is obvious that the maximum Elmore delay node is b. We obtain $t(b) = 464.2ps$.
- Edge (p, d): From Figure 5.45(c) we see that the nearest point between d and (s, a) is p, a new Steiner point. We obtain $t(d) = 326.1ps$
- Edge (s, d): Figure 5.45(d) shows the case, where node d connects to the source directly. We obtain $t(d) = 331.0ps$.

- Edge (p, c): From Figure 5.45(e) we see that the nearest point between c and (s, a) is p, a new Steiner point. We obtain $t(c) = 268.6ps$.
- Edge (s, c): From Figure 5.45(f) we obtain $t(c) = 273.5ps$.

Thus, adding (p, c) results in the minimum Elmore delay increase.

(c) Third iteration: Figure 5.46 shows the seven candidate connections we consider. Each node outside the tree can connect to either the nearest

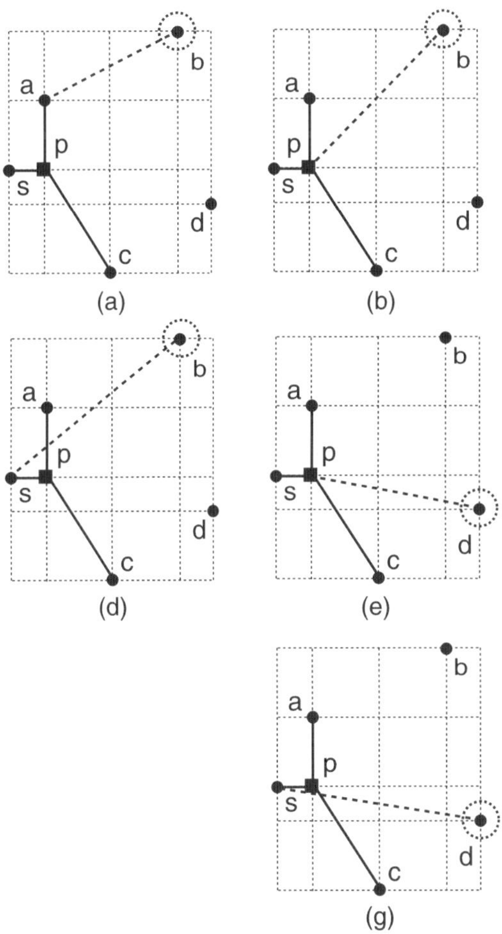

Figure 5.46. Third iteration of SERT algorithm. (**a–d**) Four ways node b can connect to the tree, (**e–g**) three ways node d can connect to the tree. (**f**) is the tree with minimum Elmore delay increase.

point on each edge in the tree via a Steiner point or the source directly.

- Edge (a, b): Figure 5.46(a) shows that the nearest point between b and (p, a) is a. It is obvious that the maximum Elmore delay node is b. We obtain $t(b) = 533.3ps$.
- Edge (p, b): Figure 5.46(b) shows that the nearest point between b and (s, p) is p. We obtain $t(b) = 579.6ps$.
- Edge (q, b): Figure 5.46(c) shows that the nearest point between b and (p, c) is q, a new Steiner point. We obtain $t(b) = 569.6ps$
- Edge (s, b): Figure 5.46(d) shows the case, where node b connects to the source directly. We obtain $t(b) = 553.7ps$ and $t(c) = 427.9ps$. Thus, the maximum Elmore delay is $t(b) = 553.7ps$.
- Edge (p, d): Figure 5.46(e) shows that node p is the nearest point for both d-(p, a) pair and d-(s, p) pair. We obtain $t(d) = 446.4ps$
- Edge (q, d): Figure 5.46(f) shows that the nearest point between d and (p, c) is q, a new Steiner point. We obtain $t(d) = 413.3ps$
- Edge (s, d): Figure 5.46(g) shows the case, where node d connects to the source directly. We obtain $t(d) = 420.5ps$ and $t(c) = 393.3ps$. Thus, the maximum Elmore delay is $t(d) = 420.5ps$.

Thus, adding (q, d) results in the minimum Elmore delay increase.

(d) Fourth iteration: Figure 5.47 shows the six candidate connections we consider. Node b can connect to either the nearest point on each edge in the tree via a Steiner point or the source directly.

- Edge (a, b): Figure 5.47(a) shows that the nearest point between b and (p, a) is a. We obtain $t(b) = 606.3ps$ and $t(d) = 557.3ps$. Thus, the maximum Elmore delay is $t(b) = 606.3ps$.
- Edge (p, b): Figure 5.47(b) shows that the nearest point between b and (s, p) is p. We obtain $t(b) = 652.5ps$ and $t(d) = 557.3ps$. Thus, the maximum Elmore delay is $t(b) = 652.5ps$.
- Edge (r, b): Figure 5.47(c) shows that the nearest point between b and (p, q) is r, a new Steiner point. We obtain $t(b) = 680.0ps$ and $t(d) = 631.0ps$. Thus, the maximum Elmore delay is $t(b) = 680.0ps$.
- Edge (q, b): Figure 5.47(d) shows that the nearest point between b and (q, c) is q. We obtain $t(b) = 833.0ps$.
- Edge (r, b): Figure 5.47(e) shows that the nearest point between b and (q, d) is r, a new Steiner point. We obtain $t(b) = 762.5ps$.

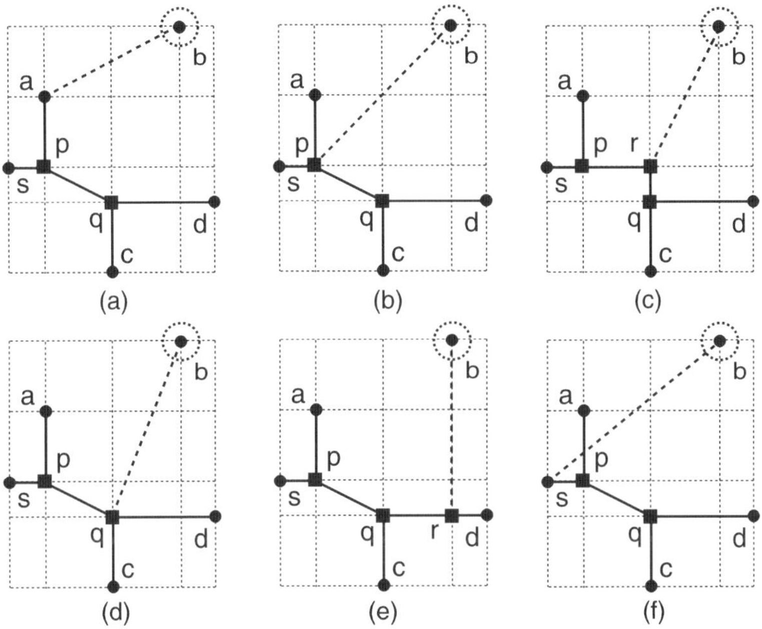

Figure 5.47. Fourth iteration of SERT algorithm. (**a–f**) Six ways node b can connect to the tree. (s) is the tree with minimum Elmore delay increase.

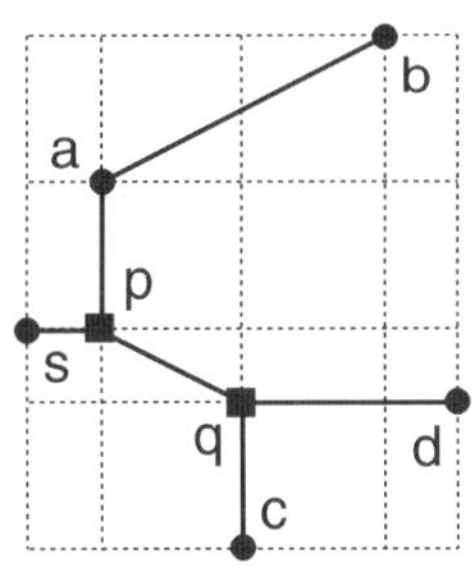

Figure 5.48. Final tree obtained by SERT algorithm with the maximum Elmore delay $t(b) = 606.3ps$.

- Edge (s, b): Figure 5.47(f) shows the case, where b connects to the source directly. We obtain $t(b) = 608.0ps$ and $t(d) = 573.2ps$. Thus, the maximum Elmore delay is $t(b) = 608.0ps$.

Thus, adding (a, b) results in the minimum Elmore delay increase.

Figure 5.48 shows the final tree built by SERT algorithm, where the maximum Elmore delay is $t(b) = 606.3ps$.

6. More Practice Problems

1. Perform the L-RST algorithm [Ho et al., 1990] on the following point set:

$$\{s(3,5), a(2,6), b(5,4), c(10,6), d(7,2), e(7,7), f(1,9)\}$$

   Node s as the source. Show that the final L-RST is stable under re-routing.

2. Perform the 1-Steiner algorithm by Borah, Owens, and Irwin [Borah et al., 1994] on the BR-MSTs shown in Figure 5.29. Can we run the L-RST algorithm [Ho et al., 1990] on these MSTs? Why or Why not?

3. Complete the remaining steps of the 1-Steiner algorithm using the MST shown in Figure 5.15(d). How does the final solution compare to the tree shown in Figure 5.17?

4. Perform the remaining step of the Borah, Owens, and Irwin algorithm [Borah et al., 1994] with the following pairs: $\{a,(b,c)\}$, $\{b,(a,c)\}$, $\{b,(c,e)\}$, $\{c,(e,f)\}$, $\{c,(b,d)\}$. (see Table 5.1)

5. Consider the following point set:

$$\{s(2,6), a(5,4), b(7,1), c(7,2), d(6,6), e(1,2)\}$$

   Use s as the source node.

   (a) Perform 1-Steiner point insertion using the "naive" method presented in [Kahng and Robins, 1992].

   (b) Perform a single pass of 1-Steiner point insertion using the Borah, Owens, and Irwin algorithm [Borah et al., 1994]

   (c) Compare the Steiner trees built by the two algorithms in terms of wire-length.

6. Consider the point set given in problem #6.

   (a) Construct BR-MSTs using the BPRIM algorithm under $\epsilon = 0$, $\epsilon = 0.5$, and $\epsilon = \infty$. Compare these trees in terms of radius and wirelength.

   (b) Construct a BR-MST using the BRBC algorithm under $\epsilon = 0.5$.

   (c) Compare the BR-MSTs built by the BPRIM and BRBC algorithms under $\epsilon = 0.5$ in terms of radius and wirelength.

7. Perform the A-tree algorithm on the problem instance shown in Figure 5.37, where the safe moves are chosen based on the reverse alphabetical order.

8. Perform the 1-Steiner algorithm by Kahng and Robins on the problem instance shown in Figure 5.37. How does the final wirelength compare to the one obtained by the A-tree algorithm shown in Figure 5.39(i)? What about the source-sink path length?

9. Perform the A-tree algorithm on the following point set:

$$\{s(0,0), a(2,6), b(5,4), c(3,5), d(2,2), e(1,3)\}$$

   where the origin is the source. Choose the safe moves based on alphabetical order.

10. Repeat the SERT algorithm practice problem shown on page 185 using the $65nm$ technology used in the ERT algorithm on page 182.

11. Consider the following point set: {(2,6), (5,4), (3,5), (7,2), (1,2)}. Use (5,4) as the source node. Use the following parameters for $65nm$ technology: unit-length resistance is $r = 0.4\,\Omega/\mu m$, unit-length capacitance is $c = 0.2fF/\mu m$, driver output resistance is $r_d = 250\,\Omega$, and input capacitance of the sinks is $50fF$.

    (a) Perform the ERT algorithm. Break ties based on alphabetical order.

    (b) Perform the SERT algorithm.

7. Probing Further

Disclaimer: The list here is meant to be representative, not comprehensive. A comprehensive survey on Steiner routing is provided in [Kahng and Robins, 1994] and in [Cong et al., 1996].

L-Shaped Steiner Routing Algorithm

The authors of [Robins and Salowe, 1994] solved the Bounded Degree Minimum Spanning Tree (BDMST) problem, where one seeks an MST with the maximum node degree bounded by a constant D. When $D = 2$, the problem becomes the well-known NP-hard Traveling Salesman Problem. $D = 8$ in rectilinear plane is polynomial-time solvable as shown by [Ho et al., 1990]. The authors of [Robins and Salowe, 1994] showed that in rectilinear plane $D \leq 3$ is NP-hard and that $D \geq 4$ is polynomial-time solvable.

The L-RST built by [Ho et al., 1990] is stable under re-routing, where any alternate path between a sink and a Steiner point does not reduce the total wirelength of the tree further. The authors of [Bozorgzadeh et al., 2001] presented an optimal algorithm which takes a Steiner tree and outputs a more flexible Steiner tree. The main idea is to identify the Steiner points that can be moved without changing the stability of the tree as well as its topology. They showed that a net with a flexible Steiner tree increases its routability.

1-Steiner Routing Algorithms

The authors of [Griffith et al., 1994] developed a straightforward, efficient (O^3) implementation of [Kahng and Robins, 1992], achieving a speedup factor of three orders of magnitude over previous $O(n^4 \log n)$ implementation. A key observation is that once we have computed an MST over the point set P, the addition of a single new point x into P can only induce a small constant number of changes between $MST(P)$ and $MST(P \cup \{x\})$. They also give a parallel implementation that achieves near-linear speedup on multiple processors.

The original 1-Steiner heuristic proposed by [Borah et al., 1994] is extended by the same authors in [Borah et al., 1997] to build routing trees with near optimal Elmore delay. The basic approach is the same in both works: replace an existing edge in the tree for another costlier edge, possibly introducing a Steiner point. However, the gain formulation used in [Borah et al., 1997] is based on Elmore delay improvement instead of wirelength reduction as in [Borah et al., 1994]. The $O(n^2)$ time complexity stays the same in the new Elmore delay-oriented heuristic [Borah et al., 1997].

The authors of [Mandoiu et al., 2000] presented an approximation algorithm for 1-Steiner problem, where a 3/2 approximation algorithm named RV [Rajagopalan and Vazirani, 1999] is adopted. The RV algorithm is designed for the metric Steiner tree problem on quasi-bipartite graphs that do not contain edges connecting pairs of Steiner vertices. The RV algorithm is built around the linear programming relaxation of an integer program formulation. Their heuristic achieves a good running time by combining an efficient implementation of the RV algorithm with simple, but powerful geometric reductions.

The authors of [Zarkesh-Ha et al., 2000] presented a stochastic wirelength distribution for global interconnects in a non-homogeneous System-On-Chip (SOC), which is useful in system-level interconnect planning. The distribution is derived using novel models for netlist, placement, and routing information, where the routing model is constructed based on 1-Steiner algorithm of [Kahng and Robins, 1992]. Through comparison with actual product data, it is shown that the proposed stochastic model successfully predicts the global net-length distribution of a heterogeneous system.

The author of [Zhou, 2004] presented on $O(n \log n)$ time and $O(n)$ space heuristic for 1-Steiner construction, which is the fastest known 1-Steiner heuristic. They combined the edge replacement heuristic of [Borah et al., 1994] and the concept of Spanning Graph in [Zhou et al., 2002]. The spanning graph is an intermediate step for the MST construction, where it is defined as a graph that contains at least one MST. This spanning graph helps reduce the number of point-edge pairs examined for tree improvements.

Bounded Radius Routing Algorithms

Prim's MST algorithm minimizes the wirelength while Dijkstra's shortest path algorithm minimizes the radius if used to construct rectilinear MST. The authors of [Alpert et al., 1995] studied the tradeoff between these min-cost and min-radius objectives and provided an algorithm named AHHK to balance the objectives. The key idea is to use a combined cost function when choosing an edge to be added to the growing tree.

The authors of [Cong and Madden, 1997b] studied the problem of routing nets with multiple sources, such as those found in signal busses. When there are multiple sources and sinks, path length minimization can be achieved by minimizing the maximum distance between any pair of nodes, which leads to diameter minimization. Their goal is to construct a minimum diameter routing tree with minimum total tree cost, as measured by a combination of maximum path length, average path length, and total tree length.

The authors of [Alpert et al., 2002] solved the buffered Steiner tree for "difficult instances", which are characterized by a large number of sinks, large variations in sink criticalities, etc. Their solution is C-tree, a two-level Steiner router that first clusters sinks with common characteristics together, constructs low-level Steiner trees for each cluster, then performs a timing-driven Steiner construction on the top-level clustering. The authors used the AHHK heuristic [Alpert et al., 1995] as the tree topology constructor.

Buffer block planning is a popular method for early stage interconnect planning. The authors of [Jiang et al., 2004] performed simultaneous floorplanning and buffer block planning for more rigorous interconnect planning. In order to accurately compute the size and location of buffer blocks, the authors performed global routing for all nets using the AHHK heuristic [Alpert et al., 1995]. This routing result is also used to satisfy timing constraints.

The Minimum Radius Minimum Cost (MRMC) routing problem seeks a routing tree with the minimum radius and wirelength. The authors of [Ho et al., 2005] proposed a fast multi-level routing considering crosstalk and performance optimization. To handle crosstalk minimization, they incorporated an intermediate stage of layer/track assignment into the multilevel routing framework. For MRMC routing, they adopted the Bounded Radius Bounded Cost (BRBC) algorithm [Cong et al., 1992] to construct performance-oriented routing trees.

A-tree Algorithm

Given a signal net with a required arrival time associated with each sink, the authors of [Okamoto and Cong, 1996] presented an efficient algorithm that finds a Steiner tree with buffer insertion and wire sizing so that the required arrival time (or timing slack) at the source is maximized. A unique contribution of this algorithm is that it performs Steiner tree construction, buffer insertion, and wire sizing simultaneously with consideration of both critical delay and total capacitance minimization. The authors combined the A-tree construction and dynamic programming based buffer insertion and wire sizing to accomplish this goal.

The authors of [Alexander and Robins, 1996] presented two Steiner arborescence based heuristics to perform performance-driven FPGA routing. Given an arbitrary weighted routing graph, their arborescence algorithms produce a Steiner tree, where all source-sink paths are the shortest possible and the total wirelength is optimized as a secondary objective. Their first graph-based Steiner arborescence heuristic named PFA is based on a path-folding strategy that overlaps and merges shortest paths in order to reduce the overall wirelength. The second arborescence heuristic named IDOM iteratively selects Steiner nodes which improve the total wirelength with respect to an optimal spanning arborescence algorithm.

The authors of [Cong et al., 1998] presented the following graph-based heuristics and exact algorithms to construct Steiner arborescence: (1) RSA/G algorithm is an efficient graph-based adaptation of the greedy RSA heuristic in [Rao et al., 1992], (2) RSA/BnB/G algorithm is an optimal exponential-time branch-and-bound variant of RSA/G, (3) RSA/DP/G algorithm is a fast implementation of RSA/BnB/G based on dynamic programming, (4) k-IDeA/G algorithm is a "scaled-down" near optimal version of RSA/BnB/G, and (5) k-IA/G algorithm is a natural dual of k-IDeA/G that implements the IDOM heuristic [Alexander and Robins, 1996]. These heuristics are shown to generate near optimal results and achieve speedups of orders of magnitude over existing algorithms.

The Minimal Cover approach presented in [Ramnath, 2003] is an approximation algorithm for minimum Rectilinear Steiner Arborescence (RSA) construction that runs in $O(n \log n)$ time. The minimal cover algorithm maintains an expanding "diamond", which initially consists of the origin. The arborescence is constructed as the diamond expands. At any intermediate stage, the arborescence contains several incomplete edges, all of which terminate at the boundary of the diamond. The edges are extended as the diamond expands. There are three kinds of events that take place as the diamond expands: (1) an incomplete edge is extended, (2) a Steiner point is introduced, and (3) a sink is encountered. This approach yields an RSA that has a wirelength no more than twice the optimal RSA.

Elmore Routing Tree Algorithms

Permutation-constrained routing tree (P-tree) [Lillis et al., 1996] is an Elmore delay-oriented Steiner tree that balances the area and delay objectives. A key idea is that routing topologies are induced by a permutation on the sinks of a net, which is quite different from the conventional "tree growth" scheme as in [Boese et al., 1995]. Each topology is then mapped into a routing tree under the area/deay tradeoff. They presented several ideas to obtain "high quality permutations", which translate to high quality routing trees. P-tree also incorporates simultaneous wire-sizing for further delay improvement.

MVERT [Hou et al., 1999] (Maximum delay Violation Elmore Routing Tree) is a timing-driven Steiner tree construction algorithm that is based on non-Hanan grid. The algorithm works in two phases, where a minimum-delay Steiner Elmore routing tree is first constructed using a minor variant of ERT algorithm [Boese et al., 1995]. A difference is that they minimize the maximum delay violation rather than the maximum delay. Then the tree is iteratively modified, using an efficient search method, to reduce its wirelength. Non-Hanan nodes are introduced during this refinement phase. The search method exploits the piecewise concavity of the delay function to arrive at a solution efficiently.

The RATS-tree algorithm [Cong et al., 2001] considers the inductance effect as well as the RC parasitics during Steiner tree routing. The proposed RLC model is more suitable than the RC-based Elmore delay model for Multi-Chip Module (MCM) global routing. RATS-tree is capable of constructing a large class of routing topologies, ranging from shortest path Steiner trees [Cong et al., 1993] to bounded radius Steiner trees [Cong et al., 1992] and Elmore Steiner trees [Boese et al., 1995]. By considering a large class of routing topologies, they are able to produce a set of topologies, providing smooth tradeoffs among signal delay, waveform, routing area, and congestion.

3D IC is a new system integration technology, where bare dies are stacked on top of each other and are vertically connected with Through-Silicon-Vias (TSVs). The authors of [Pathak and Lim, 2007] presented the first Steiner routing algorithm for 3D ICs. Their algorithm is an extension of SERT [Boese et al., 1995], where a two-variable optimization is performed to decide the connection point and TSV location for Elmore delay minimization upon each edge addition. Given a set of 3D Steiner trees, they also proposed an integer linear programming formulation to reposition the TSVs for thermal optimization.

Chapter 6

MULTI-NET ROUTING

Given a graph $G(V, E)$ that represents the routing resources such as channels and regions and a netlist NL that consists of multiple nets, where each net is a subset of vertices in V, the goal of multi-net routing is to construct routing trees of all nets in NL such that the capacity constraint specified in each edge in E is satisfied. The objective is to minimize the total wirelength, routing congestion, longest source-sink path length, etc. This chapter presents sample problems related to the following works:

- Steiner min-max tree algorithm [Chiang et al., 1990]
- Multi-commodity flow routing algorithm [Shragowitz and Keel, 1987]
- Iterative deletion algorithm [Cong and Preas, 1988]
- Yoshimura and Kuh algorithm [Yoshimura and Kuh, 1982]

The first router is a sequential router, where the nets are routed one-by-one. The others route all nets simultaneously. The first three routers are global routers, whereas the last one is a detailed router.

S.K. Lim, *Practical Problems in VLSI Physical Design Automation*

1. Steiner Min-Max Tree Algorithm

Given an undirected, edge-weighted graph $G(V, E)$ and a net n that contains a subset of nodes $D \subseteq V$, the Steiner Min-Max Tree (SMMT) of n is a Steiner tree of n, where the maximum weight among all edges in the tree is minimized. The rationale behind minimizing the maximum edge weight rather than the traditional wirelength or diameter objectives is that we try to minimize congestion (and thus improve routability) in case of routing multiple nets. If the edge weight indicates the routing usage (= congestion) of the edge, minimizing the maximum edge weight in the Steiner tree means avoiding congested spots. Unlike the NP-hard Steiner tree problem for wirelength minimization, this SMMT problem can be solved optimally with a simple and efficient algorithm named ALG-SMMT presented by Chiang, Sarrafzadeh, and Wong [Chiang et al., 1990]. However, if we desire that the wirelength is also minimized in SMMT, this problem so called MSMMT becomes NP-hard.

In case of the "sequential routing approach" for multiple-net routing, where the nets are routed one-by-one, the quality of routing heavily depends on the net order. This is because the nets that are routed earlier become obstacles for the nets to be routed. The authors of [Chiang et al., 1990] first suggested various factors to consider to obtain good orderings such as half-perimeter of the bounding box (HPBB) and the number of terminals of the net. Then they proposed a heuristic algorithm to solve the MSMMT problem, which consists of two phases, namely, SMMT and Shortest Path (SP) phase. The basic approach is to build SMMTs of the nets first and then re-route them for further wirelength minimization. The SMMT phase consists of multiple passes, where we visit the nets one-by-one and route them using their ALG-SMMT algorithm in each pass. If the wirelength of SMMT of a net is below the threshold, we accept it; otherwise, we skip it and re-route it during the next pass or the SP phase. The SP phase is also iterative, where each net is ripped-up and re-routed using the ALG-SP algorithm, which resembles Prim's MST algorithm [Prim, 1957], for wirelength minimization in each pass. The user specifies the total number of passes used for the SMMT and SP phases.

Quick Overview

During the SMMT phase, we first order the nets in an increasing order of their HPBB. Given a net n to be routed in a routing graph G, the ALG-SMMT algorithm first builds T, a minimum spanning tree (MST) of G (not n). Then we remove all "degree 1 Steiner nodes" and their incident edges from T one-by-one. Any node in T that is not a terminal of n is a Steiner node, and a degree 1 Steiner node has only one edge incident to it. Once all the degree 1 Steiner nodes are removed, T becomes an optimal SMMT. If the wirelength of T is shorter than the threshold, we accept it; otherwise we reject the SMMT.

Once we finish our SMMT routing effort for all nets, the current pass ends. The entire SMMT pass is repeated if desired. Note that the nets that we failed to route during the current pass may become routable during the next pass because the underlying routing graph may have been updated.

During the SP phase, we visit the nets in the same order as in the SMMT phase and try to re-route them for wirelength minimization. Given a net n to be re-routed in a routing graph G, we first rip-up n from G and update the edge weights in G. Let T denote the final SP-tree for n. The ALG-SP algorithm selects a source terminal, which is the closest to the geometric center of the bounding box of n, and adds it to T. We then find a terminal of n that is the closest to T and add it to T via a shortest feasible path. A path is feasible if it does not violate edge capacity constraints. We add the remaining terminals of n one-by-one until T spans all terminals. Lastly, we accept T if the wirelength or the maximum edge weight is reduced after the re-routing. Once we finish our SP re-routing effort for all nets, the current pass ends. The entire SP pass is repeated if desired. As in the case of SMMT phase, SP may give us different routing trees for the same net in different passes because the underlying routing graph may have been updated.

Practice Problem

Consider the following netlist, where the nets are sorted based on their HPBB:

$$
\begin{aligned}
n_1 &= \{(1,0),(0,3),(3,2),(3,4)\} \\
n_2 &= \{(0,2),(3,0),(4,3)\} \\
n_3 &= \{(1,1),(2,2),(4,0),(4,4)\} \\
n_4 &= \{(0,0),(2,1),(1,3),(4,1),(2,4)\} \\
n_5 &= \{(2,0),(0,4),(4,2),(3,3)\}
\end{aligned}
$$

These nets are to be routed in this order. Use a 5×5 mesh for the routing graph G, where the weight of each edge in G corresponds to the demand, i.e., current usage. The edge capacity is set to 3.

1. Perform a single pass of the SMMT-phase under $c_j = 2.0$.

   We visit the nets in the given order and apply the ALG-SMMT algorithm.

   (a) Net n_1: Figure 6.1(a) shows the 4 terminals of n_1, where HPBB is 7. The weight of edges in the routing graph is zero. Figure 6.1(b) shows a MST of the routing graph. Note that there exist many other MSTs with the same cost. Lastly, Figure 6.1(c) shows the final SMMT with the maximum edge weight of 0 and wirelength of 9. We obtain this SMMT after removing all degree-1 Steiner vertices from the initial MST. We

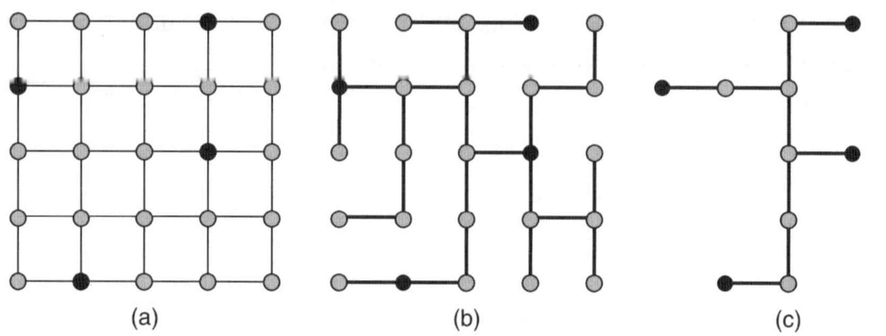

Figure 6.1. (**a**) Net n_1 with HPBB of 7, where the weight of all edges in the underlying routing graph G is initially set to zero, (**b**) a MST of G, (**c**) final SMMT with maximum edge weight of 0, and wirelength of 9. We accept this solution because $9 < 2.0 \times 7$.

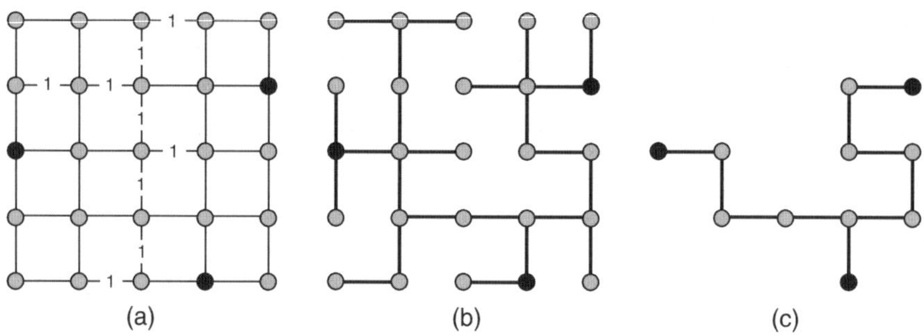

Figure 6.2. (**a**) Net n_2 with HPBB of 7, where the edge label in G denotes the current weight (no label indicates zero weight), (**b**) a MST of G, (**c**) final SMMT with maximum edge weight of 0, and wirelength of 10. We accept this solution because $10 < 2.0 \times 7$.

accept this SMMT because the wirelength is shorter than the threshold $(9 < 2.0 \times 7)$.

(b) Net n_2: Figure 6.2(a) shows the 3 terminals of n_2. Note that the edge weights in the routing graph are updated based on the routing result of n_1. Figure 6.2(b) shows a MST of the routing graph. Again this MST is not unique. Lastly, Figure 6.2(c) shows the final SMMT with the maximum edge weight of 0 and wirelength of 10. We accept this SMMT because the wirelength is shorter than the threshold $(10 < 2.0 \times 7)$.

(c) Net n_3: Figure 6.3(a) shows the 4 terminals of n_3. The routing graph now contains the routing results of n_1 and n_2. Figure 6.3(b) shows a MST of the routing graph. Lastly, Figure 6.3(c) shows the final SMMT with the maximum edge weight of 1 and wirelength of 15. We reject this SMMT because the wirelength is longer than the threshold $(15 > 2.0 \times 7)$.

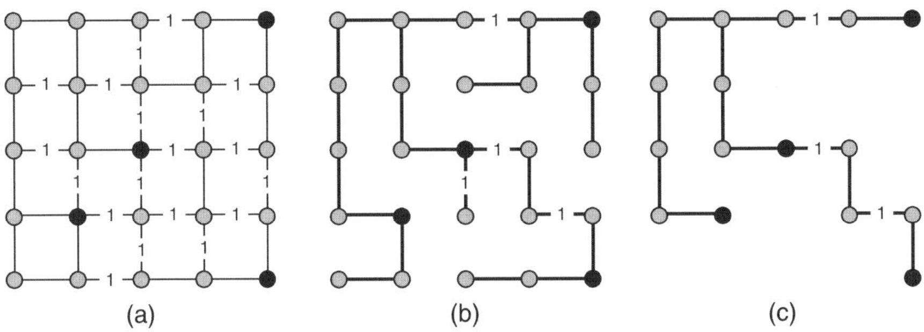

Figure 6.3. (**a**) Net n_3 with HPBB of 7, (**b**) a MST of G, (**c**) final SMMT with maximum edge weight of 1, and wirelength of 15. We reject this solution because $15 > 2.0 \times 7$.

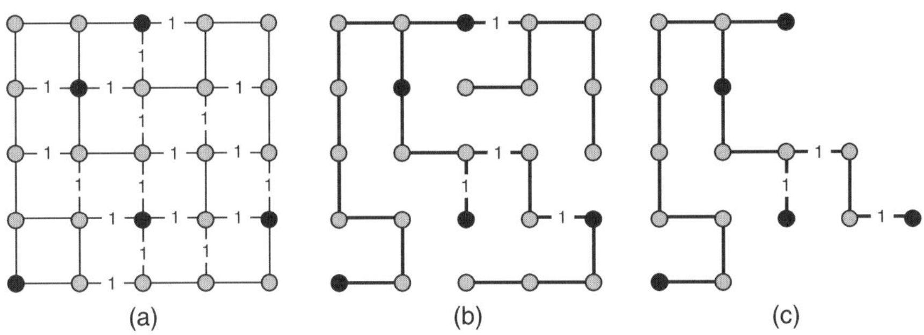

Figure 6.4. (**a**) Net n_4 with HPBB of 8, (**b**) a MST of G, (**c**) final SMMT with maximum edge weight of 1, and wirelength of 15. We accept this solution because $15 < 2.0 \times 8$.

(d) Net n_4: Figure 6.4(a) shows the 5 terminals of n_4. Note that we use the same routing graph as the one used for n_3 because the routing of n_3 has failed. Figure 6.4(b) shows a MST of the routing graph. Lastly, Figure 6.4(c) shows the final SMMT with the maximum edge weight of 1 and wirelength of 15. We accept this solution because the wirelength is shorter than the threshold $(15 < 2.0 \times 8)$.

(e) Net n_5: Figure 6.5(a) shows the 4 terminals of n_5. The routing graph now contains the routing results of n_1, n_2, and n_4. Figure 6.5(b) shows a MST of the routing graph. Lastly, Figure 6.5(c) shows the final SMMT with the maximum edge weight of 1 and wirelength of 12. We accept this SMMT because the wirelength is shorter than the threshold $(12 < 2.0 \times 8)$.

Figure 6.6(a) shows the final routing graph that contains the routing results of n_1, n_2, n_4, and n_5. The maximum demand (= current usage) among all edges is 2. Figure 6.6(b–e) show the SMMTs of the routed nets.

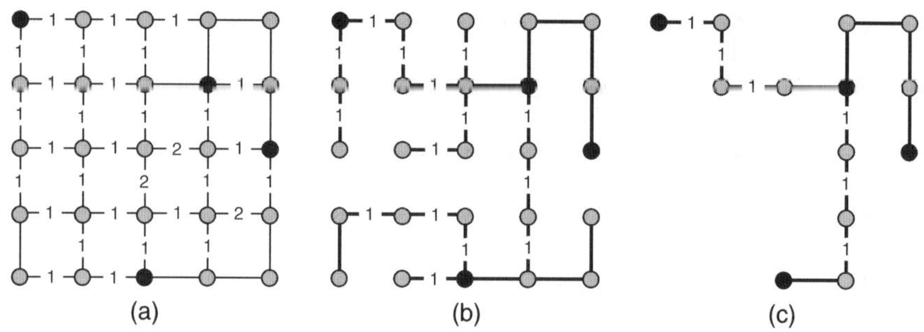

Figure 6.5. (**a**) Net n_5 with HPBB of 8, (**b**) a MST of G, (**c**) final SMMT with maximum edge weight of 1, and wirelength of 12. We accept this solution because $12 < 2.0 \times 8$.

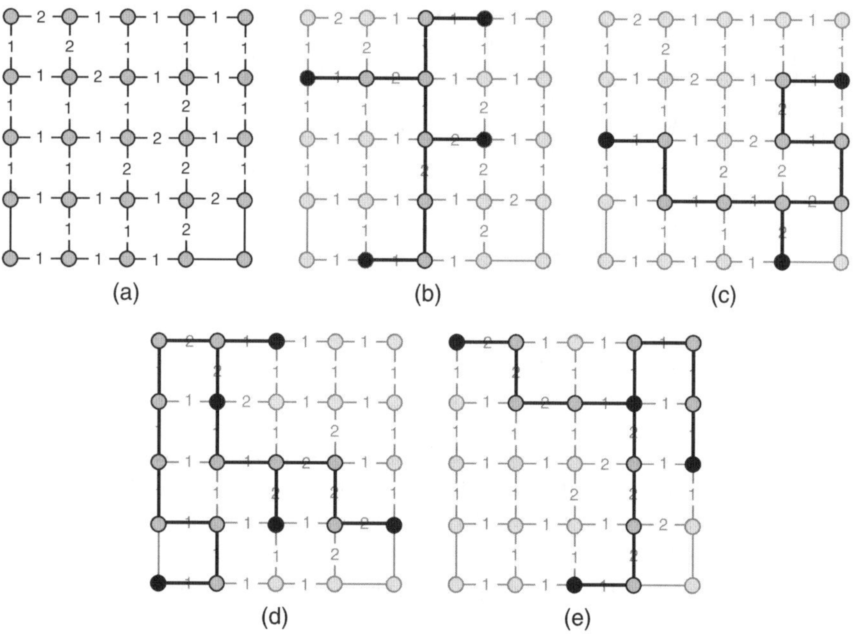

Figure 6.6. (**a**) Final routing graph, (**b**) SMMT of n_1, (**c**) SMMT of n_2, (**d**) SMMT of n_4, (**e**) SMMT of n_5. Note that n_3 routing has failed.

2. Perform a single pass of the SP-phase.

   We visit the nets in the given order and attempt to re-route them using the ALG-SP algorithm.

   (a) Net n_1: Figure 6.7(a) first shows the SMMT of n_1 we obtained from the SMMT phase. Our goal is to rip up SMMT of n_1 and re-route it using ALG-SP algorithm. Figure 6.7(b) shows the routing graph after removing SMMT of n_1. The arrow points to the source terminal $s = (3, 2)$

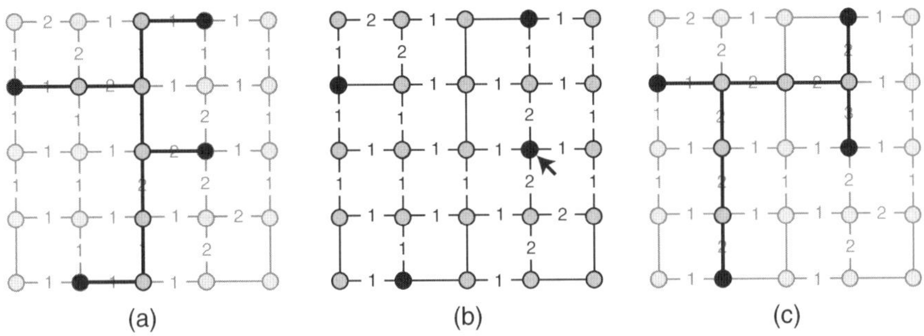

Figure 6.7. (**a**) SMMT of n_1, (**b**) routing graph after ripping up SMMT of n_1. The arrow points to the source for SP computation, (**c**) SP of n_1.

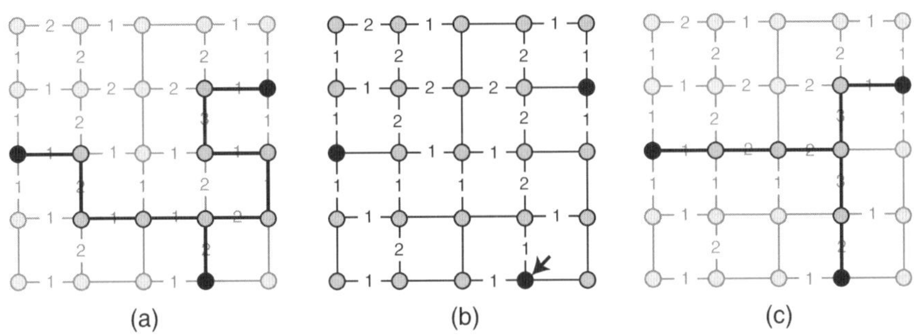

Figure 6.8. (**a**) SMMT of n_2, (**b**) routing graph after ripping up SMMT of n_2. The arrow points to the source for SP computation, (**c**) SP of n_2.

needed in ALG-SP. This terminal is chosen because it is the closest to the geometric center of the bounding box of n_1. The order of terminals added to grow s is as follows: $(3, 4)$, $(0, 3)$, $(1, 0)$. Figure 6.7(c) shows the final SP of n_1. The routing graph is updated to reflect the re-routing of n_1.

(b) Net n_2: Figure 6.8(a) shows the SMMT of n_2 we obtained from the SMMT phase. The underlying routing graph is the same as the one shown in Figure 6.7(c). Figure 6.8(b) shows the routing graph after ripping up the SMMT of n_2. We select $s = (3, 0)$ as the source terminal. The order of terminals added to grow s is as follows: $(4, 3)$, $(0, 2)$. Figure 6.8(c) shows the final SP of n_2. The routing graph is updated to reflect the re-routing of n_2.

(c) Net n_3: SMMT of n_3 does not exist due to the routing failure during the SMMT phase. Figure 6.9(b) shows the routing graph for n_3, which is the same as Figure 6.8(c). The source terminal $s = (2, 2)$. The order of terminals added to grow s is as follows: $(1, 1)$, $(4, 0)$, $(4, 4)$.

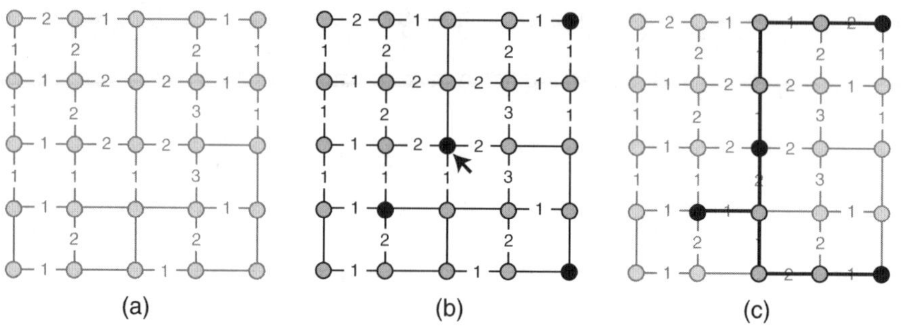

Figure 6.9. (**a**) SMMT of n_3 does not exist due to the routing failure, (**b**) routing graph for n_3. The arrow points to the source for SP computation, (**c**) SP of n_3.

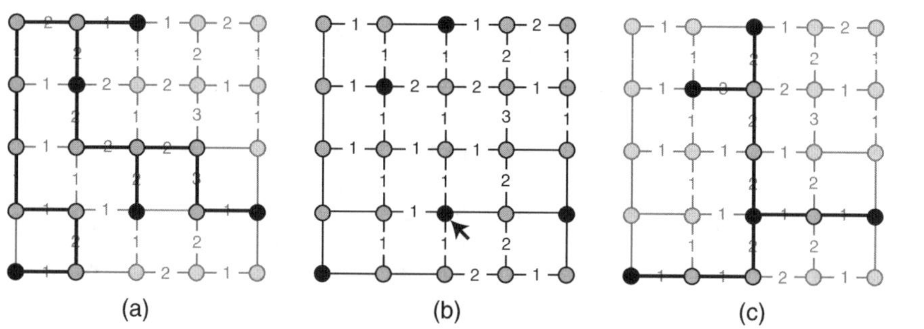

Figure 6.10. (**a**) SMMT of n_4, (**b**) routing graph after ripping up SMMT of n_4. The arrow points to the source for SP computation, (**c**) SP of n_4.

Figure 6.9(c) shows the final SP of n_3. The routing graph is updated to reflect the routing of n_3.

(d) Net n_4: Figure 6.10(a) shows the SMMT of n_4 we obtained from the SMMT phase. The underlying routing graph is the same as the one shown in Figure 6.9(c). Figure 6.10(b) shows the routing graph after ripping up the SMMT of n_4. We select $s = (2, 1)$ as the source terminal. The order of terminals added to grow s is as follows: $(4, 1)$, $(0, 0)$, $(1, 3)$, $(2, 4)$. Figure 6.10(c) shows the final SP of n_4. The routing graph is updated to reflect the re-routing of n_4.

(e) Net n_5: Figure 6.11(a) shows the SMMT of n_5 we obtained from the SMMT phase. The underlying routing graph is the same as the one shown in Figure 6.10(c). Figure 6.11(b) shows the routing graph after ripping up the SMMT of n_5. We select $s = (2, 0)$ as the source terminal. The order of terminals added to grow s is as follows: $(4, 2)$, $(3, 3)$, $(0, 4)$. Figure 6.11(c) shows the final SP of n_5. The routing graph is updated to reflect the re-routing of n_5.

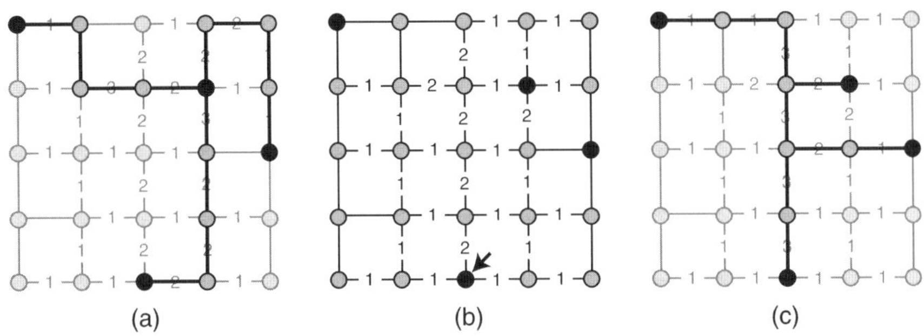

Figure 6.11. (**a**) SMMT of n_5, (**b**) routing graph after ripping up SMMT of n_5. The arrow points to the source for SP computation, (**c**) SP of n_5.

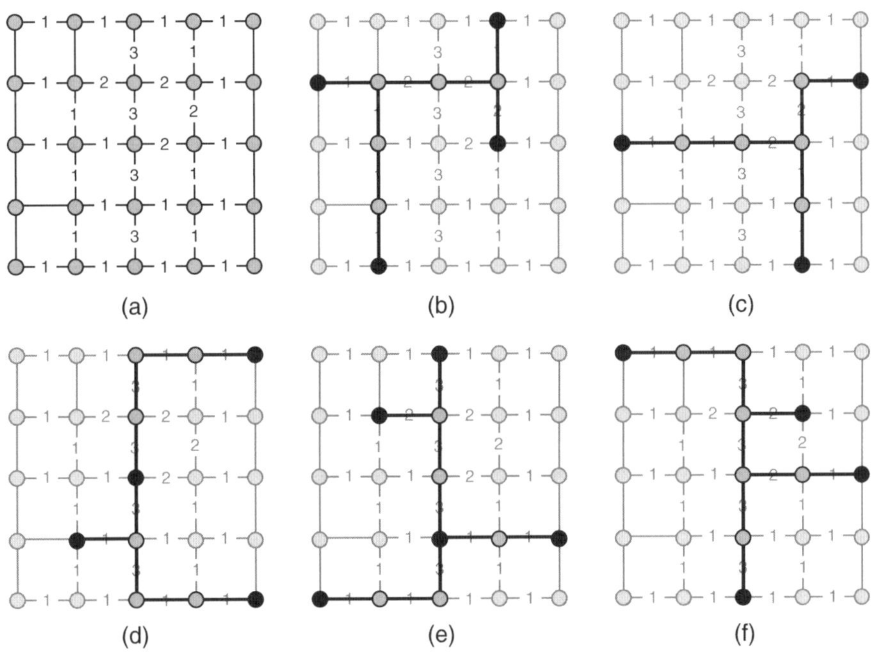

Figure 6.12. SP routing results. (**a**) Final routing graph, (**b–f**) SP of nets n_1 to n_5.

Figure 6.12(a) shows the final routing graph, where the maximum demand among all edges is 3. Figure 6.12(b–f) show the SPs of the nets n_1 to n_5.

3. Compare the results obtained by SMMT and SP phases.

   Figure 6.13 shows the routing graphs after the SMMT and SP phases. We observe that SMMT phase shows more uniform use of routing resource compared with SP phase. Table 6.1 shows the summary of SMMT and SP phases based on the results shown in Figure 6.6 and Figure 6.12. We observe

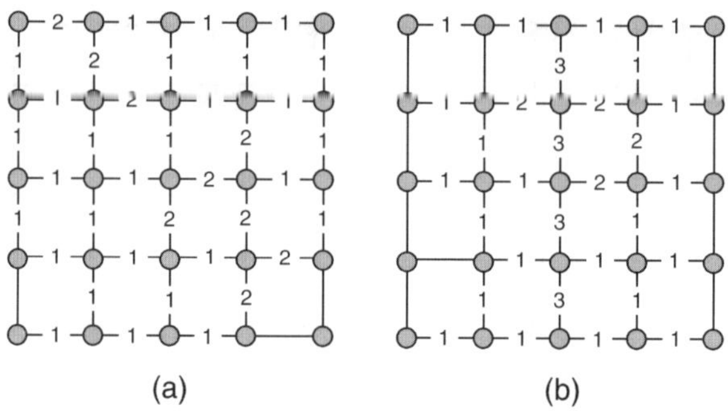

Figure 6.13. (**a**) Routing graph after SMMT phase, (**b**) routing graph after ST phase. SMMT shows more uniform use of routing resource. The edge capacity is set to 3.

Table 6.1. Summary of SMMT and SP phases based on Figures 6.6 and 6.12.

Net	SMMT phase		SP phase	
	max e-wgt	wirelength	max e-wgt	wirelength
n_1	2	9	2	8
n_2	2	10	2	7
n_3	Routing failed		3	9
n_4	2	15	3	9
n_5	2	12	3	9

that the maximum edge weights are higher in SP trees. The wirelengths, however, are shorter in SP trees. This clearly demonstrates the congestion vs wirelength tradeoff existing in multiple net routing.

2. Multi-Commodity Flow Routing Algorithm

The multi-commodity flow problem is a generalization of the network flow problem, where multiple commodities must be shipped from their own sources to the sinks on a common network. The goal is to minimize the total shipping cost. The global routing problem can be formulated as a multi-commodity flow problem, where each net is treated as a commodity, and the routing resource graph is treated as the flow network. It is well known that the multi-commodity flow problem is NP-complete and can be formulated as a linear programming (LP) problem. Thus, one can utilize an existing LP solver to solve the global routing problem. One significant benefit of the multi-commodity flow based global routing is that it routes *all* nets simultaneously, thereby overcoming the net ordering problem. However, the runtime of LP solver increases quickly with bigger size problems. Thus, the research on the multi-commodity flow problem has mostly focused on heuristics and combinatorial approximation algorithms.

It is generally believed that the work by Shragowitz and Keel [Shragowitz and Keel, 1987] is the first reported work on global routing using the multi-commodity flow model. They formulated the routing problem as integer linear programming (ILP) and solved it using a heuristic named MM (minimax) algorithm instead of using an off-the-shelf ILP solver. The objective is to minimize the total wirelength under the capacity constraints. MM algorithm repeatedly performs shortest path computation [Dijkstra, 1959] and rip-up-and-reroute [Ting and Tien, 1983]. One disadvantage of this work is that it can only handle 2-pin nets so that a multi-pin net must be decomposed into a set of two-pin connections and solved independently.

Quick Overview

Given an instance of routing problem that consists of a routing graph $G(V, E)$ and a netlist NL, we first construct the flow network $F(V, A)$. F inherits all nodes in G. For each edge $e(x, y) \in E$, we add a pair of directed arcs (x, y) and (y, x) to A. The cost of each arc represents the cost of routing using the corresponding edge in the routing graph, which is usually set to its Manhattan distance. The ILP formulation of the multi-commodity flow based routing problem consists of the objective function and three sets of constraints: demand, capacity, and integer constraints.

- Objective function: Let $x_a^k \in \{0, 1\}$ denote the integer variable for arc a with respect to net (= commodity) k. $x_a^k = 1$ means that net k uses arc a in its route. Thus, the total number of these integer variables is $|A| \times |NL|$. The objective function is constructed in such a way that we visit each arc in A and compute the weighted sum of its x-variables (see the sample problem for details).

- Demand constraint: Given a node $v \in V$ and a net $k \in NL$, the demand constraint basically states that given a node v, the total amount of flow for net k entering and leaving v equals 1 if v is the source of net k; -1 if v is the sink of net k; 0 otherwise. We set up this equation for all nets in NL for all vertices in V (see the sample problem for details).
- Capacity constraint: for each arc $a \in A$, the total usage, i.e., the sum of x_a^k for all $k \in NL$ should not exceed the capacity of the corresponding edge $e \in E$.
- Integer constraint: the x variables can only take either 0 or 1.

The MM algorithm starts with shortest path computation [Dijkstra, 1959] for all nets while ignoring capacity constraints. We call this result the solution at iteration $r = 0$. We increment r and proceed with the rest of the steps as follows:

- Step 1: we compute M_r, the maximum overflow among all arcs in the solution at iteration $r - 1$. If $M_r \leq 0$, we terminate.
- Step 2: we obtain J_r, the set of arcs with the maximum overflow in the solution at iteration $r - 1$. We also compute J_r^0, the set of arcs with maximum overflow and maximum minus 1 overflow.
- Step 3: we assign ∞ as the cost of all arcs in J_r^0.
- Step 4: we compute K_r, the set of nets that use arcs in J_r in the solution at iteration $r - 1$. We also compute K_r^0, the set of nets that use arcs in J_r^0.
- Step 5: we compute shortest paths for all nets in K_r using the new cost from step 3. We ignore capacity constraint in this case. If there is a net with non-infinity cost, we go to step 6; otherwise we go to step 9.
- Step 6: we choose k^0 among the nets with non-infinity cost from step 5, which is the net with the minimum cost increase between the old and the new routes.
- Step 7: we construct the routing solution for iteration r by using all of the routes from iteration $r - 1$ except for k^0, where we use the new route.
- Step 8: we increment r and go back to step 1.
- Step 9: given a net $k \in K_r^0$, we check to see if $x_p^k \times x_q^k = 0$ for a pair of arcs $p, q \in J_r^0$ in the solution at iteration $r - 1$. If this is true for all such pairs in J_r^0 for all nets in K_r^0, we decide that the routing problem does not have a solution; otherwise, we go to step 10.[26]

[26]The practice problem provided in this section does not require steps 9 and 10.

- Step 10 (escape procedure): we choose $k' \in K_r^0$, the net that uses the maximum number of arcs in J_r^0 in the solution at iteration $r-1$. We rip up k' and reroute it so that the original route is preserved as much as possible. This new route for k' together with old routes from iteration $r-1$ for other nets becomes the solution for iteration r. We go back to step 1.

Practice Problem

Consider the routing graph G shown in Figure 6.14(a), where the capacity of all edges is 2. Figure 6.14(b) shows the flow network.[27] Table 6.2 shows the cost of the arcs in the flow network. The following six nets are to be routed: $n_1 = \{a, l\}, n_2 = \{i, c\}, n_3 = \{d, f\}, n_4 = \{k, d\}, n_5 = \{g, h\}, n_6 = \{b, k\}$. The first node in each net is the source.

1. Set up and solve the integer linear programming (ILP) formulation of the multi-commodity flow based global routing.

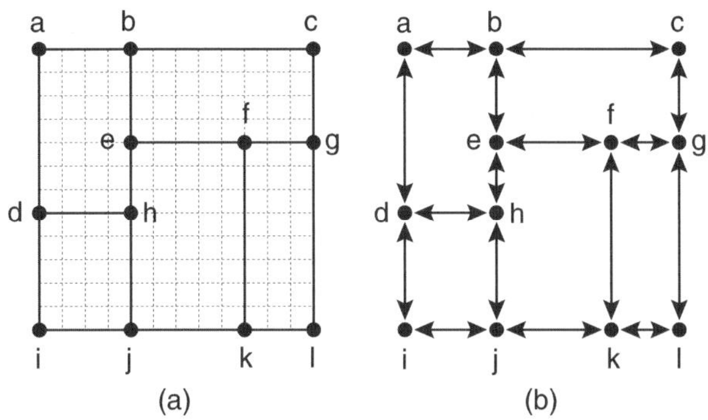

Figure 6.14. (**a**) Routing graph, where the capacity of all edges is 2, (**b**) its flow network.

Table 6.2. List of the arcs in the flow network.

Arc	Cost	Arc	Cost	Arc	Cost	Arc	Cost
(a,b)	4	(b,a)	4	(b,c)	8	(c,b)	8
(d,h)	4	(h,d)	4	(e,f)	5	(f,e)	5
(f,g)	3	(g,f)	3	(i,j)	4	(j,i)	4
(j,k)	5	(k,j)	5	(k,l)	3	(l,k)	3
(a,d)	7	(d,a)	7	(d,i)	5	(i,d)	5
(b,e)	4	(e,b)	4	(e,h)	3	(h,e)	3
(h,j)	5	(j,h)	5	(f,k)	8	(k,f)	8
(c,g)	4	(g,c)	4	(g,l)	8	(l,g)	8

[27]This directed graph, where bi-directional edges are used for each channel, is preferred over an undirected graph. The number of constraints is smaller with directed graphs because absolute functions are not necessary in its ILP.

First, we compute the objective function as follows: Let x_e^k denote the integer variable for arc e with respect to net (= commodity) k. The total number of these x-variables is $16 \times 2 \times 6 = 192$. The objective function is as follows:

$$\begin{aligned}
&4(x_{a,b}^1 + \cdots + x_{a,b}^6) + 4(x_{b,a}^1 + \cdots + x_{b,a}^6) + 8(x_{b,c}^1 + \cdots + x_{b,c}^6) \\
&+ 8(x_{c,b}^1 + \cdots + x_{c,b}^6) + 4(x_{d,h}^1 + \cdots + x_{d,h}^6) + 4(x_{h,d}^1 + \cdots + x_{h,d}^6) \\
&+ 5(x_{e,f}^1 + \cdots + x_{e,f}^6) + 5(x_{f,e}^1 + \cdots + x_{f,e}^6) + 3(x_{f,g}^1 + \cdots + x_{f,g}^6) \\
&+ 3(x_{g,f}^1 + \cdots + x_{g,f}^6) + 4(x_{i,j}^1 + \cdots + x_{i,j}^6) + 4(x_{j,i}^1 + \cdots + x_{j,i}^6) \\
&+ 5(x_{j,k}^1 + \cdots + x_{j,k}^6) + 5(x_{k,j}^1 + \cdots + x_{k,j}^6) + 3(x_{k,l}^1 + \cdots + x_{k,l}^6) \\
&+ 3(x_{l,k}^1 + \cdots + x_{l,k}^6) + 7(x_{a,d}^1 + \cdots + x_{a,d}^6) + 7(x_{d,a}^1 + \cdots + x_{d,a}^6) \\
&+ 5(x_{d,i}^1 + \cdots + x_{d,i}^6) + 5(x_{i,d}^1 + \cdots + x_{i,d}^6) + 4(x_{b,e}^1 + \cdots + x_{b,e}^6) \\
&+ 4(x_{e,b}^1 + \cdots + x_{e,b}^6) + 3(x_{e,h}^1 + \cdots + x_{e,h}^6) + 3(x_{h,e}^1 + \cdots + x_{h,e}^6) \\
&+ 5(x_{h,j}^1 + \cdots + x_{h,j}^6) + 5(x_{j,h}^1 + \cdots + x_{j,h}^6) + 8(x_{f,k}^1 + \cdots + x_{f,k}^6) \\
&+ 8(x_{k,f}^1 + \cdots + x_{k,f}^6) + 4(x_{c,g}^1 + \cdots + x_{c,g}^6) + 4(x_{g,c}^1 + \cdots + x_{g,c}^6) \\
&+ 8(x_{g,l}^1 + \cdots + x_{g,l}^6) + 8(x_{l,g}^1 + \cdots + x_{l,g}^6) \qquad (6.1)
\end{aligned}$$

Second, we compute the demand constraints as follows: Let z_v^k denote the demand variable for node v with respect to net (= commodity) k. The total number of these z-variables is $12 \times 6 = 72$. All z-variables are initialized to zero. We then obtain non-zero demand variables by examining the netlist. From net $n_1 = \{a, l\}$, we have $z_a^1 = 1$, $z_l^1 = -1$. From net $n_2 = \{i, c\}$, we have $z_i^2 = 1$, $z_c^2 = -1$. From net $n_3 = \{d, f\}$, we have $z_d^3 = 1$, $z_f^3 = -1$. From net $n_4 = \{k, d\}$, we have $z_k^4 = 1$, $z_d^4 = -1$. From net $n_5 = \{g, h\}$, we have $z_g^5 = 1$, $z_h^5 = -1$. From net $n_6 = \{b, k\}$, we have $z_b^6 = 1$, $z_k^6 = -1$. Next, we visit the nodes in the flow network and obtain the demand constraints as follows:

- Node a: node a is the source of net n_1. Thus,

$$x_{a,b}^1 + x_{a,d}^1 - x_{b,a}^1 - x_{d,a}^1 = 1 \qquad (6.2)$$
$$x_{a,b}^2 + x_{a,d}^2 - x_{b,a}^2 - x_{d,a}^2 = 0 \qquad (6.3)$$
$$x_{a,b}^3 + x_{a,d}^3 - x_{b,a}^3 - x_{d,a}^3 = 0 \qquad (6.4)$$
$$x_{a,b}^4 + x_{a,d}^4 - x_{b,a}^4 - x_{d,a}^4 = 0 \qquad (6.5)$$
$$x_{a,b}^5 + x_{a,d}^5 - x_{b,a}^5 - x_{d,a}^5 = 0 \qquad (6.6)$$
$$x_{a,b}^6 + x_{a,d}^6 - x_{b,a}^6 - x_{d,a}^6 = 0 \qquad (6.7)$$

- Node b: node b is the source of net n_6. Thus,

$$x^1_{b,a} + x^1_{b,e} + x^1_{b,c} - x^1_{a,b} - x^1_{e,b} - x^1_{c,b} = 0 \tag{6.8}$$
$$x^2_{b,a} + x^2_{b,e} + x^2_{b,c} - x^2_{a,b} - x^2_{e,b} - x^2_{c,b} = 0 \tag{6.9}$$
$$x^3_{b,a} + x^3_{b,e} + x^3_{b,c} - x^3_{a,b} - x^3_{e,b} - x^3_{c,b} = 0 \tag{6.10}$$
$$x^4_{b,a} + x^4_{b,e} + x^4_{b,c} - x^4_{a,b} - x^4_{e,b} - x^4_{c,b} = 0 \tag{6.11}$$
$$x^5_{b,a} + x^5_{b,e} + x^5_{b,c} - x^5_{a,b} - x^5_{e,b} - x^5_{c,b} = 0 \tag{6.12}$$
$$x^6_{b,a} + x^6_{b,e} + x^6_{b,c} - x^6_{a,b} - x^6_{e,b} - x^6_{c,b} = 1 \tag{6.13}$$

- Node c: node c is the sink of net n_2. Thus,

$$x^1_{c,b} + x^1_{c,g} - x^1_{b,c} - x^1_{g,c} = 0 \tag{6.14}$$
$$x^2_{c,b} + x^2_{c,g} - x^2_{b,c} - x^2_{g,c} = -1 \tag{6.15}$$
$$x^3_{c,b} + x^3_{c,g} - x^3_{b,c} - x^3_{g,c} = 0 \tag{6.16}$$
$$x^4_{c,b} + x^4_{c,g} - x^4_{b,c} - x^4_{g,c} = 0 \tag{6.17}$$
$$x^5_{c,b} + x^5_{c,g} - x^5_{b,c} - x^5_{g,c} = 0 \tag{6.18}$$
$$x^6_{c,b} + x^6_{c,g} - x^6_{b,c} - x^6_{g,c} = 0 \tag{6.19}$$

- Node d: node d is the source of net n_3 and the sink of net n_4. Thus,

$$x^1_{d,a} + x^1_{d,h} + x^1_{d,i} - x^1_{a,d} - x^1_{h,d} - x^1_{i,d} = 0 \tag{6.20}$$
$$x^2_{d,a} + x^2_{d,h} + x^2_{d,i} - x^2_{a,d} - x^2_{h,d} - x^2_{i,d} = 0 \tag{6.21}$$
$$x^3_{d,a} + x^3_{d,h} + x^3_{d,i} - x^3_{a,d} - x^3_{h,d} - x^3_{i,d} = 1 \tag{6.22}$$
$$x^4_{d,a} + x^4_{d,h} + x^4_{d,i} - x^4_{a,d} - x^4_{h,d} - x^4_{i,d} = -1 \tag{6.23}$$
$$x^5_{d,a} + x^5_{d,h} + x^5_{d,i} - x^5_{a,d} - x^5_{h,d} - x^5_{i,d} = 0 \tag{6.24}$$
$$x^6_{d,a} + x^6_{d,h} + x^6_{d,i} - x^6_{a,d} - x^6_{h,d} - x^6_{i,d} = 0 \tag{6.25}$$

- Node e: node e is not involved with any net. Thus,

$$x^1_{e,b} + x^1_{e,f} + x^1_{e,h} - x^1_{b,e} - x^1_{f,e} - x^1_{h,e} = 0 \tag{6.26}$$
$$x^2_{e,b} + x^2_{e,f} + x^2_{e,h} - x^2_{b,e} - x^2_{f,e} - x^2_{h,e} = 0 \tag{6.27}$$
$$x^3_{e,b} + x^3_{e,f} + x^3_{e,h} - x^3_{b,e} - x^3_{f,e} - x^3_{h,e} = 0 \tag{6.28}$$
$$x^4_{e,b} + x^4_{e,f} + x^4_{e,h} - x^4_{b,e} - x^4_{f,e} - x^4_{h,e} = 0 \tag{6.29}$$
$$x^5_{e,b} + x^5_{e,f} + x^5_{e,h} - x^5_{b,e} - x^5_{f,e} - x^5_{h,e} = 0 \tag{6.30}$$
$$x^6_{e,b} + x^6_{e,f} + x^6_{e,h} - x^6_{b,e} - x^6_{f,e} - x^6_{h,e} = 0 \tag{6.31}$$

- Node f: node f is the sink of net n_3. Thus,

$$x^1_{f,e} + x^1_{f,k} + x^1_{f,g} - x^1_{e,f} - x^1_{k,f} - x^1_{g,f} = 0 \quad (6.32)$$
$$x^2_{f,e} + x^2_{f,k} + x^2_{f,g} - x^2_{e,f} - x^2_{k,f} - x^2_{g,f} = 0 \quad (6.33)$$
$$x^3_{f,e} + x^3_{f,k} + x^3_{f,g} - x^3_{e,f} - x^3_{k,f} - x^3_{g,f} = -1 \quad (6.34)$$
$$x^4_{f,e} + x^4_{f,k} + x^4_{f,g} - x^4_{e,f} - x^4_{k,f} - x^4_{g,f} = 0 \quad (6.35)$$
$$x^5_{f,e} + x^5_{f,k} + x^5_{f,g} - x^5_{e,f} - x^5_{k,f} - x^5_{g,f} = 0 \quad (6.36)$$
$$x^6_{f,e} + x^6_{f,k} + x^6_{f,g} - x^6_{e,f} - x^6_{k,f} - x^6_{g,f} = 0 \quad (6.37)$$

- Node g: node g is the source of net n_5. Thus,

$$x^1_{g,c} + x^1_{g,f} + x^1_{g,l} - x^1_{c,g} - x^1_{f,g} - x^1_{l,g} = 0 \quad (6.38)$$
$$x^2_{g,c} + x^2_{g,f} + x^2_{g,l} - x^2_{c,g} - x^2_{f,g} - x^2_{l,g} = 0 \quad (6.39)$$
$$x^3_{g,c} + x^3_{g,f} + x^3_{g,l} - x^3_{c,g} - x^3_{f,g} - x^3_{l,g} = 0 \quad (6.40)$$
$$x^4_{g,c} + x^4_{g,f} + x^4_{g,l} - x^4_{c,g} - x^4_{f,g} - x^4_{l,g} = 0 \quad (6.41)$$
$$x^5_{g,c} + x^5_{g,f} + x^5_{g,l} - x^5_{c,g} - x^5_{f,g} - x^5_{l,g} = 1 \quad (6.42)$$
$$x^6_{g,c} + x^6_{g,f} + x^6_{g,l} - x^6_{c,g} - x^6_{f,g} - x^6_{l,g} = 0 \quad (6.43)$$

- Node h: node h is the sink of net n_5. Thus,

$$x^1_{h,e} + x^1_{h,d} + x^1_{h,j} - x^1_{e,h} - x^1_{d,h} - x^1_{j,h} = 0 \quad (6.44)$$
$$x^2_{h,e} + x^2_{h,d} + x^2_{h,j} - x^2_{e,h} - x^2_{d,h} - x^2_{j,h} = 0 \quad (6.45)$$
$$x^3_{h,e} + x^3_{h,d} + x^3_{h,j} - x^3_{e,h} - x^3_{d,h} - x^3_{j,h} = 0 \quad (6.46)$$
$$x^4_{h,e} + x^4_{h,d} + x^4_{h,j} - x^4_{e,h} - x^4_{d,h} - x^4_{j,h} = 0 \quad (6.47)$$
$$x^5_{h,e} + x^5_{h,d} + x^5_{h,j} - x^5_{e,h} - x^5_{d,h} - x^5_{j,h} = -1 \quad (6.48)$$
$$x^6_{h,e} + x^6_{h,d} + x^6_{h,j} - x^6_{e,h} - x^6_{d,h} - x^6_{j,h} = 0 \quad (6.49)$$

- Node i: node i is the source of net n_2. Thus,

$$x^1_{i,d} + x^1_{i,j} - x^1_{d,i} - x^1_{j,i} = 0 \quad (6.50)$$
$$x^2_{i,d} + x^2_{i,j} - x^2_{d,i} - x^2_{j,i} = 1 \quad (6.51)$$
$$x^3_{i,d} + x^3_{i,j} - x^3_{d,i} - x^3_{j,i} = 0 \quad (6.52)$$
$$x^4_{i,d} + x^4_{i,j} - x^4_{d,i} - x^4_{j,i} = 0 \quad (6.53)$$
$$x^5_{i,d} + x^5_{i,j} - x^5_{d,i} - x^5_{j,i} = 0 \quad (6.54)$$
$$x^6_{i,d} + x^6_{i,j} - x^6_{d,i} - x^6_{j,i} = 0 \quad (6.55)$$

- Node j: node j is not involved with any net. Thus,

$$x_{j,i}^1 + x_{j,h}^1 + x_{j,k}^1 - x_{i,j}^1 - x_{h,j}^1 - x_{k,j}^1 = 0 \quad (6.56)$$
$$x_{j,i}^2 + x_{j,h}^2 + x_{j,k}^2 - x_{i,j}^2 - x_{h,j}^2 - x_{k,j}^2 = 0 \quad (6.57)$$
$$x_{j,i}^3 + x_{j,h}^3 + x_{j,k}^3 - x_{i,j}^3 - x_{h,j}^3 - x_{k,j}^3 = 0 \quad (6.58)$$
$$x_{j,i}^4 + x_{j,h}^4 + x_{j,k}^4 - x_{i,j}^4 - x_{h,j}^4 - x_{k,j}^4 = 0 \quad (6.59)$$
$$x_{j,i}^5 + x_{j,h}^5 + x_{j,k}^5 - x_{i,j}^5 - x_{h,j}^5 - x_{k,j}^5 = 0 \quad (6.60)$$
$$x_{j,i}^6 + x_{j,h}^6 + x_{j,k}^6 - x_{i,j}^6 - x_{h,j}^6 - x_{k,j}^6 = 0 \quad (6.61)$$

- Node k: node k is the source of net n_4 and the sink of net n_6. Thus,

$$x_{k,j}^1 + x_{k,f}^1 + x_{k,l}^1 - x_{j,k}^1 - x_{f,k}^1 - x_{l,k}^1 = 0 \quad (6.62)$$
$$x_{k,j}^2 + x_{k,f}^2 + x_{k,l}^2 - x_{j,k}^2 - x_{f,k}^2 - x_{l,k}^2 = 0 \quad (6.63)$$
$$x_{k,j}^3 + x_{k,f}^3 + x_{k,l}^3 - x_{j,k}^3 - x_{f,k}^3 - x_{l,k}^3 = 0 \quad (6.64)$$
$$x_{k,j}^4 + x_{k,f}^4 + x_{k,l}^4 - x_{j,k}^4 - x_{f,k}^4 - x_{l,k}^4 = 1 \quad (6.65)$$
$$x_{k,j}^5 + x_{k,f}^5 + x_{k,l}^5 - x_{j,k}^5 - x_{f,k}^5 - x_{l,k}^5 = 0 \quad (6.66)$$
$$x_{k,j}^6 + x_{k,f}^6 + x_{k,l}^6 - x_{j,k}^6 - x_{f,k}^6 - x_{l,k}^6 = -1 \quad (6.67)$$

- Node l: node l is the sink of net n_1. Thus,

$$x_{l,k}^1 + x_{l,g}^1 - x_{k,l}^1 - x_{g,l}^1 = -1 \quad (6.68)$$
$$x_{l,k}^2 + x_{l,g}^2 - x_{k,l}^2 - x_{g,l}^2 = 0 \quad (6.69)$$
$$x_{l,k}^3 + x_{l,g}^3 - x_{k,l}^3 - x_{g,l}^3 = 0 \quad (6.70)$$
$$x_{l,k}^4 + x_{l,g}^4 - x_{k,l}^4 - x_{g,l}^4 = 0 \quad (6.71)$$
$$x_{l,k}^5 + x_{l,g}^5 - x_{k,l}^5 - x_{g,l}^5 = 0 \quad (6.72)$$
$$x_{l,k}^6 + x_{l,g}^6 - x_{k,l}^6 - x_{g,l}^6 = 0 \quad (6.73)$$

Third, we compute the capacity constraints as follows:

$$x_{a,b}^1 + \cdots x_{a,b}^6 + x_{b,a}^1 + \cdots x_{b,a}^6 \leq 2 \quad (6.74)$$
$$x_{b,c}^1 + \cdots + x_{b,c}^6 + x_{c,b}^1 + \cdots + x_{c,b}^6 \leq 2 \quad (6.75)$$
$$x_{d,h}^1 + \cdots + x_{d,h}^6 + x_{h,d}^1 + \cdots + x_{h,d}^6 \leq 2 \quad (6.76)$$
$$x_{e,f}^1 + \cdots + x_{e,f}^6 + x_{f,e}^1 + \cdots + x_{f,e}^6 \leq 2 \quad (6.77)$$
$$x_{f,g}^1 + \cdots + x_{f,g}^6 + x_{g,f}^1 + \cdots + x_{g,f}^6 \leq 2 \quad (6.78)$$
$$x_{i,j}^1 + \cdots + x_{i,j}^6 + x_{j,i}^1 + \cdots + x_{j,i}^6 \leq 2 \quad (6.79)$$

$$x_{j,k}^1 + \cdots + x_{j,k}^6 + x_{k,j}^1 + \cdots + x_{k,j}^6 \le 2 \quad (6.80)$$

$$x_{k,l}^1 + \cdots + x_{k,l}^6 + x_{l,k}^1 + \cdots + x_{l,k}^6 \le 2 \quad (6.81)$$

$$x_{a,d}^1 + \cdots + x_{a,d}^6 + x_{d,a}^1 + \cdots + x_{d,a}^6 \le 2 \quad (6.82)$$

$$x_{d,i}^1 + \cdots + x_{d,i}^6 + x_{i,d}^1 + \cdots + x_{i,d}^6 \le 2 \quad (6.83)$$

$$x_{b,e}^1 + \cdots + x_{b,e}^6 + x_{e,b}^1 + \cdots + x_{e,b}^6 \le 2 \quad (6.84)$$

$$x_{e,h}^1 + \cdots + x_{e,h}^6 + x_{h,e}^1 + \cdots + x_{h,e}^6 \le 2 \quad (6.85)$$

$$x_{h,j}^1 + \cdots + x_{h,j}^6 + x_{j,h}^1 + \cdots + x_{j,h}^6 \le 2 \quad (6.86)$$

$$x_{f,k}^1 + \cdots + x_{f,k}^6 + x_{k,f}^1 + \cdots + x_{k,f}^6 \le 2 \quad (6.87)$$

$$x_{c,g}^1 + \cdots + x_{c,g}^6 + x_{g,c}^1 + \cdots + x_{g,c}^6 \le 2 \quad (6.88)$$

$$x_{g,l}^1 + \cdots + x_{g,l}^6 + x_{l,g}^1 + \cdots + x_{l,g}^6 \le 2 \quad (6.89)$$

Fourth, the integer constraints are as follows:

$$x_e^k \in \{0, 1\}, \ \forall e, k \quad (6.90)$$

Lastly, the ILP formulation is as follows:
Minimize (6.1)
Under constraints (6.2)–(6.73), (6.74)–(6.89), and (6.90).

We use GLPK package [FSF, 2006] to solve this ILP and obtain the following results (runtime was less than a second on a Linux machine with 2.5 GHz CPU)[28]:

- Minimum cost: 108 (this is equal to the total wirelength of the six nets).
- Non-zero x-variables: $x_{d,h}^1$, $x_{j,k}^1$, $x_{k,l}^1$, $x_{a,d}^1$, $x_{h,j}^1$, $x_{a,b}^2$, $x_{b,c}^2$, $x_{i,d}^2$, $x_{d,a}^2$, $x_{d,h}^3$, $x_{e,f}^3$, $x_{h,e}^3$, $x_{k,j}^4$, $x_{i,d}^4$, $x_{j,i}^4$, $x_{f,e}^5$, $x_{g,f}^5$, $x_{e,h}^5$, $x_{b,c}^6$, $x_{g,f}^6$, $x_{f,k}^6$, $x_{c,g}^6$.
- Path for $n_1 = \{a, l\}$: we have $x_{a,d}^1$, $x_{d,h}^1$, $x_{h,j}^1$, $x_{j,k}^1$, $x_{k,l}^1$ as the non-zero variables. Thus, the path is $a \to d \to h \to j \to k \to l$. The wirelength is 24.
- Path for $n_2 = \{i, c\}$: we have $x_{i,d}^2$, $x_{d,a}^2$, $x_{a,b}^2$, $x_{b,c}^2$ as the non-zero variables. Thus, the path is $i \to d \to a \to b \to c$. The wirelength is 24.
- Path for $n_3 = \{d, f\}$: we have $x_{d,h}^3$, $x_{h,e}^3$, $x_{e,f}^3$ as the non-zero variables. Thus, the path is $d \to h \to e \to f$. The wirelength is 12.

[28]The source file for this integer linear programming formulation is available for download at: `http://users.ece.gatech.edu/limsk/book`

- Path for $n_4 = \{k, d\}$: we have $x^4_{k,j}$, $x^4_{j,i}$, $x^4_{i,d}$ as the non-zero variables. Thus, the path is $k \to j \to i \to d$. The wirelength is 14.
- Path for $n_5 = \{g, h\}$: we have $x^5_{g,f}$, $x^5_{f,e}$, $x^5_{e,h}$ as the non-zero variables. Thus, the path is $g \to f \to e \to h$. The wirelength is 11.
- Path for $n_6 = \{b, k\}$: we have $x^6_{b,c}$, $x^6_{c,g}$, $x^6_{g,f}$, $x^6_{f,k}$ as the non-zero variables. Thus, the path is $b \to c \to g \to f \to k$. The wirelength is 23.

Figure 6.15 shows the routing results for the six nets. Note that we obtained optimal routing for all nets except for n_6. Figure 6.15 also shows the corresponding routing channel usage.

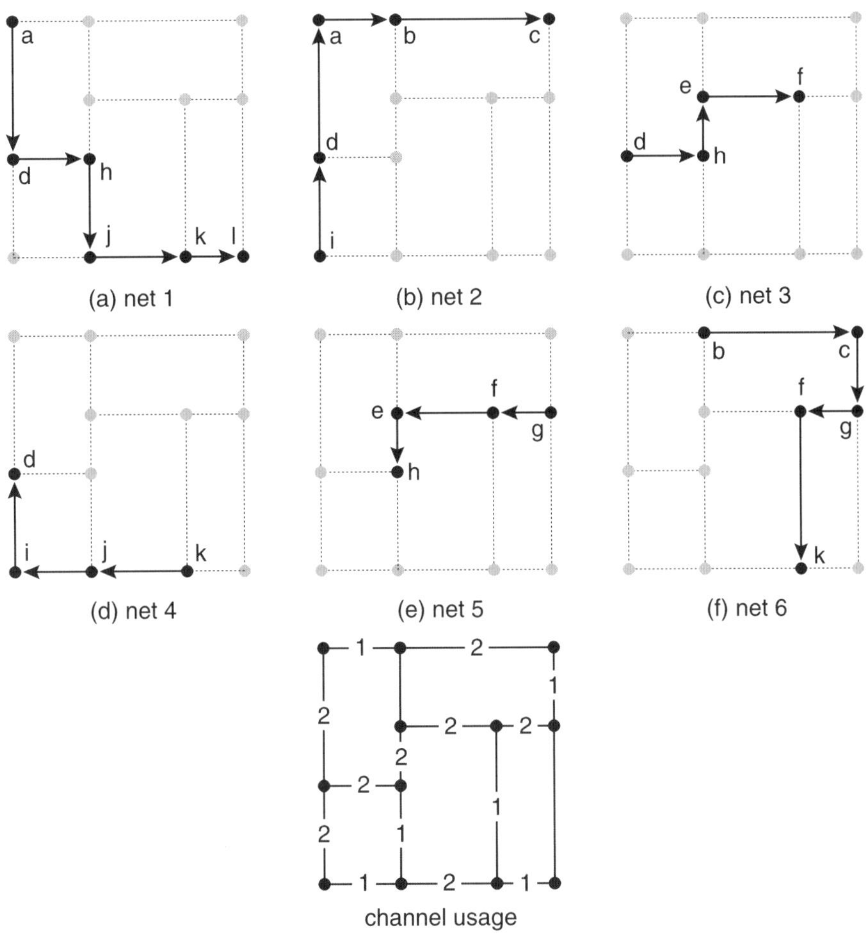

Figure 6.15. Final result of ILP-based multi-commodity flow global routing.

2. Perform the shortest path based MM heuristic. Break ties so that the number of turns (= vias) is minimized.

   The initial step is to find shortest paths for all nets while ignoring capacity constraints. Figure 6.16 shows the shortest paths and the channel usage. Under the channel capacity of two, an overflow occurs on two channels: $c(d, i)$ and $c(e, f)$. Note that we distinguish arcs and channels because we represent a channel $c(x, y)$ by a pair of arcs (x, y) and (y, x). The rest of the steps of the algorithm proceeds as follows:

   (a) Iteration 1: the following steps are performed for iteration $r = 1$:

   - Step 1: from Figure 6.16 the maximum level of overflow is $M_1 = 3 - 2 = 1$. Since $M_1 > 0$, we proceed.

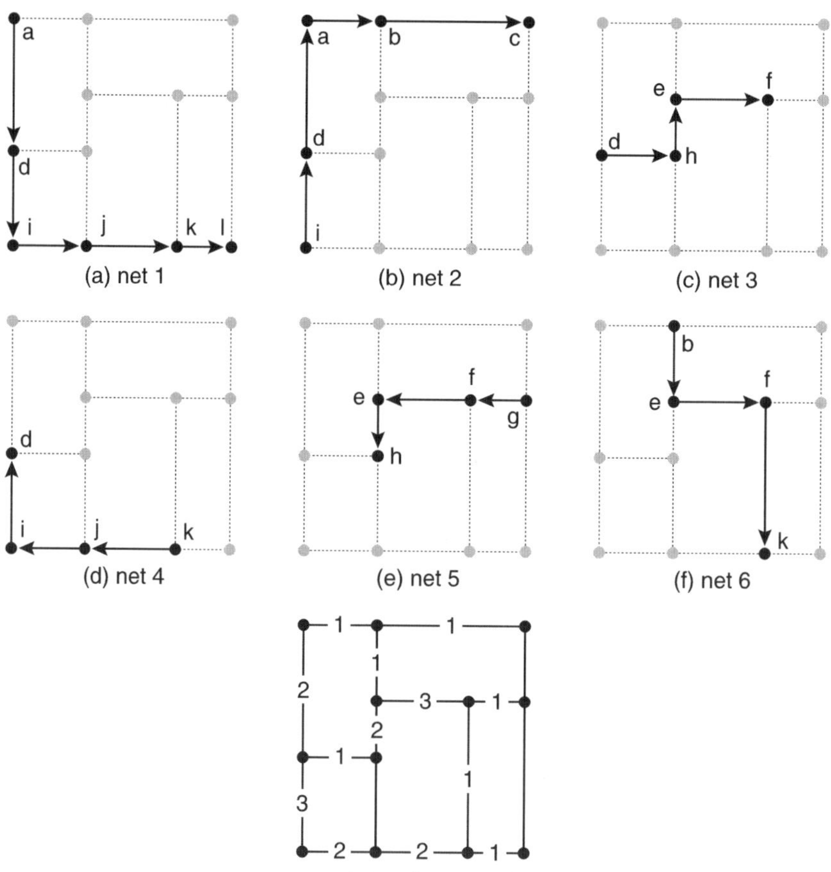

Figure 6.16. Initial step of MM algorithm: shortest paths for the nets. Note that some paths are not unique. Two channels have overflow.

- Step 2: from Figure 6.16 we obtain the set of channels with overflow of M_1, which is $J_1 = \{c(d,i), c(e,f)\}$. Also, we obtain the set of channels with overflow of M_1 and $M_1 - 1$, which is $J_1^0 = \{c(a,d), c(e,h), c(i,j), c(j,k), c(d,i), c(e,f)\}$.
- Step 3: we assign ∞ as the cost of all channels in J_1^0.
- Step 4: from Figure 6.16 we obtain the set of nets that use channels in J_1, which is $K_1 = \{n_1, n_2, n_3, n_4, n_5, n_6\}$. Also, we obtain the set of nets that use channels in J_1^0, which in this case is $K_1^0 = K_1$.
- Step 5: we obtain shortest paths for all nets in K_1 using the new cost from step 3. Figure 6.17 shows the results. We see that the cost of n_2, n_3, n_4, and n_5 are infinity, whereas n_1 has the cost of 24, and n_6 has the cost of 23. Since there are two nets with non-infinity cost, we move to step 6.
- Step 6: the old (= Figure 6.16) and new (= Figure 6.17) cost of n_1 are 24 and 24, respectively. The old and the new cost of n_6 are 17 and 23. Thus, we choose $k^0 = n_1$ as the net with minimum cost increase.
- Step 7: we keep the new path (= Figure 6.17) for $k^0 = n_1$ and retrieve old paths from Figure 6.16 for other nets. Figure 6.18 shows this result.

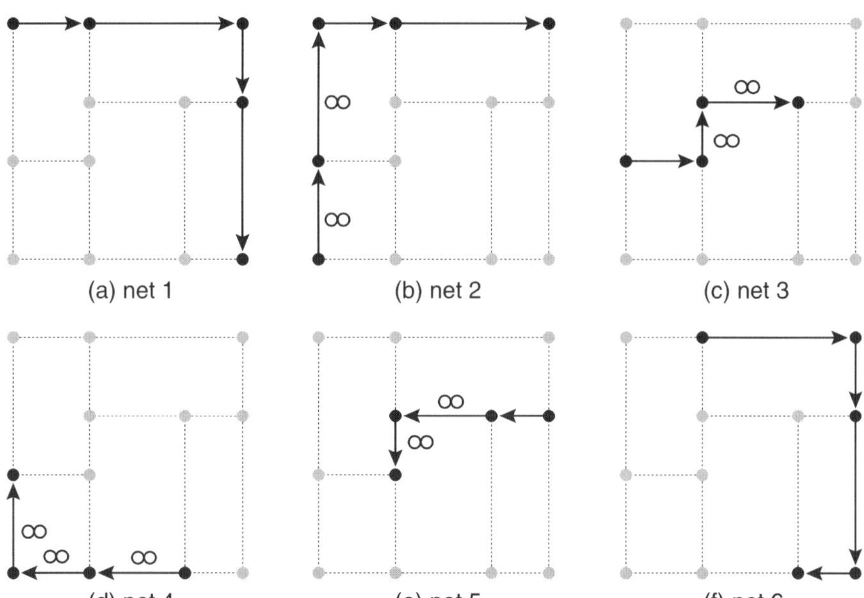

Figure 6.17. Iteration 1 of MM algorithm: shortest paths for the nets under new cost. The cost of n_2, n_3, n_4, n_5 are infinity, n_1 is 24, and n_6 is 23.

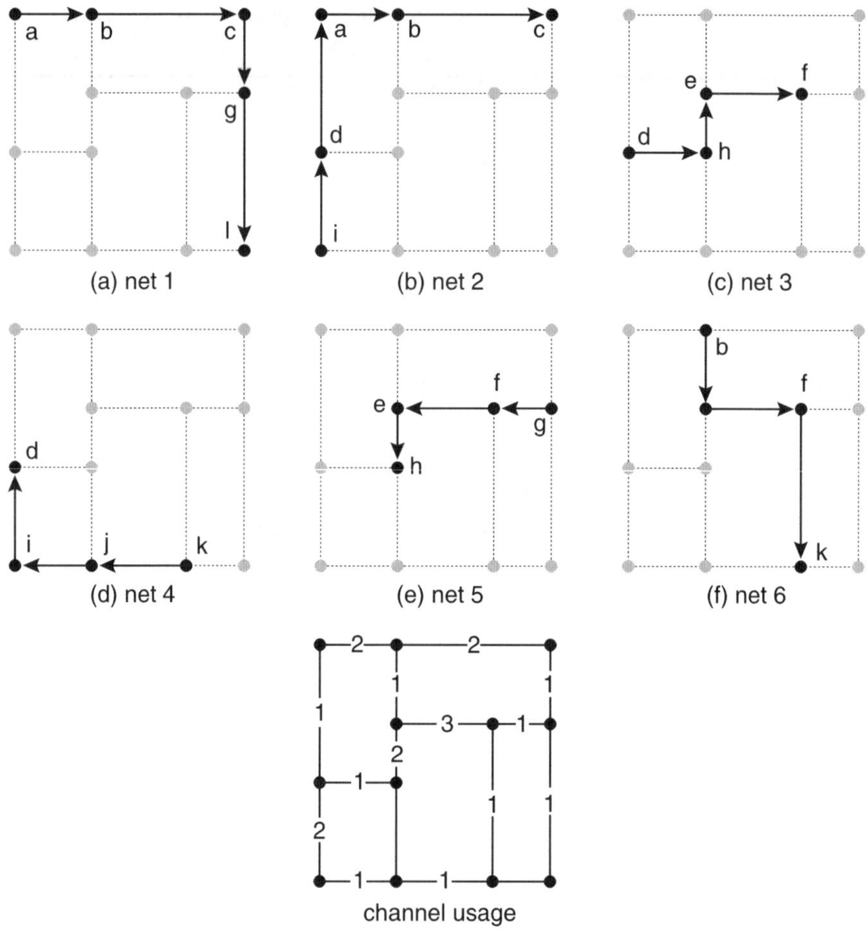

Figure 6.18. Final result of iteration 1. Net 1 is rerouted compared with Figure 6.16.

(b) Iteration 2: the following steps are performed for iteration $r = 2$:

- Step 1: from Figure 6.18, we compute $M_2 = 3 - 2 = 1$. Since $M_2 > 0$, we proceed.
- Step 2: Figure 6.18 we obtain the set of channels with overflow of M_2, which is $J_2 = \{c(e, f)\}$. Also, we obtain the set of channels with overflow of M_2 and $M_2 - 1$, which is

$$J_2^0 = \{c(a, b), c(b, c), c(e, h), c(d, i), c(e, f)\}$$

- Step 3: we assign ∞ as the cost of all channels in J_2^0.

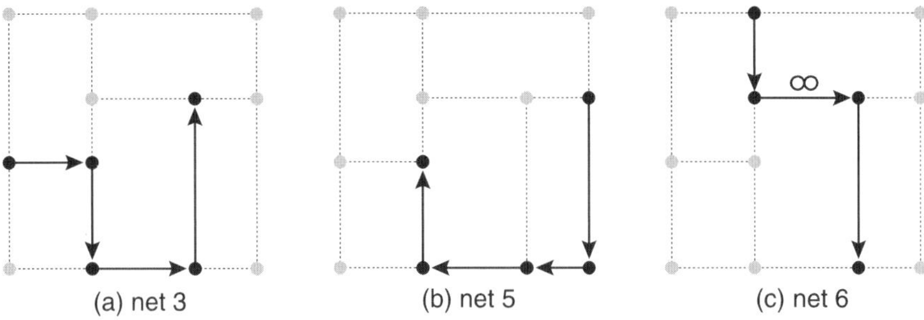

Figure 6.19. Iteration 2 of MM algorithm: shortest paths for the nets under new cost. The cost of n_3 is 22, n_5 is 21, and n_6 is infinity.

- Step 4: From figure 6.18 we obtain the set of nets that use channels in J_2, which is $K_2 = \{n_3, n_5, n_6\}$. Also, we obtain the set of nets that use channels in J_2^0, which is $K_2^0 = \{n_1, n_2, n_3, n_4, n_5, n_6\}$.
- Step 5: we obtain shortest paths for all nets in K_2 using the new cost from step 3. Figure 6.19 shows the results. We see that the cost of n_6 is infinity, whereas n_3 is 22, and n_5 is 21. Since there are two nets with non-infinity cost, we move to step 6.
- Step 6: the old (= Figure 6.18) and new (= Figure 6.19) cost of n_3 are 12 and 22, respectively. The old and the new cost of n_5 are 11 and 21. Both nets have the same cost increase, so we randomly break the tie and choose $k^0 = n_3$.
- Step 7: we keep the new path for $k^0 = n_3$ and retrieve old paths from Figure 6.18 for other nets. Figure 6.20 shows this result. Since there exists no overflow in this routing result, the algorithm terminates.

The wirelength of the nets are $wl(n_1) = 24$, $wl(n_2) = 24$, $wl(n_3) = 22$, $wl(n_4) = 14$, $wl(n_5) = 11$, and $wl(n_6) = 17$. Thus, the total wirelength is 112.

3. Compare the results obtained by ILP and MM methods.

   The wirelength cost is lower with ILP (108 vs 112). In case of ILP approach shown in Figure 6.15, net n_6 is sub-optimal. In case of MM algorithm shown in Figure 6.20, net n_3 is sub-optimal.

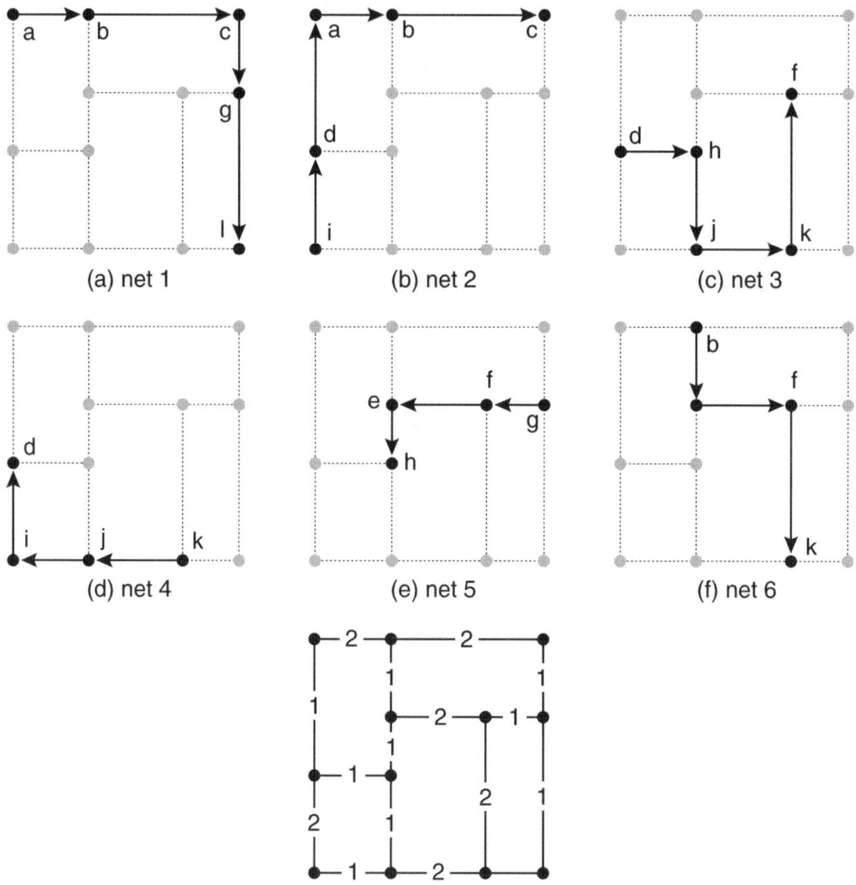

Figure 6.20. Final result of iteration 2. Net 3 is rerouted compared with Figure 6.18. MM algorithm terminates because there is no overflow.

3. Iterative Deletion Algorithm

One of the well known routing algorithms that overcome the net ordering problem of the sequential routing approach is iterative deletion [Cong and Preas, 1988]. Originally proposed for standard cell global routing, iterative deletion algorithm constructs routing trees for *all* nets simultaneously. The algorithm consists of two phases, namely feedthrough insertion and net segment computation. During the feedthrough insertion phase, feedthroughs are inserted for the nets that connect cells in non-adjacent rows. These feedthroughs not only allow a net to cross a row but also help reduce the routing channel density. The way the algorithm builds the routing trees and inserts feedthroughs for all nets simultaneously is by utilizing the so called net connection graph $G(V, E)$. This graph is the union of the complete graphs of all individual nets. Each edge in E is given a weight based on its x-coordinate span and its requirement for feedthroughs. The edge with the minimum weight, i.e., an edge that is the shortest in x-direction and requires none or minimum number of feedthroughs is chosen to be added to grow the routing trees. Since this edge is chosen from a global list of edges, all routing trees have an equal chance to grow. The width of rows keep increasing from feedthrough insertion, thereby causing the x-span of the affected edges to change as well. Thus, the weight of the affected edges are dynamically updated. Once we find the spanning forest for all nets, the first phase terminates.

The iterative deletion phase is used to refine the routing result obtained from the feedthrough insertion phase. A net segment is an edge that connects a pair of pins in the same net. Determining the net segments for a net is, therefore, constructing a routing tree for the net. Instead of *adding* edges in E to obtain a spanning forest, however, the iterative deletion phase computes the spanning forest by *deleting* the edges in E. Each edge is given a weight that is based on the routing density so that deleting the edges with larger weights helps better detect congested regions. Again all nets are given an equal chance to converge to their final routing trees because the algorithm maintains the list of *all* candidate edges to be included in the spanning forest. Compared to the $O(n)$ number of edges to be added during the first phase, the number of edges to be deleted during the second phase is $O(n^2)$. In order to cope with this higher complexity, the authors of [Cong and Preas, 1988] proposed a way to simplify the initial net connection graph.

Quick Overview

Given a standard cell placement along with the netlist, we first build the net connection graph $G(V, E)$ by merging the complete graphs for all individual nets. We then compute the weight of each edge in E as follows:

$$w(e(i,j)) = |x_i - x_j| + K \cdot \sum_{e \cap R_k \neq \emptyset} width(R_k)$$

where x_i is the x-location of pin i, K is the user-specified constant, R_k is the row k, and $width(R_k)$ is the current width of row R_k. Our goal is to obtain a set of spanning trees for all nets, i.e., a spanning forest F for the entire netlist. This is done by adding edges in E in an increasing order of their weight. Given an edge e under consideration, we first see if adding e creates a cycle in the current forest. If so, we discard it; otherwise, we add e to F and insert feedthroughs if needed. If feedthroughs are inserted, we update the width of the rows that accept the feedthroughs as well as the weight of all edges intersecting with these rows. Once E is re-sorted based on the updated edge weights, we insert the next edge.

Since all the necessary feedthroughs are inserted during the first phase, we build the net connection graph for the iterative deletion phase by forming complete graphs among the pins in the same channel. We then simplify this graph further to obtain the simplified net connection graph $G'(E', V')$ by removing edges that connect non-adjacent pins in the same channel. We compute the weight of edges in E' as follows:

$$w(e) = d(e)/d(C_e)$$

where $d(e)$ is the maximum density over e, and $d(C_e)$ is the density of the channel that e belongs to. The density of a given column in a channel is the number of edges that either pass through, begin at, or end at that column. We compute $d(e)$ by computing the maximum density among all columns that e spans. $d(C_e)$ is the maximum density over the entire channel that e belongs to. We delete the edge with the maximum weight in E' and update the weight of all edges in the same channel. If this edge disconnects any pin from the net, we do not delete it. The iterative deletion phase terminates once we obtain the final spanning forest.

Practice Problem

Consider the standard placement shown in 6.21. We are to route the following nets: $n_1 = \{b, c, g, h, i, k\}$, $n_2 = \{a, d, e, f, j\}$. Assume that the width is 2 and the height is 3 for the gates and feedthroughs. The gates are to be shifted to the right upon feedthrough insertion. Assume that each cell has a built-in feedthrough (= vertical edge within each cell).

1. Perform the feedthrough insertion phase with $K = 0.5$. Break ties in lexicographical order. Place feedthroughs right below the top gate.

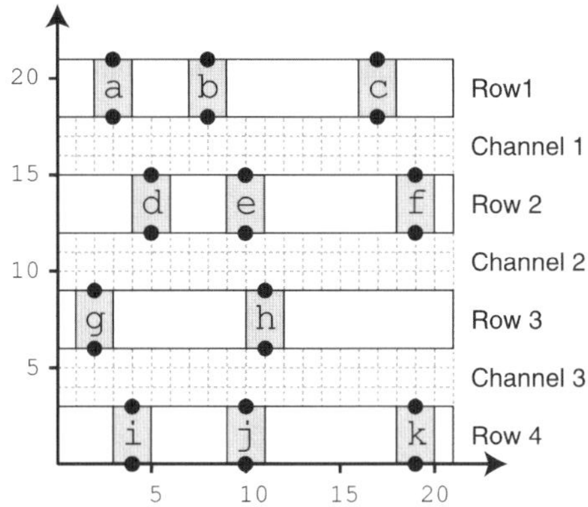

Figure 6.21. A standard cell placement with four rows and three channels.

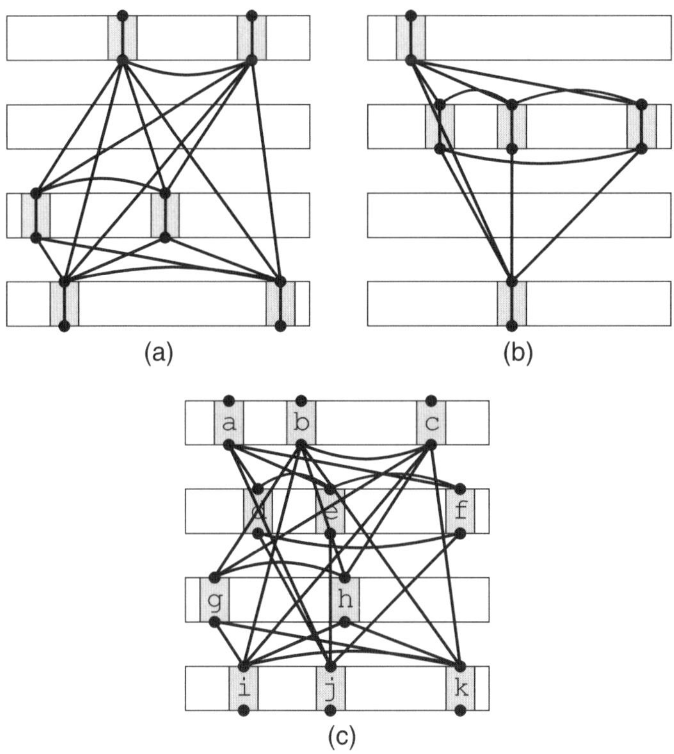

Figure 6.22. (**a**) Net connection graph for $n_1 = \{b, c, g, h, i, k\}$, (**b**) net connection graph for $n_2 = \{a, d, e, f, j\}$, (**c**) net connection graph for the entire netlist.

The first step is to build the net connection graph for the netlist. Figure 6.22(a–b) show the net connection graphs for n_1 and n_2. These are merged to form the net connection graph G for the netlist as shown in Figure

Table 6.3. A sorted list of the edges in the net connection graph shown in Figure 6.22(c). We list the intersecting rows for each edge under the R_i column. Ties are broken based on lexicographical order.

Edge	$\|x_i - x_j\|$	R_i	$w(e)$	Action
(a, d)	2	–	$2 + 0.5(0) = 2$	Added
(g, i)	2	–	$2 + 0.5(0) = 2$	Added
(d, e)	5	–	$5 + 0.5(0) = 5$	Added
(a, e)	7	–	$7 + 0.5(0) = 7$	Cycle
(h, i)	7	–	$7 + 0.5(0) = 7$	Added
(h, k)	8	–	$8 + 0.5(0) = 8$	Added
(b, c)	9	–	$9 + 0.5(0) = 9$	Added
(e, f)	9	–	$9 + 0.5(0) = 9$	Added
(g, h)	9	–	$9 + 0.5(0) = 9$	Cycle
(e, j)	0	R_3	$0 + 0.5(21) = 10.5$	
(b, h)	3	R_2	$3 + 0.5(21) = 13.5$	
(d, f)	14	–	$14 + 0.5(0) = 14$	
(i, k)	15	–	$15 + 0.5(0) = 15$	
(d, j)	5	R_3	$5 + 0.5(21) = 15.5$	
(a, f)	16	–	$16 + 0.5(0) = 16$	
(b, g)	6	R_2	$6 + 0.5(21) = 16.5$	
(c, h)	6	R_2	$6 + 0.5(21) = 16.5$	
(g, k)	17	–	$17 + 0.5(0) = 17$	
(f, j)	9	R_3	$9 + 0.5(21) = 19.5$	
(c, k)	2	R_2, R_3	$2 + 0.5(21 + 21) = 23$	
(b, i)	4	R_2, R_3	$4 + 0.5(21 + 21) = 25$	
(c, g)	15	R_2	$15 + 0.5(21) = 25.5$	
(a, j)	7	R_2, R_3	$7 + 0.5(21 + 21) = 28$	
(b, k)	11	R_2, R_3	$11 + 0.5(21 + 21) = 32$	
(c, i)	13	R_2, R_3	$13 + 0.5(21 + 21) = 34$	

6.22(c). Table 6.3 shows a sorted list of the edges in G based on an increasing order of their weights. Since n_1 connects 6 gates, we need five edges in its routing tree. Similarly, we need four edges for n_2. We construct the spanning forest F by adding the nine edges as follows:

(a) First seven edges: The first edge in the list is (a, d). Note that adding this edge to F does not create a cycle. In addition, edge weight update is not necessary because feedthrough is not inserted. Thus, we can add the next edge (g, i) using the same sorted list. We skip (a, e) and (g, h) because they create cycles. Figure 6.23(a) shows the spanning forest after adding the first seven edges.

(b) Eighth edge: We see from Table 6.3 that the next edge to be inserted is (e, j). This edge does not create a cycle but intersects with row 3, which requires a feedthrough in row 3. Figure 6.23(b) shows the spanning

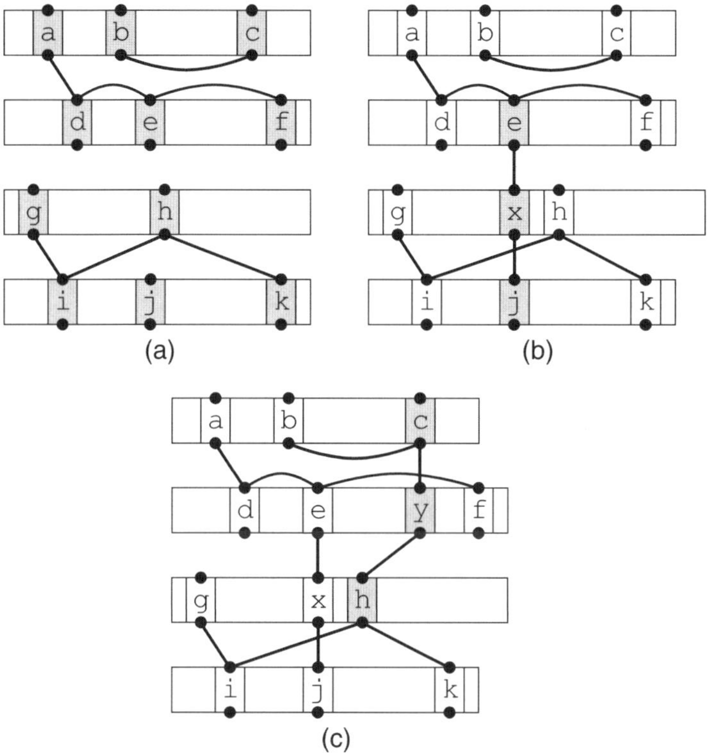

Figure 6.23. (**a**) After adding the first seven edges, (**b**) after adding the eighth edge (e, j) and its feedthrough x, (**c**) after adding the ninth edge (c, h) and its feedthrough y, which corresponds to the final spanning forest.

forest after adding (e, j) and its feedthrough x. This causes an update on all edges that intersect with row 3 plus the edges that are incident to h, which is shifted during the feedthrough insertion. Table 6.4 shows the edge list before and after adding (e, j). We note that there are nine edges that are affected by the feedthrough insertion.

(c) Ninth edge: Table 6.5 shows the updated edge list. The first edge in the sorted list (d, f) creates a cycle. Thus, we insert (c, h). This edge requires a feedthrough in row 2. Figure 6.23(c) shows the final spanning forest after adding (c, h) and its feedthrough y. No more edge weight update is necessary because the final spanning forest is obtained.

Figure 6.24 shows the final routing trees for n_1 and n_2 after feedthrough insertion.

2. Build the simplified net connection graph G'.

Table 6.4. Before and after adding the eighth edge (e, j) that creates a feedthrough in row 3. Gate h is shifted during the feedthrough insertion. We update the weight of nine edges that are affected by the feedthrough insertion.

Before		**After**			
Edge	$w(e)$	$\|x_i - x_j\|$	R_i	$w(e)$	Action
(e, j)	10.5				Added
(b, h)	13.5	5	R_2	$5 + 0.5(21) = 15.5$	Updated
(d, f)	14	14	–	$14 + 0.5(0) = 14$	
(i, k)	15	15	–	$15 + 0.5(0) = 15$	
(d, j)	15.5	5	R_3	$5 + 0.5(23) = 16.5$	Updated
(a, f)	16	16	–	$16 + 0.5(0) = 16$	
(b, g)	16.5	6	R_2	$6 + 0.5(21) = 16.5$	
(c, h)	16.5	4	R_2	$4 + 0.5(21) = 14.5$	Updated
(g, k)	17	17	–	$17 + 0.5(0) = 17$	
(f, j)	19.5	9	R_3	$9 + 0.5(23) = 20.5$	Updated
(c, k)	23	2	R_2, R_3	$2 + 0.5(21 + 23) = 24$	Updated
(b, i)	25	4	R_2, R_3	$4 + 0.5(21 + 23) = 26$	Updated
(c, g)	25.5	15	R_2	$15 + 0.5(21) = 25.5$	
(a, j)	28	7	R_2, R_3	$7 + 0.5(21 + 23) = 29$	Updated
(b, k)	32	11	R_2, R_3	$11 + 0.5(21 + 23) = 33$	Updated
(c, i)	34	13	R_2, R_3	$13 + 0.5(21 + 23) = 35$	Updated

Table 6.5. Adding the ninth edge (c, h) that creates a feedthrough in row 2. Gate f is shifted during the feedthrough insertion. No more update is necessary because the spanning forest is completed.

Edge	$w(e)$	Action
(d, f)	14	Cycle
(c, h)	14.5	Added
(i, k)	15	
(b, h)	15.5	
(a, f)	16	
(b, g)	16.5	
(d, j)	16.5	
(g, k)	17	
(f, j)	20.5	
(c, k)	24	
(c, g)	25.5	
(b, i)	26	
(a, j)	29	
(b, k)	33	
(c, i)	35	

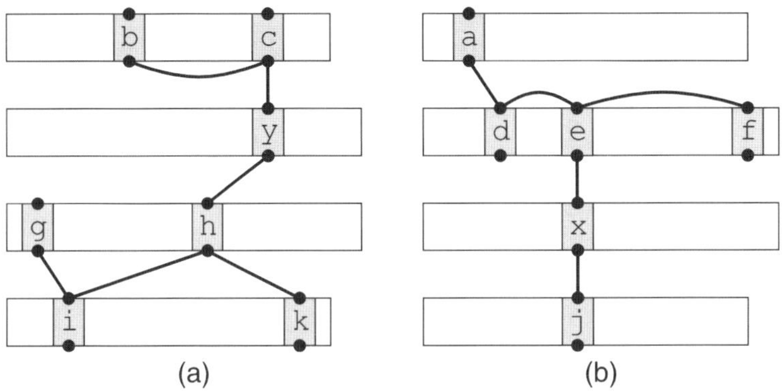

Figure 6.24. Feedthrough insertion result. (**a**) Final routing tree for n_1, (**b**) final routing tree for n_2.

Figure 6.25 shows the simplified net connection graphs, which are derived based on the routing trees shown in Figure 6.24. We form cliques among the pins in the same channel and remove the edges that connect non-adjacent pins (shown in dotted lines). Figure 6.25(c) shows the simplified net connection graph G' of the entire netlist.

3. Compute the density of the channels in G'.

   Recall that the density of a given column in a channel is the number of edges that either pass through, begin at, or end at that column. From Figure 6.26, we see that the density of channel 1 is 4, channel 2 is 6, and channel 3 is 2.

4. Perform Iterative Deletion to determine the net segments. Break ties based on the following factors in this order: (1) delete edges with longer x-span $(= |x_i - x_j|)$, (2) delete edges with higher edge density $d(e)$, (3) delete edges from the bottom-most channel, (4) delete edges with higher lexicographical order.

   We perform iterative deletion as follows until we obtain the final spanning forest. We do not delete the edges that cause the pins that belong to the same net to be disconnected.

   (a) First edge: Table 6.6 shows the sorted list of the edges in G'. We delete (x, f) from channel 2. The density of channel 2 reduces to 5, causing the weight of all edges in channel 2 to be updated. Figure 6.27(a) shows the net connection graph after deleting (x, f). Table 6.7 shows how the edge weights are updated after the deletion.

   (b) Second edge: Table 6.8 shows the updated list of the edges in G'. We delete (g, h) from channel 2. The density of channel 2 reduces to 4,

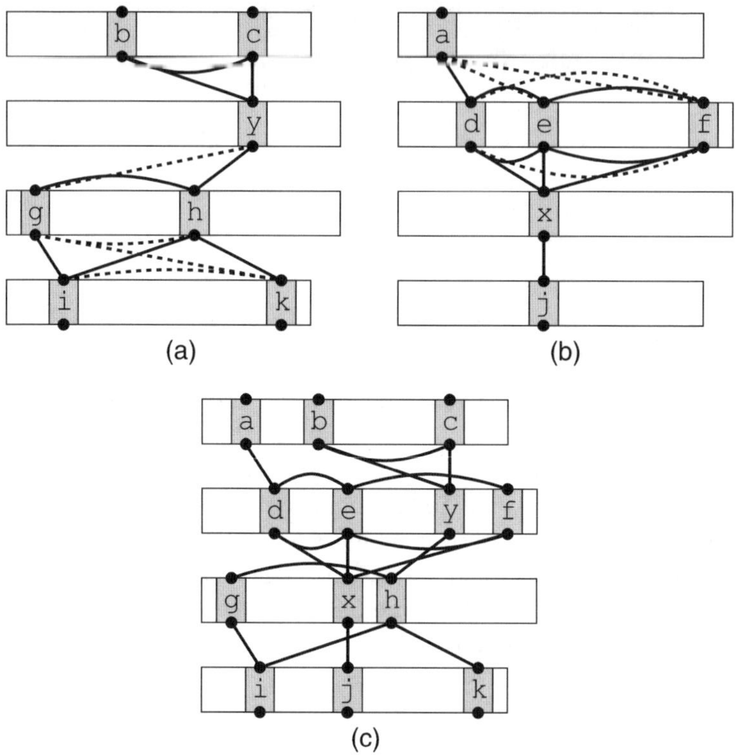

Figure 6.25. (**a**) Simplified net connection graph for n_1, where the edges to be removed are shown in dotted lines, (**b**) simplified net connection graph for n_2, (**c**) simplified net connection graph for the netlist.

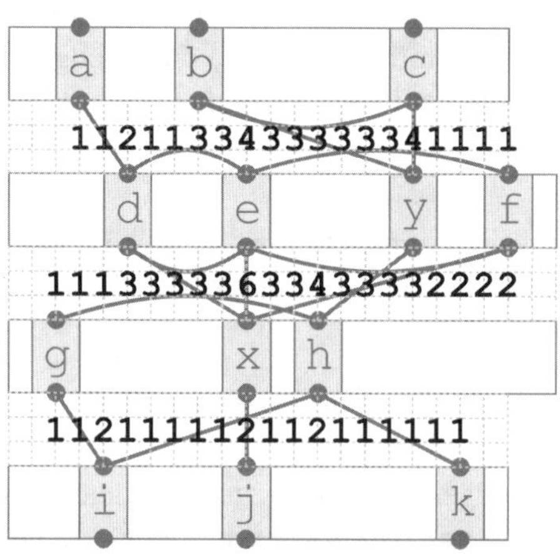

Figure 6.26. Density at each column in each channel. The density of channels 1, 2, and 3 are 4, 6, and 2, respectively.

Table 6.6. Sorted list of the edges in the simplified net connection graph shown in Figure 6.25(c).

Edge	x-span	$d(e)$	C_i	$w(e)$
(x, f)	11	6	C_2	$6/6 = 1$
(g, h)	11	6	C_2	$6/6 = 1$
(e, f)	11	6	C_2	$6/6 = 1$
(e, f)	11	4	C_1	$4/4 = 1$
(b, y)	9	4	C_1	$4/4 = 1$
(b, c)	9	4	C_1	$4/4 = 1$
(h, i)	9	2	C_3	$2/2 = 1$
(h, k)	6	2	C_3	$2/2 = 1$
(d, x)	5	6	C_2	$6/6 = 1$
(d, e)	5	6	C_2	$6/6 = 1$
(d, e)	5	4	C_1	$4/4 = 1$
(g, i)	2	2	C_3	$2/2 = 1$
(e, x)	0	6	C_2	$6/6 = 1$
(c, y)	0	4	C_1	$4/4 = 1$
(x, j)	0	2	C_3	$2/2 = 1$
(h, y)	4	4	C_2	$4/6 = 0.67$
(a, d)	2	2	C_1	$2/4 = 0.5$

causing the weight of all edges in channel 2 to be updated. Figure 6.27(b) shows the net connection graph after deleting (g, h). Table 6.8 shows how the edge weights are updated after the deletion.

(c) Third edge: Table 6.9 shows the updated list of the edges in G'. We delete (e, f) from channel 2. The density of channel 2 reduces to 3, causing the weight of all edges in channel 2 to be updated. Figure 6.27(c) shows the net connection graph after deleting (e, f). Table 6.9 shows how the edge weights are updated after the deletion.

(d) Fourth edge: Table 6.10 shows the updated list of the edges in G'. Note that deleting (e, f) isolates node f. Thus, we skip it and delete (b, y) instead. The density of channel 1 reduces to 3, causing the weight of all edges in channel 1 to be updated. Figure 6.27(d) shows the net connection graph after deleting (b, y). Table 6.10 shows how the edge weights are updated after the deletion.

(e) Fifth edge: Table 6.11 shows the updated list of the edges in G'. Note that we skip the first three edges because deleting them will disconnect the nodes that belong to the same net. Thus, we delete (d, x) instead. The density of channel 2 reduces to 2, causing the weight of all edges in channel 2 to be updated. Figure 6.27(e) shows the net connection graph after deleting (d, x). Table 6.11 shows how the edge weights are updated after the deletion.

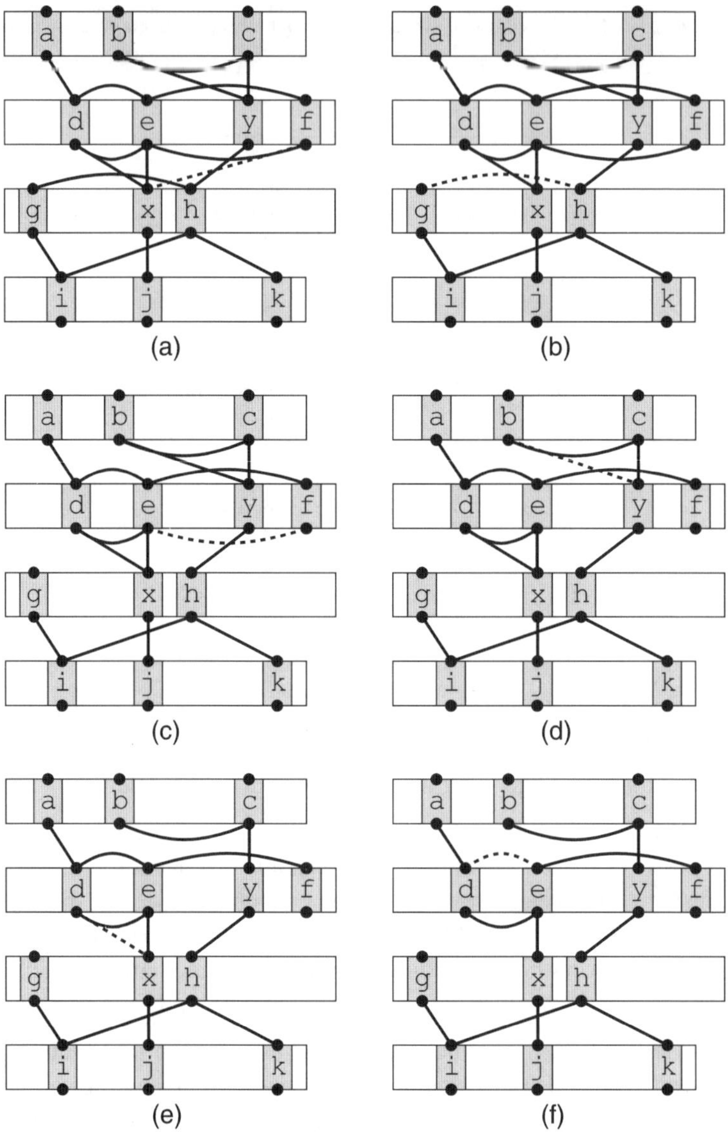

Figure 6.27. Iterative deletion. (**a**) Deleting (x, f), (**b**) deleting (g, h), (**c**) deleting (e, f) (in channel 2), (**d**) deleting (b, y), (**e**) deleting (d, x), (**f**) deleting (d, e) (in channel 1).

(f) Sixth edge: Table 6.12 shows the updated list of the edges in G'. We delete (d, e) from channel 1 and obtain the final spanning forest.

Figure 6.28 shows the final routing trees for n_1 and n_2 after iterative deletion.

Table 6.7. Deleting the first edge (x, f). The density of channel 2 reduces to 5, causing the weight of all edges in channel 2 to be updated.

	Before				**After**	
Edge	$w(e)$	x-span	$d(e)$	C_i	$w(e)$	Action
(x, f)	1	11	6	C_2	$6/6 = 1$	Deleted
(g, h)	1	11	5	C_2	$5/5 = 1$	Updated
(e, f)	1	11	5	C_2	$5/5 = 1$	Updated
(e, f)	1	11	4	C_1	$4/4 = 1$	
(b, y)	1	9	4	C_1	$4/4 = 1$	
(b, c)	1	9	4	C_1	$4/4 = 1$	
(h, i)	1	9	2	C_3	$2/2 = 1$	
(h, k)	1	6	2	C_3	$2/2 = 1$	
(d, x)	1	5	5	C_2	$5/5 = 1$	Updated
(d, e)	1	5	5	C_2	$5/5 = 1$	Updated
(d, e)	1	5	4	C_1	$4/4 = 1$	
(g, i)	1	2	2	C_3	$2/2 = 1$	
(e, x)	1	0	5	C_2	$5/5 = 1$	Updated
(c, y)	1	0	4	C_1	$4/4 = 1$	
(x, j)	1	0	2	C_3	$2/2 = 1$	
(h, y)	0.67	4	3	C_2	$3/5 = 0.6$	Updated
(a, d)	0.5	2	2	C_1	$2/4 = 0.5$	

Table 6.8. Deleting the second edge (g, h). The density of channel 2 reduces to 4, causing the weight of all edges in channel 2 to be updated.

	Before				**After**	
Edge	$w(e)$	x-span	$d(e)$	C_i	$w(e)$	Action
(g, h)	1	11	5	C_2	$5/5 = 1$	Deleted
(e, f)	1	11	4	C_2	$4/4 = 1$	Updated
(e, f)	1	11	4	C_1	$4/4 = 1$	
(b, y)	1	9	4	C_1	$4/4 = 1$	
(b, c)	1	9	4	C_1	$4/4 = 1$	
(h, i)	1	9	2	C_3	$2/2 = 1$	
(h, k)	1	6	2	C_3	$2/2 = 1$	
(d, x)	1	5	4	C_2	$4/4 = 1$	Updated
(d, e)	1	5	4	C_2	$4/4 = 1$	Updated
(d, e)	1	5	4	C_1	$4/4 = 1$	
(g, i)	1	2	2	C_3	$2/2 = 1$	
(e, x)	1	0	4	C_2	$4/4 = 1$	Updated
(c, y)	1	0	4	C_1	$4/4 = 1$	
(x, j)	1	0	2	C_3	$2/2 = 1$	
(h, y)	0.6	4	2	C_2	$2/4 = 0.5$	Updated
(a, d)	0.5	2	2	C_1	$2/4 = 0.5$	

Table 6.9. Deleting the third edge (e, f) in channel 2. The density of channel 2 reduces to 3, causing the weight of all edges in channel 2 to be updated.

	Before				After	
Edge	$w(e)$	x-span	$d(e)$	C_i	$w(e)$	Action
(e, f)	1	11	4	C_2	$4/4 = 1$	Deleted
(e, f)	1	11	4	C_1	$4/4 = 1$	
(b, y)	1	9	4	C_1	$4/4 = 1$	
(b, c)	1	9	4	C_1	$4/4 = 1$	
(h, i)	1	9	2	C_3	$2/2 = 1$	
(h, k)	1	6	2	C_3	$2/2 = 1$	
(d, x)	1	5	3	C_2	$3/3 = 1$	Updated
(d, e)	1	5	4	C_1	$4/4 = 1$	
(d, e)	1	5	3	C_2	$3/3 = 1$	Updated
(g, i)	1	2	2	C_3	$2/2 = 1$	
(e, x)	1	0	3	C_2	$3/3 = 1$	Updated
(c, y)	1	0	4	C_1	$4/4 = 1$	
(x, j)	1	0	2	C_3	$2/2 = 1$	
(h, y)	0.5	4	1	C_2	$1/3 = 0.33$	Updated
(a, d)	0.5	2	2	C_1	$2/4 = 0.5$	

Table 6.10. Deleting the fourth edge (b, y). Note that deleting (e, f) results in isolation of node f. The density of channel 1 reduces to 3, causing the weight of all edges in channel 1 to be updated.

	Before				After	
Edge	$w(e)$	x-span	$d(e)$	C_i	$w(e)$	Action
(e, f)	1	11	4	C_1	$4/4 = 1$	Skip
(b, y)	1	9	4	C_1	$4/4 = 1$	Deleted
(b, c)	1	9	3	C_1	$3/3 = 1$	Updated
(h, i)	1	9	2	C_3	$2/2 = 1$	
(h, k)	1	6	2	C_3	$2/2 = 1$	
(d, x)	1	5	3	C_2	$3/3 = 1$	
(d, e)	1	5	3	C_1	$3/3 = 1$	Updated
(d, e)	1	5	3	C_2	$3/3 = 1$	
(g, i)	1	2	2	C_3	$2/2 = 1$	
(c, y)	1	0	3	C_1	$3/3 = 1$	Updated
(e, x)	1	0	3	C_2	$3/3 = 1$	
(x, j)	1	0	2	C_3	$2/2 = 1$	
(a, d)	0.5	2	2	C_1	$2/3 = 0.67$	Updated
(h, y)	0.33	4	1	C_2	$1/3 = 0.33$	

Table 6.11. Deleting the fifth edge (d, x). The density of channel 2 reduces to 2, causing the weight of all edges in channel 2 to be updated.

	Before				**After**	
Edge	$w(e)$	x-span	$d(e)$	C_i	$w(e)$	Action
(b, c)	1	9	3	C_1	$3/3 = 1$	Skip
(h, i)	1	9	2	C_3	$2/2 = 1$	Skip
(h, k)	1	6	2	C_3	$2/2 = 1$	Skip
(d, x)	1	5	3	C_2	$3/3 = 1$	Deleted
(d, e)	1	5	2	C_2	$2/2 = 1$	Updated
(d, e)	1	5	3	C_1	$3/3 = 1$	
(g, i)	1	2	2	C_3	$2/2 = 1$	
(e, x)	1	0	2	C_2	$2/2 = 1$	Updated
(c, y)	1	0	3	C_1	$3/3 = 1$	
(x, j)	1	0	2	C_3	$2/2 = 1$	
(a, d)	0.67	2	2	C_1	$2/3 = 0.67$	
(h, y)	0.33	4	1	C_2	$1/2 = 0.5$	Updated

Table 6.12. Deleting the sixth edge (d, e) from channel 1. No more edge deletion is necessary.

Edge	$w(e)$	Action
(d, e)	1	Removed
(d, e)	1	
(g, i)	1	
(c, y)	1	
(x, j)	1	
(e, x)	1	
(a, d)	0.67	
(h, y)	0.5	

5. Compare the routing results obtained by feedthrough insertion and iterative deletion using their net connection graphs.

   Figure 6.29 shows the comparison between the net connection graphs obtained after feedthrough insertion and iterative deletion. The density of channel 1 is lower in the case of iterative deletion.

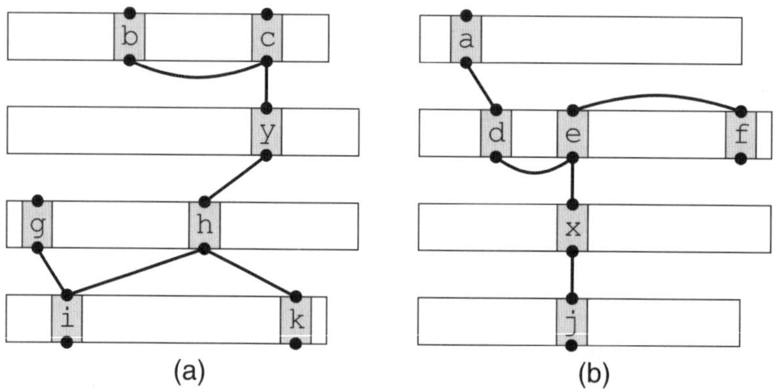

Figure 6.28. (**a**) Final routing tree for n_1 after iterative deletion, (**b**) final routing tree for n_2.

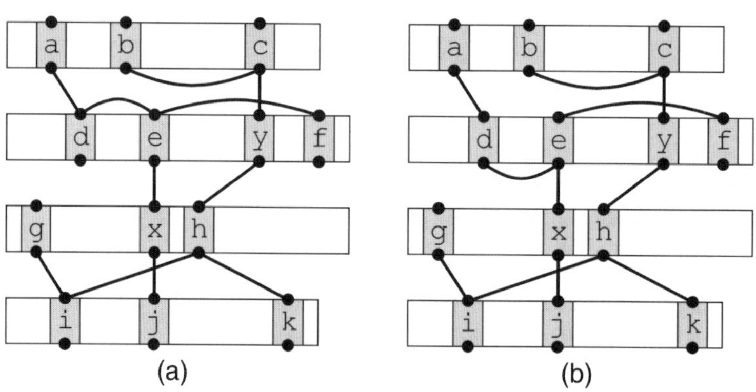

Figure 6.29. (**a**) Net connection graph after feedthrough insertion, (**b**) net connection graph after iterative deletion. The density of channel 1 is lower in (**b**) (= 3 vs 2).

4. Yoshimura and Kuh Algorithm

Channel routing is another important multi-net routing problem, where the pins are arranged into two rows and connected using the space in between the rows (= channel). The goal is to connect the pins from the same nets using rectilinear lines within the channel so that the height of the channel is minimized. Secondary goals include the minimization of via, wirelength, crosstalk noise, etc. Two layers of metal are given usually, one for horizontal wires and the other for vertical wires. The channel routing problem is used in standard cell detail routing.

The first known algorithm for channel routing is named Left Edge (LE) algorithm [Hashimoto and Stevens, 1971]. One important follow-up work is the Constrained LE (CLE) algorithm by [Perskey et al., 1976], where the pins under vertical constraints are handled efficiently (to be discussed later). The "net merging" method proposed by Yoshimura and Kuh [Yoshimura and Kuh, 1982] is another significant improvement over the basic LE, where some subsets of nets are merged before the routing starts. This pre-process of merging nets is proved to be effective in reducing the channel height further compared with the original/constraned LE. Once the merging step is done, any channel routing algorithm such as CLE algorithm can be used to route the nets.

Quick Overview

Given a channel routing problem instance, we first construct the horizontal constraint graph (HCG) and vertical constraint graph (VCG). Each net becomes a node in HCG, and an undirected edge (x, y) exists if the horizonal span of nets x and y overlap at some column(s). In case of VCG, each node represents a net, and a directed edge (x, y) exists if net x is on the top row and y is in the bottom row, and both x and y are located in the same column. If nets x and y are connected in HCG, x and y cannot be assigned to the same track (= intermediate row in the channel) due to the horizontal overlap. Likewise, if net x is connected to net y in VCG, x needs to be assigned to a track above y due to the vertical overlap at the column in which nets x and y are co-located.

The next step is to construct the zone representation, where each zone corresponds to a maximal clique in the HCG. The horizonal span of each net is represented in terms of the zones it spans, e.g., net 1 spans zone 2 and 3. We then visit a pair of neighboring zones from left to right. Given a pair of zones z and $z + 1$, we obtain two sets of nets L and R, where L is the set of nets ending at zone z or before z. R is the set of nets that begin at zone $z + 1$. We then obtain a subset of nets $P \subseteq L$ and another subset $Q \subseteq R$ so that any net in P and any net in Q are not on the same path in the VCG. This is done to make sure cycles are not formed from net merging. If $|P| < |Q|$, we switch the names of the subsets so that $|P| \geq |Q|$ always. Two heuristics are proposed in

[Yoshimura and Kuh, 1982] to choose a pair of nets for merging, one from P and another from Q. We only utilize their "simple" heuristic in this book. The goal of this heuristic is to find a pair of nets for merging so that the increase in the longest path length in VCG is minimized after the merging. We first find $m^* \in Q$ that maximizes

$$f(m) = K \cdot \{u(m) + d(m)\} + \max\{u(m), d(m)\}$$

where K is the user-specified constant. To compute $u(x)$ and $u(y)$, we add a source and a sink to the VCG so that the source connects to all nodes with no incoming edge, and all nodes with no outgoing edge connects to the sink. Then, $u(x)$ denotes the length of the longest source-to-x path, and $d(x)$ denotes the length of the longest x-to-sink path. Next, we find $n^* \in P$ that minimizes

$$g(n, m) = K \cdot h(n, m) - \{\sqrt{u(m) \cdot u(n)} + \sqrt{d(m) \cdot d(n)}\}$$

where $n \in P$, $m \in Q$, and

$$\begin{aligned} h(n, m) &= \max\{u(n), u(m)\} + \max\{d(n), d(m)\} \\ &\quad - \max\{u(n) + d(n), u(m) + d(m)\} \end{aligned}$$

We then merge n^* and m^* and remove them from P and Q. This selection continues until Q is empty. Note that we do not update VCG until the net pair selection is completed for a given pair of zones. The net merging algorithm terminates if all zone pairs are processed.

Once the net merging step is completed, we update the VCG and zone representation to reflect the merging. In the Constrained Left Edge (CLE) algorithm, we assign nets from top to bottom tracks and left to right columns in the same track. Starting with the top-most track, we first choose a subset of nets X so that these nets do not have any incoming edges in the VCG. If there are multiple nets in X, we choose the one, say x_1, that begins at the left-most column. We then remove x_1 from X and assign it to the first track. If X is not empty, we choose the next net, say x_2, so that it does not cause horizontal overlap with x_1, i.e., edge (x_1, x_2) does not exist in the HCG. In addition, x_2 should begin at the left-most column among all nets in X. We then assign x_2 to the same track and remove it from X. Once all nets in X are examined, we remove the nets that are assigned to the current track from VCG and go to the next track and repeat the track assignment. The CLE algorithm terminates once we assign all nets to the tracks.

Practice Problem

Consider the following two-layer channel routing problem.

$$\text{TOP} = [1, 1, 4, 2, 3, 4, 3, 6, 5, 8, 5, 9]$$
$$\text{BOT} = [2, 3, 2, 0, 5, 6, 4, 7, 6, 9, 8, 7]$$

1. Construct the constraint graphs (HCG and VCG) and the zone representation.

   Table 6.13 shows the horizontal span of the nets as well as the zones defined by the spans. Figure 6.30 shows the horizontal and vertical constraint graphs. Lastly, Figure 6.31 shows the zone representation. Note that each zone corresponds to a maximal clique in the HCG. Also note that the VCG is acyclic, requiring no dogleg to break the cycles. The size of maximum clique in the HCG is 4, and the length of the longest path in the VCG is 6. Thus, the lower bound of track usage is 6, the maximum of these two values.

Table 6.13. Horizonal span of the nets and their zones.

	Zone 1		Zone 2		Zone 3			Zone 4		Zone 5		
Net	c1	c2	c3	c4	c5	c6	c7	c8	c9	c10	c11	c12
1	1	1										
2	2	2	2	2								
3		3	3	3	3	3	3					
4			4	4	4	4	4					
5					5	5	5	5	5	5	5	
6						6	6	6	6			
7								7	7	7	7	7
8										8	8	
9										9	9	9

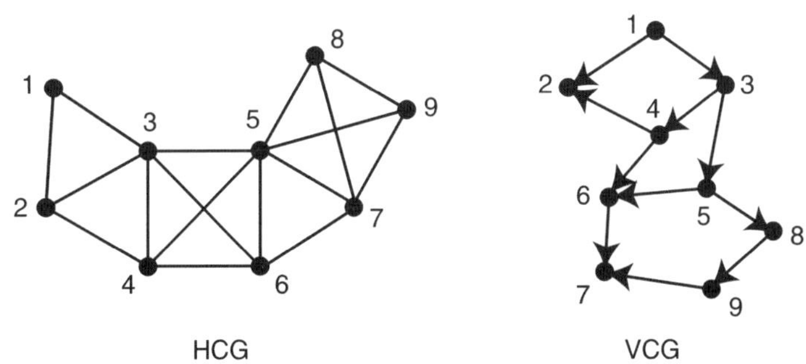

Figure 6.30. Constraint graphs.

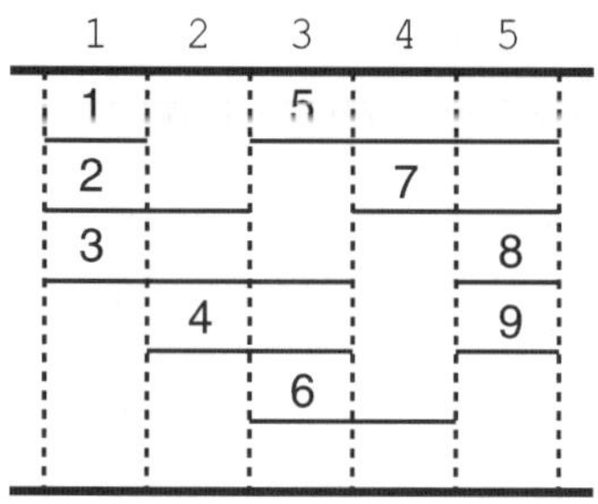

Figure 6.31. Zone representation.

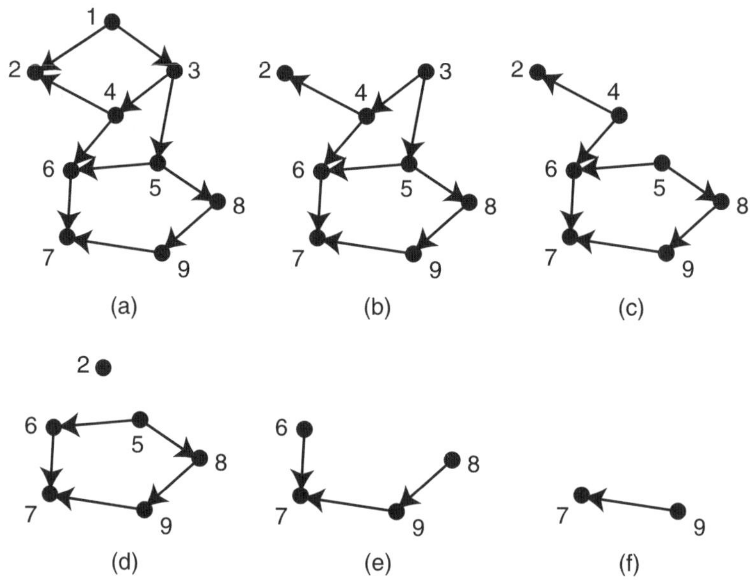

Figure 6.32. VCG after track assignment. (**a**) Initial VCG, (**b**) after assigning net 1, (**c**) after assigning net 3, (**d**) after assigning net 4, (**e**) after assigning nets 2 and 5, (**f**) after assigning nets 6 and 8.

2. Perform the Constrained Left Edge (CLE) algorithm [Perskey et al., 1976].

   We assign the nets to the tracks as follows:

   - Track 1: we can only assign net 1 because it is the only node with no incoming edge in the VCG shown in Figure 6.32(a).
   - Track 2: we can only assign net 3 because it is the only node with no incoming edge in the VCG shown in Figure 6.32(b).
   - Track 3: the VCG shown in Figure 6.32(c) shows two nets with no incoming edge: nets 4 and 5. We assign net 4 to track 3 first because

the starting column of net 4 is to the left of the starting column of net 5. Next, we decide that we cannot assign net 5 in the same track due to the edge between nets 4 and 5 in the HCG shown in Figure 6.30, i.e., horizontal overlap will occur.

- Track 4: the VCG shown in Figure 6.32(d) shows two nets with no incoming edge: nets 2 and 5. We assign net 2 to track 4 first because the starting column of net 2 is to the left of the starting column of net 5. Since there is no edge between nets 2 and 5 in the HCG, we assign net 5 to track 4 as well.
- Track 5: the VCG shown in Figure 6.32(e) shows two nets with no incoming edge: nets 6 and 8. We assign net 6 to track 5 first. There is no edge between nets 6 and 8 in the HCG, so we assign net 8 to track 5 as well.
- Track 6: we can only assign net 9 because it is the only node with no incoming edge in the VCG shown in Figure 6.32(f).
- Track 7: we assign net 7, the last node remaining in the VCG.

The final track usage is 7. Figure 6.33 shows the final channel routing result.

3. Perform net merging using the "simple method" given in [Yoshimura and Kuh, 1982]. Assume $K = 100$.

   We visit the zones shown in Figure 6.31 from the left-most position:

   (a) Zones 1 and 2: From Figure 6.31, we see that net 1 ends at zone 1, so $L = \{1\}$. In addition, net 4 begins at zone 2, so $R = \{4\}$. Since 1 and 4 are on the same path in the VCG shown in Figure 6.30, no merging is possible between nets 1 and 4.

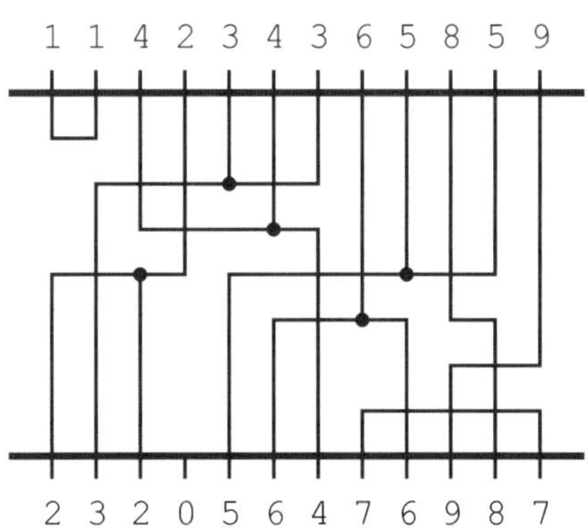

Figure 6.33. Final routing after constrained LE algorithm is applied on the original routing problem.

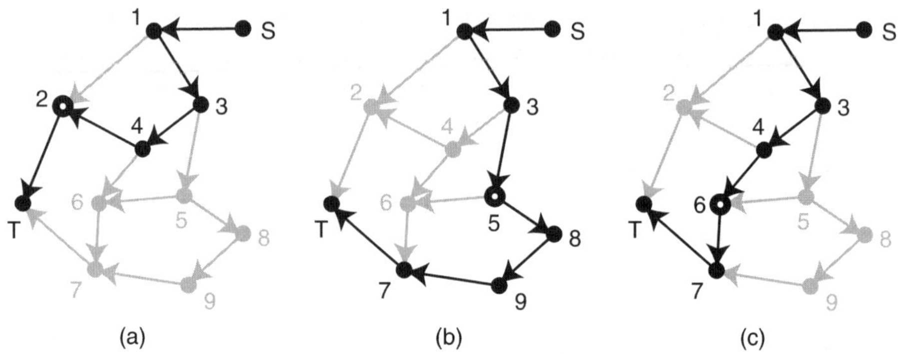

Figure 6.34. Computation of $u(x)$ and $d(x)$ for nets 2, 5, and 6. **(a)** $u(2) = 4$, $d(2) = 1$, **(b)** $u(5) = 3$, $d(5) = 4$, **(c)** $u(6) = 4$, $d(6) = 2$.

(b) Zones 2 and 3: we start with $L = \{1\}$ from the previous iteration. We note from Figure 6.31 that net 2 ends at zone 2, so $L = \{1, 2\}$. In addition, nets 5 and 6 begin at zone 3, so $R = \{5, 6\}$. The net pairs $(1, 5)$ and $(1, 6)$ are on the same path in the VCG shown in Figure 6.30. Thus, only the pairs $(2, 5)$ and $(2, 6)$ can be merged. We perform net pair selection as follows:

(i) We form $P = \{5, 6\}$ and $Q = \{2\}$.

(ii) We compute $u(x)$ and $d(x)$ values for nets 2, 5, and 6 as shown in Figure 6.34. We get $u(2) = 4$, $d(2) = 1$, $u(5) = 3$, $d(5) = 4$, $u(6) = 4$, and $d(6) = 2$.

(iii) Since there is only one net in Q, it is trivial to see that $m^* = 2$.

(iv) We compute $h(n, 2)$ and $g(n, 2)$ for each $n \in P$ as follows:

$$\begin{aligned} h(5,2) &= \max\{u(5), u(2)\} + \max\{d(5), d(2)\} \\ &\quad - \max\{u(5) + d(5), u(2) + d(2)\} = 1 \\ h(6,2) &= \max\{u(6), u(2)\} + \max\{d(6), d(2)\} \\ &\quad - \max\{u(6) + d(6), u(2) + d(2)\} = 0 \end{aligned}$$

Thus,

$$\begin{aligned} g(5,2) &= 100 \cdot h(5,2) - \{\sqrt{u(2) \cdot u(5)} + \sqrt{d(2) \cdot d(5)}\} \\ &= 94.5 \\ g(6,2) &= 100 \cdot h(6,2) - \{\sqrt{u(2) \cdot u(6)} + \sqrt{d(2) \cdot d(6)}\} \\ &= -5.4 \end{aligned}$$

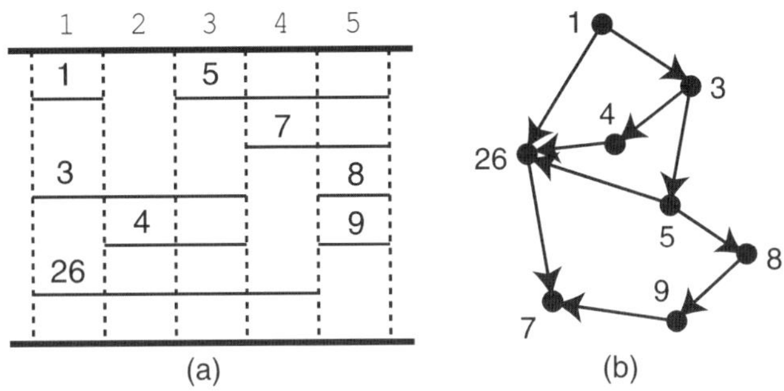

Figure 6.35. Result after merging net $(2, 6)$. (**a**) Zone representation, (**b**) VCG.

Since $g(5, 2) > g(6, 2)$, we see that $n^* = 6$. Thus, we merge net $n^* = 6$ and $m^* = 2$.

(v) We remove net 2 from Q and 6 from P. Since Q is empty, we are done.

We remove net 2 from L because it is merged. Thus, $L = \{1\}$. Figure 6.35 shows the updated zone representation and VCG after merging net 2 and 6.

(c) Zones 3 and 4: we start with $L = \{1\}$ from the previous iteration. From the new zone representation shown in Figure 6.35(a), we note that nets 3 and 4 end at zone 3, so $L = \{1, 3, 4\}$. In addition, net 7 begin at zone 4, so $R = \{7\}$. From the new VCG shown in Figure 6.35(b), we see that all nets in L and R are on the same path. Thus, no net merging is possible.

(d) Zones 4 and 5: we start with $L = \{1, 3, 4\}$ from the previous iteration. From the zone representation shown in Figure 6.35(a), we note that net 26 ends at zone 4, so $L = \{1, 3, 4, 26\}$. In addition, nets 8 and 9 begin at zone 5, so $R = \{8, 9\}$. From the VCG shown in Figure 6.35(b), we see that the pairs $(4, 8)$, $(4, 9)$, $(26, 8)$, and $(26, 9)$ can be merged because the nets in each pair are not on the same path. We perform net pair selection as follows:

(i) We form $P = \{4, 26\}$ and $Q = \{8, 9\}$.

(ii) We compute $u(x)$ and $d(x)$ values for nets 4, 26, 8, and 9 as shown in Figure 6.36. We get $u(4) = 3$, $d(4) = 3$, $u(26) = 4$, $d(26) = 2$, $u(8) = 4$, $d(8) = 3$, $u(9) = 5$, and $d(9) = 2$.

(iii) We compute $f(m)$ for each $m \in Q$ as follows:

$$f(8) = 100\{u(8) + d(8)\} + \max\{u(8), d(8)\} = 704$$
$$f(9) = 100\{u(9) + d(9)\} + \max\{u(9), d(9)\} = 705$$

Thus, we choose $m^* = 9$.

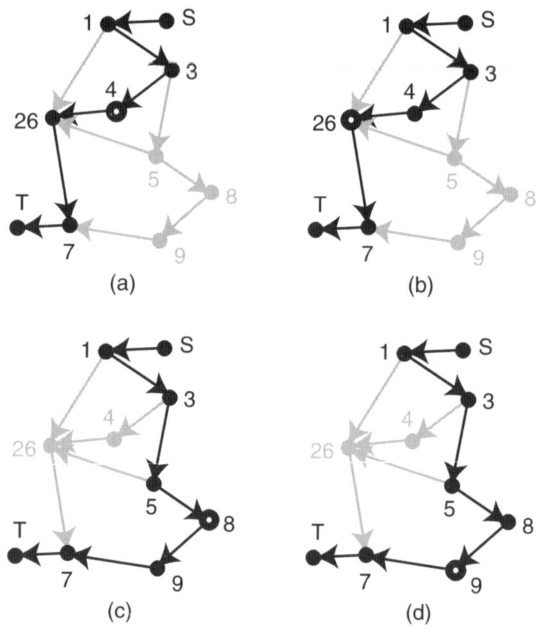

Figure 6.36. Computation of $u(x)$ and $d(x)$ for nets 4, 26, 8, and 9. (**a**) $u(4) = 3$, $d(4) = 3$, (**b**) $u(26) = 4$, $d(26) = 2$, (**c**) $u(8) = 4$, $d(8) = 3$, (**d**) $u(9) = 5$, $d(9) = 2$.

(iv) We compute $h(n, 9)$ and $g(n, 9)$ for each $n \in P$ as follows:

$$\begin{aligned} h(4,9) &= \max\{u(4), u(9)\} + \max\{d(4), d(9)\} \\ &\quad - \max\{u(4) + d(4), u(9) + d(9)\} = 1 \\ h(26,9) &= \max\{u(26), u(9)\} + \max\{d(26), d(9)\} \\ &\quad - \max\{u(26) + d(26), u(9) + d(9)\} = 0 \end{aligned}$$

Thus,

$$\begin{aligned} g(4,9) &= 100 \cdot h(4,9) - \{\sqrt{u(9) \cdot u(4)} + \sqrt{d(9) \cdot d(4)}\} \\ &= 93.7 \\ g(26,9) &= 100 \cdot h(26,9) - \{\sqrt{u(9) \cdot u(26)} + \sqrt{d(9) \cdot d(26)}\} \\ &= -6.5 \end{aligned}$$

Since $g(4,9) > g(26,9)$, we see that $n^* = 26$. Thus, we merge net $n^* = 26$ and $m^* = 9$.

(v) We remove net 26 from P and 9 from Q. Since Q is not empty, we repeat the whole process. At this point, $P = \{4\}$ and $Q = \{8\}$.

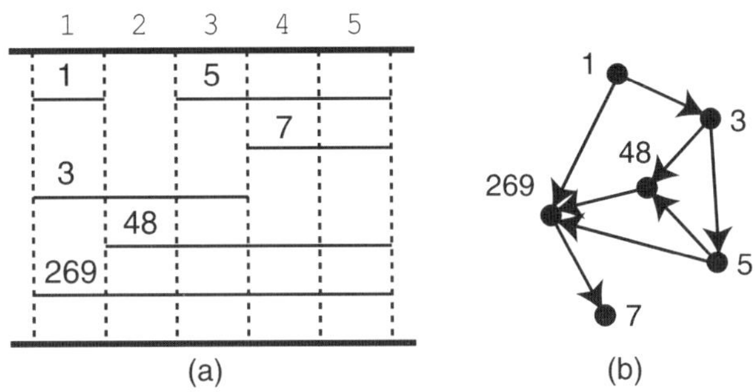

Figure 6.37. Result after merging nets $(26, 9)$ and $(8, 4)$. (**a**) Zone representation, (**b**) VCG.

Thus, it is trivial to see that $m^* = 8$, and $n^* = 4$. This means we merge nets 8 and 4 and remove them from P and Q. Since there is no more element left in P and Q, we are done.

Figure 6.37 shows the updated zone representation and VCG after merging nets $(26, 9)$ and $(8, 4)$.

4. Perform the Constrained Left Edge (CLE) algorithm [Perskey et al., 1976] on the net merging result and draw the channel routing result. Assign tracks from top to bottom and left to right.

   We assign the nets to the tracks as follows:

   - Track 1: we can only assign net 1 because it is the only node with no incoming edge in the VCG shown in Figure 6.38(a).
   - Track 2: we can only assign net 3 because it is the only node with no incoming edge in the VCG shown in Figure 6.38(b).
   - Track 3: we can only assign net 5 because it is the only node with no incoming edge in the VCG shown in Figure 6.38(c).
   - Track 4: we can only assign net 48 because it is the only node with no incoming edge in the VCG shown in Figure 6.38(d).
   - Track 5: we can only assign net 269 because it is the only node with no incoming edge in the VCG shown in Figure 6.38(e).
   - Track 6: we assign net 7, the last node remaining in the VCG.

   From the initial VCG in Figure 6.38(a), we see that the longest path length is 6, which is the lower bound of track usage. We obtain the same track usage as this lower bound. Figure 6.39 shows the final routing result. Note that this result is better than the result shown in Figure 6.33.

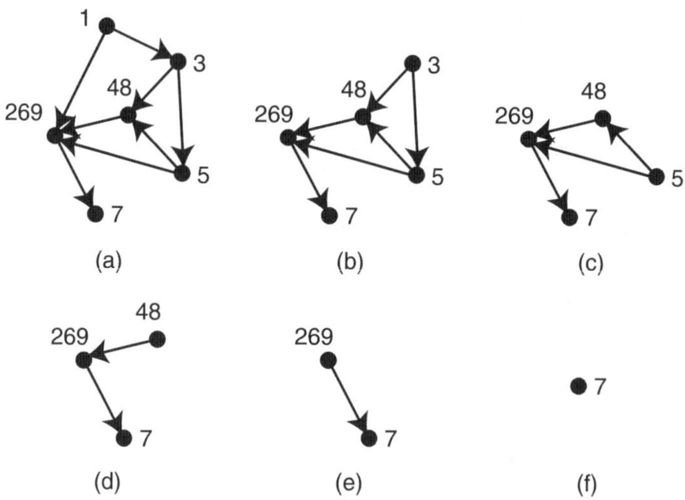

Figure 6.38. VCGs after track assignment. (**a**) Initial VCG, (**b**) after assigning net 1, (**c**) after assigning net 3, (**d**) after assigning net 5, (**e**) after assigning net 48, (**f**) after assigning net 269.

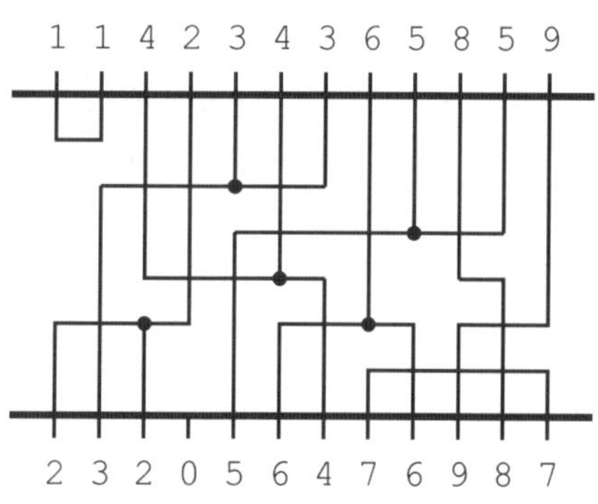

Figure 6.39. Final routing after performing constrained LE algorithm on top of net merging result.

5. More Practice Problems

1. Consider the following netlist:

$$
\begin{aligned}
n_1 &= \{(1,0),(0,3),(3,2),(3,3)\} \\
n_2 &= \{(0,0),(2,1),(1,3),(3,1),(2,3)\} \\
n_3 &= \{(0,2),(3,1),(2,2)\} \\
n_4 &= \{(1,1),(2,2),(3,0),(3,3)\}
\end{aligned}
$$

   Use a 4×4 mesh for the routing graph, where the weight of each edge corresponds to the current usage. The edge capacity is set to 3.

   (a) Perform a single pass of the SMMT-phase of [Chiang et al., 1990] under $c_j = 2.0$.

   (b) Perform a single pass of the SP-phase of [Chiang et al., 1990].

   (c) Compare the results obtained by the SMMT and SP phases.

2. Consider the following routing graph G shown in Figure 6.40, where the capacity of all edges is 2. The following six nets are to be routed: $n_1 = \{a, i\}$, $n_2 = \{a, h\}$, $n_3 = \{d, c\}$, $n_4 = \{c, h\}$, $n_5 = \{g, b\}$, $n_6 = \{i, d\}$. The first node in each net is the source.

   (a) Set up and solve the integer linear programming (ILP) formulation of the multi-commodity flow based global routing.

   (b) Perform the shortest path based MM heuristic. Break ties so that the number of turns (= vias) is minimized.

   (c) Compare the results obtained by ILP and MM methods.

3. Consider the standard placement shown in 6.41. We are to route the following nets: $n_1 = \{a, c, d, e, h\}$, $n_2 = \{b, e, f, g\}$. Assume that the width is 2 and the height is 3 for the gates and feedthroughs. The gates are to be

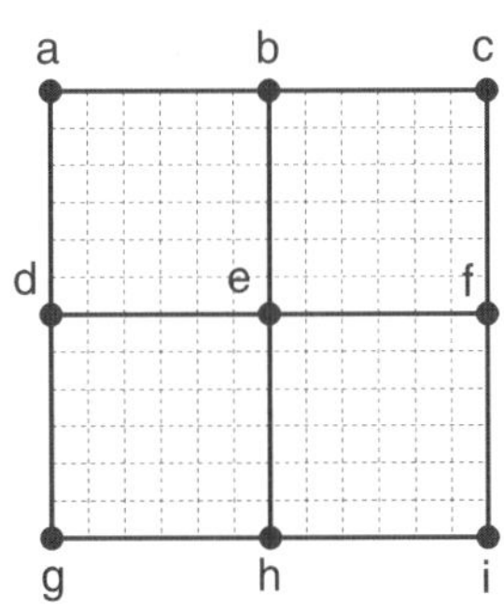

Figure 6.40. Routing graph for multi-commodity flow based routing.

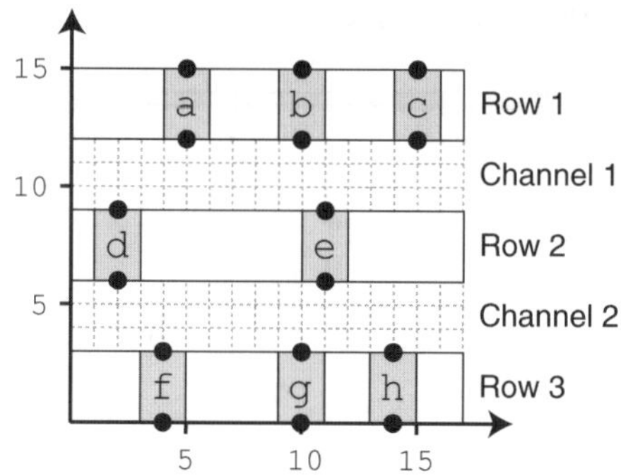

Figure 6.41. A standard cell placement with three rows.

shifted to the right upon feedthrough insertion. Assume that each cell has a built-in feedthrough (= vertical edge within each cell).

(a) Perform the feedthrough insertion phase of [Cong and Preas, 1988] with $K = 0.5$. Break ties in lexicographical order. Place feedthroughs right below the top gate.

(b) Build the simplified net connection graph G', and compute the density of the channels in G'.

(c) Perform Iterative Deletion [Cong and Preas, 1988] to determine the net segments. Break ties based on the following factors in this order: (1) delete edges with longer x-span ($= |x_i - x_j|$), (2) delete edges with higher edge density $d(e)$, (3) delete edges from the bottom-most channel, (4) delete edges with higher lexicographical order.

(d) Compare the routing results obtained by feedthrough insertion and iterative deletion using their net connection graphs.

4. Consider the following two-layer channel routing problem.

$$\text{TOP} = [1, 2, 1, 4, 2, 5, 4, 8, 8, 6]$$
$$\text{BOT} = [3, 4, 3, 5, 2, 6, 6, 7, 5, 7]$$

(a) Construct the constraint graphs (HCG and VCG) and the zone representation.

(b) Perform the Constrained Left Edge (CLE) algorithm [Perskey et al., 1976] on the original routing problem and draw the channel routing result.

(c) Perform net merging using the "simple" heuristic given in [Yoshimura and Kuh, 1982]. Assume $K = 100$.

(d) Perform the Constrained Left Edge (CLE) algorithm [Perskey et al., 1976] on the net merging result and draw the channel routing result.

6. Probing Further

Disclaimer: The list here is meant to be representative, not comprehensive. A comprehensive survey on multi-net routing algorithms is provided in [Hu and Sapatnekar, 2001].

Steiner Min-Max Tree Algorithm

The Steiner Min-Max Tree formulation [Chiang et al., 1990] is extended to Weighted Rectilinear Steiner Tree (WRST) problem in [Chiang et al., 1994]. WRST problem seeks Steiner trees with minimum weighted wirelength. The goal is to simultaneously minimize congestion and wirelength. The routing of each net is done on a routing graph that is the union of the Hanan grid and the grid formed by extending the boundaries of each routing region. The weight of the routing graph represents the congestion of the corresponding routing regions. The weights for all regions are updated after each net is routed.

The authors of [Changfan et al., 2000] presented a methodology for timing optimization at the post-routing stage. A post-routing logic optimization is performed based on incremental placement and routing characterization techniques. With incremental placement during logic optimization, timing can be evaluated using the accurate parasitics from placement. The authors performed global routing using WRST algorithm [Chiang et al., 1994] to predict the routing region usage and congestion.

The authors of [Alpert et al., 2003a] presented a method to pre-plan buffers and interconnects in the early stage of layout generation. Both buffers and wires are considered simultaneously because wire routes determine buffer requirements and buffer locations constrain the wire routes. Their four-stage heuristic consists of initial tree construction, congestion reduction, buffer assignment, and final post-processing. The congestion reduction is accomplished by rip-up-and-reroute using SMMT approach [Chiang et al., 1990].

The authors of [Su et al., 2004] presented an early-stage interconnect planning methodology that simultaneously considers signal wires and power grid wires under the congestion constraints. The authors start with initial Steiner trees and improve the routing congestion with SMMT-based rip-up-and-reroute [Chiang et al., 1990]. This is followed by power-grid optimization that includes wire removal in non-critical regions and wire-sizing that considers the voltage-drop and current-density constraints.

Multi-Commodity Flow Routing Algorithm

The authors of [Carden et al., 1996] developed the first multi-commodity flow based global router with a theoretical bound from the optimal solution. They applied the two-terminal multi-commodity fractional flow algorithm by [Shahrokhi and Matula, 1990], followed by randomized rounding to obtain integer solutions to the multi-terminal multi-commodity problem. Instead of the min-cost multi-commodity flow formulation used in [Shragowitz and Keel, 1987], the authors adopted the concurrent multi-commodity flow formulation. The goal is to maximize routability by minimizing edge congestion as opposed to the conventional techniques which usually seek to minimize wirelength.

TIGER [Hong et al., 1997] is a multi-commodity flow based timing-driven global router for gate array and standard cell layout designs. The timing-driven global routing problem is formulated as a multi-terminal, multi-commodity network flow problem with integer flows under the timing constraints. In order to handle multi-terminal nets, a primal and dual approach is adopted to obtain a fractional routing solution, and a heuristic process is used to integerize the fractional solution. A critical path based timing analysis is used to guarantee the satisfaction of timing constraints.

The authors of [Albrecht, 2001] showed how the approximation algorithms by [Garg and Konemann, 1998] with extensions due to [Fleischer, 2000] for the multi-commodity flow problem can be used to solve the linear programming relaxation of the global routing problem. This approach is shown to be effective in minimizing the maximum congestion, and evenly distributing routing resource usage. Their approach is also shown to improve signal integrity from the extra spacing between wires.

Buffer block planning [Cong et al., 1999a] is an early interconnect planning method that reserves a set of regions in the layout for buffer insertion. The authors of [Dragan et al., 2000] formulated a multi-commodity flow based routing problem that makes use of the given buffer block planning solution. Their method routes nets using the available buffer blocks such that the required upper and lower bounds on buffer intervals—as well as wirelength upper bounds per net—are satisfied. The authors used the approximation method of [Fleischer, 2000] to overcome the runtime limit of multi-commodity flow computation.

UTACO [Jing et al., 2004] is a timing and congestion-driven standard cell global router based on multi-commodity flow. The authors adopt a shadow price mechanism to incorporate timing and congestion objectives into one unified objective function. The multi-commodity flow is expressed by a linear programming formulation as a primal problem. They convert the primal problem into a dual formulation using the shadow price as the variables, where the shadow price of a net represents the sum of its congestion price and timing price.

Iterative Deletion Algorithm

The author of [Cong, 1991] presented an algorithm that combines pin assignment global routing. The algorithm is based on two key theorems: the channel pin assignment theorem and the block boundary decomposition theorem.

According to these two theorems, one only needs to generate a coarse pin assignment and global routing solution. Iterative deletion [Cong and Preas, 1988] is used to assign the pins of all nets to block boundaries and determine the global routes of the nets. The exact pin locations and global routing topology can be determined optimally later by a linear time algorithm.

DECIMATE [Cong and Madden, 1997a] is a performance-driven standard cell global router that is based on iterative deletion. The authors use A-tree [Cong et al., 1993] and 1-Steiner tree [Kahng and Robins, 1992; Borah et al., 1994] to construct performance-driven routing tree topologies. Iterative deletion is then applied to non-timing critical nets to minimize channel density and congestion. DECIMATE constructs simplified net connection graph for each non-timing critical net and removes all redundant edges based on a weight function that represents routing congestion.

The concept of iterative deletion is used for circuit partitioning in [Madden, 1999]. As with the routing problems, they begin with a redundant initial partitioning solution. Unlike most move-based algorithms such as [Fiduccia and Mattheyses, 1982], in which a vertex is assigned to a single random partition, they assign each vertex to multiple random partitions. These redundant assignments in the initial partitioning solution are then iteratively removed in a greedy manner, until a final non-redundant solution is produced. Iterative improvement algorithms pursue moves that appear the "best," while iterative deletion algorithms eliminate moves that appear the "worst."

Global routing determines the set of routing regions used in a routing tree. Pseudo-pin assignment (PPA) is a step that determines the location of wire crossings on the routing region boundaries. The authors of [Chang and Cong, 2001] presented a PPA algorithm with crosstalk noise control in multi-layer grid-less general area routing. The entire step is divided into coarse PPA and detailed PPA. Iterative deletion is used during the coarse PPA step, where the pins are assigned to "intervals" on the region boundaries. Crosstalk and wirelength minimization is considered during this step.

Simultaneous shield insertion and net ordering (SINO) is an effective way to mitigate RLC crosstalk noise in routing solution. The authors of [Xiong and He, 2005] presented a global router that considers SINO for RLC crosstalk minimization. The key algorithm phase is the global routing synthesis with shield reservation and minimization based on pre-routing shield estimation. Iterative deletion is performed to reduce congestion, which in turn mitigates crosstalk noise among the routed nets. Shield insertion is also considered during the iterative deletion process.

Yoshimura and Kuh Algorithm

A hierarchical channel routing algorithm is presented in [Burstein and Pelavin, 1983], where the formulation is based on the reduction of the original two-layer $m \times n$ grid routing to a set of $2 \times n$ grid routings. The routing region is recursively bisected in horizontal direction into smaller super cells, and the channel routing is performed in terms of the super cells at each level. A key part of this algorithm is to refine the routing solution from one hierarchical level to a lower level after each bisection. The goal is to minimize congestion and wirelength during this process.

The authors of [Cong et al., 1988] extended two-layer channel routing problem to three and four layers. Instead of building a three- or four-layer solutions directly, they take advantage of existing two-layer routers and develop a method to transform a two-layer routing solution systematically into a three- or four-layer solution. The basic idea is to distribute the tracks evenly on the two horizontal layers: net 1 is assigned to track 1 on H-layer 1, net 2 to track 1 on H-layer 2, net 3 to track 2 on H-layer 1, net 4 to track 2 on H-layer 2, etc.

In order to further reduce the track usage, channel routing is often done "over the cell", the area above the top boundary and below the bottom boundary of the channel. A common approach to the over-the-cell channel routing problem is to divide the problem into three steps: (1) routing over the cells, (2) choosing net segments, and (3) routing within the channel. The authors of [Cong and Liu, 1990] provided an optimal algorithm for step (1), and an efficient heuristic for step (2). Step (3) can be carried out using a conventional channel router.

The authors of [Gao and Liu, 1996] considered crosstalk minimization in channel routing. They proposed an algorithm that utilizes existing channel routing algorithms and improves upon the routing results by permuting the routing tracks. Permutation of the tracks changes the relative positions of the horizontal wire segments and the lengths of the vertical wire segments in the routing solution, which in turn change the values of the crosstalk in the nets. A mixed integer linear programming (ILP) formulation and effective procedures for reducing the number of variables and constraints in the mixed ILP formulation are presented.

The author of [Sapatnekar, 2000] studied the impact of crosstalk on delay and presented a crosstalk-aware channel router. The router takes an initial channel routing solution that minimizes the number of tracks, and modifies it to reduce the crosstalk-induced delay, while leaving the number of tracks in the channel unchanged. A $O(n \log n)$ algorithm is developed to determine the effect of crosstalk on delay, which is fast enough to be used in optimization process. Compared with [Gao and Liu, 1996], this approach incorporates delay in the objective function. It also performs permutation of track "segments" instead of full tracks, thereby increasing the flexibility of routing solutions.

References

Ababei, C. and Bazargan, K. (2003). Timing minimization by statistical timing hMetis-based partitioning. *Int. Conf. on VLSI Design*, pages 58–63.

Adya, S., Chaturvedi, S., Roy, J. A., Papa, D. A., and Markov, I. (2004). Unification of partitioning, placement and floorplanning. *Proc. IEEE Int. Conf. on Computer-Aided Design*, pages 12–17.

Adya, S. and Markov, I. (2003). Fixed-outline floorplanning: enabling hierarchical design. *IEEE Trans. on VLSI Systems*, 11(6):1120–1135.

Albrecht, C. (2001). Global routing by new approximation algorithms for multicommodity flow. *IEEE Trans. on Computer-Aided Design of Integrated Circuits and Systems*, 20(5):622–632.

Alexander, M. and Robins, G. (1996). New performance-driven FPGA routing algorithms. *IEEE Trans. on Computer-Aided Design of Integrated Circuits and Systems*, 15(12):1505–1517.

Alpert, C., Gandham, G., M. Hrkic, J. Hu, Kahng, A. B., Lillis, J., Liu, B., Quay, S., Sapatnekar, S., and Sullivan, A. (2002). Buffered steiner trees for difficult instances. *IEEE Trans. on Computer-Aided Design of Integrated Circuits and Systems*, 21(1):3–14.

Alpert, C., Hu, J., Sapatnekar, S., and Villarrubia, P. (2003a). A practical methodology for early buffer and wire resource allocation. *IEEE Trans. on Computer-Aided Design of Integrated Circuits and Systems*, 22(5):573–583.

Alpert, C., Hu, T. C., Huang, J., Kahng, A. B., and Karger, D. (1995). Prim-dijkstra tradeoffs for improved performance-driven routing tree design. *IEEE Trans. on Computer-Aided Design of Integrated Circuits and Systems*, 14(7):890–896.

Alpert, C., Huang, J. H., and Kahng, A. B. (1998). Multilevel circuit partitioning. *IEEE Trans. on Computer-Aided Design of Integrated Circuits and Systems*, 17(8):655–667.

Alpert, C. and Kahng, A. B. (1995a). Multiway partitioning via geometric embeddings, orderings, and dynamic programming. *IEEE Trans. on Computer-Aided Design of Integrated Circuits and Systems*, 14(11):1342–1358.

Alpert, C. and Kahng, A. B. (1995b). Recent directions in netlist partitioning: a survey. *Integration, the VLSI Journal*, 19(1–2):1–81.

Alpert, C., Nam, Gi-Joon, and Villarrubia, P. G. (2003b). Effective free space management for cut-based placement via analytical constraint generation. *IEEE Trans. on Computer-Aided Design of Integrated Circuits and Systems*, 22(10):1343–1353.

Alpert, C. and Yao, S. (1995). Spectral partitioning: the more eigenvectors, the better. *Proc. ACM Design Automation Conf.*, pages 195–200.

Awerbuch, B., Baratz, A., and Peleg, D. (1990). Cost-sensitive analysis of communication protocols. In *Proc. ACM Symp. Principles of Distributed Computing*, pages 177–187.

Bazargan, K., Kim, S., and Sarrafzadeh, M. (1999). Nostradamus: a floorplanner of uncertain designs. *IEEE Trans. on Computer-Aided Design of Integrated Circuits and Systems*, 18(4):389–397.

Boese, K., Kahng, A. B., McCoy, B., and Robins, G. (1995). Near-optimal critical sink routing tree constructions. *IEEE Trans. on Computer-Aided Design of Integrated Circuits and Systems*, 14(12):1417–1436.

Borah, M., Owens, R., and Irwin, M. (1994). An edge-based heuristic for Steiner Routing. *IEEE Trans. on Computer-Aided Design of Integrated Circuits and Systems*, 13(12):1563–1568.

Borah, M., Owens, R., and Irwin, M. (1997). A fast algorithm for minimizing the elmore delay to identified critical sinks. *IEEE Trans. on Computer-Aided Design of Integrated Circuits and Systems*, 16(7):753–759.

Bozorgzadeh, E., Kastner, R., and Sarrafzadeh, M. (2001). Creating and exploiting flexibility in steiner trees. *Proc. ACM Design Automation Conf.*, pages 195–198.

Breuer, M. A. (1977). A Class of Min-cut Placement Algorithms. In *Proc. ACM Design Automation Conf.*, pages 284–290.

Burstein, M. and Pelavin, R. (1983). Hierarchical channel router. *Proc. ACM Design Automation Conf.*, pages 591–597.

Caldwell, A., Kahng, A. B., and Markov, I. (2000a). Can recursive bisection alone produce routable placements? *Proc. ACM Design Automation Conf.*, pages 477–482.

Caldwell, A., Kahng, A. B., and Markov, I. (2000b). Iterative partitioning with varying node weights. *VLSI Design*, 11(3):249–258.

Caldwell, A., Kahng, A. B., and Markov, I. (2000c). Optimal partitioners and end-case placers for standard-cell layout. *IEEE Trans. on Computer-Aided Design of Integrated Circuits and Systems*, 19(11):1304–1313.

Carden, R., Li, J., and Cheng, C.-K. (1996). A global router with a theoretical bound on the optimal solution. *IEEE Trans. on Computer-Aided Design of Integrated Circuits and Systems*, 15(2):208–216.

Chan, P. K., Schlag, M., and Zien, J. (1994). Spectral K-way ratio-cut partitioning and clustering. *IEEE Trans. on Computer-Aided Design of Integrated Circuits and Systems*, 13(9):1088–1096.

Chan, T., Cong, J., Kong, T., Shinnerl, J., and Sze, K. (2003). An enhanced multilevel algorithm for circuit placement. *Proc. IEEE Int. Conf. on Computer-Aided Design*, pages 299–306.

Chandy, J., Kim, S., Ramkumar, B., Parkes, S., and Banerjee, P. (1997). An evaluation of parallel simulated annealing strategies with application to standard cell placement. *IEEE Trans. on Computer-Aided Design of Integrated Circuits and Systems*, 16(4):398–410.

Chang, C.-C. and Cong, J. (2001). Pseudopin assignment with crosstalk noise control. *IEEE Trans. on Computer-Aided Design of Integrated Circuits and Systems*, 20(5):598–611.

Chang, C.-C., Cong, J., Pan, Z., and Yuan, X. (2003). Multilevel global placement with congestion control. *IEEE Trans. on Computer-Aided Design of Integrated Circuits and Systems*, 22(4):395–409.

Changfan, C., Hsu, Y.-C., and Tsai, F.-S. (2000). Timing optimization on routed designs with incremental placement and routing characterization. *IEEE Trans. on Computer-Aided Design of Integrated Circuits and Systems*, 19(2):188–196.

Chen, D. and Cong, J. (2004). DAOmap: a depth-optimal area optimization mapping algorithm for FPGA designs. In *Proc. IEEE Int. Conf. on Computer-Aided Design*, pages 752–759.

Chen, P. and Kuh, E. S. (2000). Floorplan sizing by linear programming approximation. *Proc. ACM Design Automation Conf.*

Chen, T. and Fan, M. (1998). On convex formulation of the floorplan area minimization problem. *Proc. Int. Symp. on Physical Design*, pages 124–128.

Chen, T.-C. and Chang, Y.-W. (2007). *Packing Floorplan Representation.* Physical Design Handbook, CRC Press. Edited by C. Alpert, S. Sapatnekar, and D. Mehta, Boca Raton, FL.

Chiang, C., Sarrafzadeh, M., and Wong, C. K. (1990). Global routing based on Steiner min-max trees. *IEEE Trans. on Computer-Aided Design of Integrated Circuits and Systems*, 9(12):1318–1325.

Chiang, C., Wong, C. K., and Sarrafzadeh, M. (1994). A weighted steiner tree-based global router with simultaneous length and density minimization. *IEEE Trans. on Computer-Aided Design of Integrated Circuits and Systems*, 13(12):1461–1469.

Cong, J. (1991). Pin assignment with global routing for general cell designs. *IEEE Trans. on Computer-Aided Design of Integrated Circuits and Systems*, 10(11):1401–1412.

Cong, J. and Ding, Y. (1992). An Optimal Technology Mapping Algorithm for Delay Optimization in Lookup-Table Based FPGA Designs. In *Proc. IEEE Int. Conf. on Computer-Aided Design*, pages 48–53.

Cong, J. and Ding, Y. (1994). On area/depth trade-off in LUT-based FPGA technology mapping. *IEEE Trans. on VLSI Systems*, 2(2):137–148.

Cong, J. and Ding, Y. (1996). Combinational logic synthesis for LUT based field programmable gate arrays. *ACM Trans. on Design Automation of Electronics Systems*, 1(2):145–204.

Cong, J., He, L., Koh, C.-K., and Madden, P. (1996). Performance optimization of VLSI interconnect layout. *Integration, the VLSI Journal*, 21(1–2):1–94.

Cong, J. and Hwang, Y. (1995). Simultaneous depth and area minimization in LUT-based FPGA mapping. In *Proc. Int. Symp. on Field Programmable Gate Arrays*, pages 68–74.

Cong, J., Kahng, A. B., and Leung, K.-S. (1998). Efficient algorithms for the minimum shortest path Steiner arborescence problem with applications to VLSI physical design. *IEEE Trans. on Computer-Aided Design of Integrated Circuits and Systems*, 17(1):24–39.

Cong, J., Kahng, A. B., Robins, G., Sarrafzadeh, M., and Wong, C. K. (1992). Provably good performance-driven global routing. *IEEE Trans. on Computer-Aided Design of Integrated Circuits and Systems*, 11(6):739–752.

Cong, J., Koh, C.-K., and Madden, P. (2001). Interconnect layout optimization under higher order RLC model for MCM designs. *IEEE Trans. on Computer-Aided Design of Integrated Circuits and Systems*, 20(12):1455–1463.

Cong, J., Kong, T., and Pan, D. (1999a). Buffer block planning for interconnect-driven floorplanning. *Proc. IEEE Int. Conf. on Computer-Aided Design*, pages 358–363.

Cong, J., Leung, K., and Zhou, D. (1993). Performance-driven interconnect design based on distributed RC delay model. In *Proc. ACM Design Automation Conf.*, pages 14–18.

Cong, J., Li, H., and Wu, C. (1999b). Simultaneous circuit partitioning/clustering with retiming for performance optimization. In *Proc. ACM Design Automation Conf.*, pages 460–465.

Cong, J. and Lim, S. K. (1998). Multiway partitioning with pairwise movement. In *Proc. IEEE Int. Conf. on Computer-Aided Design*, pages 512–516.

Cong, J. and Lim, S. K. (2004). Edge separability based circuit clustering with application to multi-level circuit partitioning. *IEEE Trans. on Computer-Aided Design of Integrated Circuits and Systems*, 23(3):346–357.

Cong, J., Lim, S. K., and Wu, C. (2000). Performance driven multi-level and multiway partitioning with retiming. *Proc. ACM Design Automation Conf.*, pages 274–279.

Cong, J. and Liu, C. L. (1990). Over-the-cell channel routing. *IEEE Trans. on Computer-Aided Design of Integrated Circuits and Systems*, 9(4):408–418.

Cong, J. and Madden, P. (1997a). Performance driven global routing for standard cell design. In *Proc. Int. Symp. on Physical Design*, pages 73–80.

Cong, J. and Madden, P. (1997b). Performance-driven routing with multiple sources. *IEEE Trans. on Computer-Aided Design of Integrated Circuits and Systems*, 16(4):410–419.

Cong, J., Nataneli, G., Romesis, M., and Shinnerl, J. (2004a). An area-optimality study of floorplanning. In *Proc. Int. Symp. on Physical Design*, pages 78–83.

Cong, J. and Preas, B. (1988). A new algorithm for standard cell global routing. *Proc. IEEE Int. Conf. on Computer-Aided Design*, pages 80–83.

Cong, J. and Romesis, M. (2001). Performance-driven multi-level Clustering with application to hierarchical FPGA mapping. *Proc. ACM Design Automation Conf.*, 1:58113–297.

Cong, J., Romesis, M., and Shinnerl, J. R. (2006). Fast floorplanning by look-ahead enabled recursive bipartitioning. *IEEE Trans. on Computer-Aided Design of Integrated Circuits and Systems*, 25(9):1719–1732.

Cong, J., Shinnerl, J., Xie, M., Kong, T., and Yuan, X. (2005). Large-scale circuit placement. *ACM Trans. on Design Automation of Electronics Systems*, 10(2):389–430.

Cong, J., Wei, Jie, and Zhang, Yan (2004b). A thermal-driven floorplanning algorithm for 3D ICs. In *Proc. IEEE Int. Conf. on Computer-Aided Design*, pages 306–313.

Cong, J., Wong, D. F., and Liu, C. L. (1988). A new approach to three- or four-layer channel routing. *IEEE Trans. on Computer-Aided Design of Integrated Circuits and Systems*, 7(10):1094–1104.

Dijkstra, E. (1959). A note on two problems in connexion with graphs. *Numerische Mathematik*, 1(1):269–271.

Doll, K., Johannes, F., and Antreich, K. (1994). Iterative placement improvement by network flow methods. *IEEE Trans. on Computer-Aided Design of Integrated Circuits and Systems*, pages 1190–1200.

Dragan, F., Kahng, A. B., Mandoiu, I., Muddu, S., and Zelikovsky, A. (2000). Provably good global buffering using an available buffer block plan. *Proc. IEEE Int. Conf. on Computer-Aided Design*, pages 104–109.

Dunlop, A. and Kernighan, B. (1985). A procedure for placement of standard-cell VLSI circuits. *IEEE Trans. on Computer-Aided Design of Integrated Circuits and Systems*, 4(1):92–98.

Dutt, S. and Deng, W. (1996a). VLSI circuit partitioning by cluster-removal using iterative improvement techniques. In *Proc. IEEE Int. Conf. on Computer-Aided Design*, pages 194–200.

Dutt, S. and Deng, W. (1996b). VLSI circuit partitioning by cluster-removal using iterative improvement techniques. *Proc. IEEE Int. Conf. on Computer-Aided Design*, pages 194–200.

Eisenmann, H. and Johannes, F. (1998). Generic global placement and floorplanning. *Proc. ACM Design Automation Conf.*, pages 269–274.

Ekpanyapong, M., Minz, J., Watewai, T., Lee, H. S., and Lim, S. K. (2004). Profile-guided microarchitectural floorplanning for deep submicron processor design. *Proc. ACM Design Automation Conf.*, pages 634–639.

Elmore, W. (1948). The transient response of damped linear networks with particular regard to wideband amplifiers. *Journal of Applied Physics*, pages 55–63.

Fiduccia, C. and Mattheyses, R. (1982). A linear time heuristic for improving network partitions. In *Proc. ACM Design Automation Conf.*, pages 175–181.

Fleischer, L. (2000). Approximating fractional multicommodity flow independent of the number of commodities. *SIAM Journal of Discrete Math.*, 13(4):505–520.

Ford, L. and Fulkerson, D. (1962). *Flows in Networks*. Princeton University Press, Princeton, NJ.

FSF, Free Software Foundation (2006). GLPK (GNU Linear Programming Kit).

Gao, T. and Liu, C. L. (1996). Minimum crosstalk channel routing. *IEEE Trans. on Computer-Aided Design of Integrated Circuits and Systems*, 15(5):465–474.

Garg, N. and Konemann, J. (1998). Faster and simpler algorithms for multicommodity flow and other fractional packing problems. *Proc. Annual Symposium on Foundations of Computer Science*, pages 300–309.

Georgakopoulos, G. and Papadimitriou, C. (1987). The 1-Steiner tree problem. *Journal of Algorithms*, 8:122–130.

Griffith, J., Robins, G., Salowe, J., and Zhang, T. (1994). Closing the gap: near-optimal steiner trees in polynomial time. *IEEE Trans. on Computer-Aided Design of Integrated Circuits and Systems*, 13(11):1351–1365.

Hagen, L., Huang, D., and Kahng, A. B. (1997). On implementation choices for iterative improvement partitioning algorithms. *IEEE Trans. on Computer-Aided Design of Integrated Circuits and Systems*, 16(10):1199–1205.

Hagen, L. and Kahng, A. B. (1992). Fast spectral methods for ratio cut partitioning and clustering. *IEEE Trans. on Computer-Aided Design of Integrated Circuits and Systems*, 11(9):1074–1085.

Hall, K. (1970). An r-dimensional quadratic placement algorithm. *Management Science*, 17:219–229.

Hanan, M. (1966). On Steiner's problem with rectilinear distance. *SIAM Journal on Applied Math*, 14(2):255–265.

Hashimoto, A. and Stevens, S. (1971). Wire routing by optimizing channel assignment within large aperatures. In *Proc. ACM Design Automation Conf.*, pages 155–169.

Hauck, S. and Borriello, G. (1997). An evaluation of bipartitioning techniques. *IEEE Trans. on Computer-Aided Design of Integrated Circuits and Systems*, 16(8):849–866.

Ho, J.-M., Vijayan, G., and Wong, C. K. (1989). A new approach to the rectilinear Steiner tree problem. In *Proc. ACM Design Automation Conf.*, pages 161–166.

Ho, J.-M., Vijayan, G., and Wong, C. K. (1990). New algorithms for the rectilinear Steiner tree problem. *IEEE Trans. on Computer-Aided Design of Integrated Circuits and Systems*, 9(2):185–193.

Ho, T.-Y., Chang, Y.-W., Chen, S.-J., and Lee, D.-T. (2005). Crosstalk- and performance-driven multilevel full-chip routing. *IEEE Trans. on Computer-Aided Design of Integrated Circuits and Systems*, 24(6):869–878.

Hong, X., Xue, T., Huang, J., Cheng, C.-K., and Kuh, E. (1997). TIGER: an efficient timing-driven global router for gate array and standard cell layout design. *IEEE Trans. on Computer-Aided Design of Integrated Circuits and Systems*, 16(11):1323–1331.

Hou, H., Hu, J., and Sapatnekar, S. (1999). Non-hanan routing. *IEEE Trans. on Computer-Aided Design of Integrated Circuits and Systems*, 18(4):436–444.

Hu, Bo and Marek-Sadowska, M. (2005). Multilevel fixed-point-addition-based VLSI placement. *IEEE Trans. on Computer-Aided Design of Integrated Circuits and Systems*, 24(8):1188–1203.

Hu, J. and Sapatnekar, S. (2001). A survey on multi-net global routing for integrated circuits. *Integration, the VLSI Journal*, 31(1):1–49.

Hur, S.-W. and Lillis, J. (2000). Mongrel: hybrid techniques for standard cell placement. *Proc. IEEE Int. Conf. on Computer-Aided Design*, pages 165–170.

Hwang, C. and Pedram, M. (2005). PMP: performance-driven multilevel partitioning by aggregating the preferred signal directions of I/O conduits. *Proc. Asia and South Pacific Design Automation Conf.*, pages 428–431.

Hwang, F. (1976). On Steiner minimal trees with rectilinear distance. *SIAM Journal of Applied Math*, 30(1):104–114.

Hwang, F., Richards, D., and Winter, P. (1992). *The Steiner Tree Problem (Annals of Discrete Mathematics)*. North-Holland, The Netherlands.

Hwang, L. and El Gamal, A. (1995). Min-cut replication in partitioned networks. *IEEE Trans. on Computer-Aided Design of Integrated Circuits and Systems*, 14(1):96–106.

Ihler, E., Wagner, D., and Wager, F. (1993). Modeling hypergraph by graphs with the same min-cut properties. *Info. Proc. Letter*, 45:171–175.

Jiang, I., Chang, Y.-W., Jou, J.-Y., and Chao, K.-Y. (2004). Simultaneous floor plan and buffer-block optimization. *IEEE Trans. on Computer-Aided Design of Integrated Circuits and Systems*, 23(5):694–703.

Jing, T., Hong, X.-L., Xu, J.-Y., Bao, H.-Y., Cheng, C.-K., and Gu, J. (2004). UTACO: a unified timing and congestion optimization algorithm for standard cell global routing. *IEEE Trans. on Computer-Aided Design of Integrated Circuits and Systems*, 23(3):358–365.

Kahng, A. B. (2000). Classical floorplanning harmful? In *Proc. Int. Symp. on Physical Design*, pages 207–213.

Kahng, A. B. and Reda, S. (2006). Wirelength minimization for min-cut placements via placement feedback. *IEEE Trans. on Computer-Aided Design of Integrated Circuits and Systems*, 25(7):1301–1312.

Kahng, A. B. and Robins, G. (1992). A new class of iterative Steiner tree heuristics with good performance. *IEEE Trans. on Computer-Aided Design of Integrated Circuits and Systems*, 11(7):893–902.

Kahng, A. B. and Robins, G. (1994). *On Optimal Interconnections for VLSI*. Kluwer, Boston, MA.

Kahng, A. B. and Wang, Q. (2005). Implementation and extensibility of an analytic placer. *IEEE Trans. on Computer-Aided Design of Integrated Circuits and Systems*, 24(5):734–747.

Karypis, G., Aggarwal, R., Kumar, V., and Shekhar, S. (1997). Multilevel hypergraph partitioning: application in VLSI domain. In *Proc. ACM Design Automation Conf.*, pages 526–529.

Karypis, G. and Kumar, V. (1999). Multilevel k-way hypergraph partitioning. *Proc. ACM Design Automation Conf.*, pages 343–348.

Kernighan, B. and Lin, S. (1970). An efficient heuristic procedure for partitioning of electrical circuits. *Bell System Technical Journal*, pages 291–307.

Kirkpatrick, S., Gelatt, C., and Vecchi, M. (1983). Optimization by simulated annealing. *Science*, 220(4598):671–680.

Kleinhans, J., Sigl, G., Johannes, F., and Antreich, K. (1991). GORDIAN: VLSI placement by quadratic programming and slicing optimization. *IEEE Trans. on Computer-Aided Design of Integrated Circuits and Systems*, 10(3):356–365.

Krishnamurthy, B. (1984). An improved min-cut algorithm for partitioning VLSI networks. *IEEE Trans. on Computers*, 33(5):438–446.

Kruskal, J. (1956). On the shortest spanning subtree of a graph and the traveling salesman problem. *Proc. of the American Mathematical Society*, 7(1):48–50.

Kukimoto, Y., Brayton, R. K., and Sawkar, P. (1998). Delay-optimal technology mapping by DAG covering. In *Proc. ACM Design Automation Conf.*, pages 348–351.

Lai, M. and Wong, D. F. (2001). Slicing tree is a complete floorplan representation. *Proc. Design, Automation and Test in Europe*, pages 228–232.

Li, J., Lillis, J., Liu, L.-T., and Cheng, C.-K. (1996). New spectral linear placement and clustering approach. *Proc. ACM Design Automation Conf.*, pages 88–93.

Lillis, J., Cheng, C.-K., Lin, T.-T. Y., and Ho, C.-Y. (1996). New performance driven routing techniques with explicit area/delay tradeoff and simultaneous wire sizing. *Proc. ACM Design Automation Conf.*, pages 395–400.

Lin, J.-M. and Chang, Y.-W. (2001). TCG: A transitive closure graph-based representation for nonslicing floorplans. In *Proc. ACM Design Automation Conf.*, pages 764–769.

Lin, J.-M. and Chang, Y.-W. (2004). TCG-S: orthogonal coupling of P*-admissible representations for general floorplans. *IEEE Trans. on Computer-Aided Design of Integrated Circuits and Systems*, 23(6):968–980.

Liu, H. and Wong, D. F. (1998). Network-flow-based multiway partitioning with area and pin constraints. *IEEE Trans. on Computer-Aided Design of Integrated Circuits and Systems*, 17(1):50–59.

Liu, L. T., Kuo, M. T., Cheng, C. K., and Hu, T. C. (1995). A replication cut for two-way partitioning. *IEEE Trans. on Computer-Aided Design of Integrated Circuits and Systems*, 14(5):623–630.

Madden, P. (1999). Partitioning by iterative deletion. In *Proc. Int. Symp. on Physical Design*, pages 83–89.

Mak, W. K. and Young, E. (2003). Temporal logic replication for dynamically reconfigurable FPGA partitioning. *IEEE Trans. on Computer-Aided Design of Integrated Circuits and Systems*, 22(7):952–959.

Mandoiu, I., Vazirani, V., and Ganley, J. (2000). A new heuristic for rectilinear steiner trees. *IEEE Trans. on Computer-Aided Design of Integrated Circuits and Systems*, 19(10):1129–1139.

Mishchenko, Alan, Cho, Sungmin, Chatterjee, Satrajit, and Brayton, Robert (2007). Combinational and sequential mapping with priority cuts. In *Proc. IEEE Int. Conf. on Computer-Aided Design*, pages 354–361.

Moh, T.-S., Chang, T.-S., and Hakimi, S. (1996). Globally optimal floorplanning for a layout problem. *IEEE Trans. on Circuits and Systems I: Fundamental Theory and Applications*, 43(9):713–720.

Murata, H., Fujiyoshi, K., Nakatake, S., and Kajitani, Y. (1995). Rectangle packing based module placement. In *Proc. IEEE Int. Conf. on Computer-Aided Design*, pages 472–479.

Murata, H. and Kuh, E. S. (1998). Sequence-pair based placement method for hard/soft/preplaced modules. In *Proc. Int. Symp. on Physical Design*, pages 167–172.

Nam, Gi-Joon and Cong, Jason (2007). *Modern Circuit Placement: Best Practices and Results*. Springer, New York.

Okamoto, T. and Cong, J. (1996). Buffered steiner tree construction with wire sizing for interconnect layout optimization. *Proc. IEEE Int. Conf. on Computer-Aided Design*, pages 44–49.

Ou, S.-L. and Pedram, M. (2000). Timing-driven placement based on partitioning with dynamic cut-net control. *Proc. ACM Design Automation Conf.*, pages 472–476.

Pan, P., Karandikar, A., and Liu, C. L. (1998). Optimal clock period clustering for sequential circuits with retiming. *IEEE Trans. on Computer-Aided Design of Integrated Circuits and Systems*, 17(6):489–498.

Pan, P. and Liu, C. L. (1992). Area minimization for general floorplans. *Proc. IEEE Int. Conf. on Computer-Aided Design*, pages 606–609.

Pathak, M. and Lim, S. K. (2007). Thermal-aware steiner routing for 3D stacked ICs. *Proc. IEEE Int. Conf. on Computer-Aided Design*, pages 205–211.

Perskey, A., Deutch, D., and Schweikert, D. (1976). LTX - a system for the directed automatic design of LSI circuits. In *Proc. ACM Design Automation Conf.*, pages 399–407.

Prim, R. (1957). Shortest connection networks and some generalizations. *Bell Systems Technical Journal*, 36:1389–1401.

Rafiq, F., Chrzanowska-Jeske, M., Yang, H. H., Jeske, M., and Sherwani, N. (2003). Integrated floorplanning with buffer/channel insertion for bus-based designs. *IEEE Trans. on Computer-Aided Design of Integrated Circuits and Systems*, 22(6):730–741.

Rajagopalan, S. and Vazirani, V. (1999). On the bidirected cut relaxation for the metric steiner tree problem. *Proc. ACM-SIAM Symposium on Discrete Algorithms*, pages 742–751.

Rajaraman, R. and Wong, D.F. (1995). Optimal clustering for delay minimization. *IEEE Trans. on Computer-Aided Design of Integrated Circuits and Systems*, 14(12):1490–1495.

Ramnath, S. (2003). New approximations for the rectilinear steiner arborescence problem. *IEEE Trans. on Computer-Aided Design of Integrated Circuits and Systems*, 22(7):859–869.

Ranjan, A., Bazargan, K., Ogrenci, S., and Sarrafzadeh, M. (2001). Fast floorplanning for effective prediction and construction. *IEEE Trans. on VLSI Systems*, 9(2):341–351.

Rao, S. K., Sadayappan, P., Hwang, F. K., and Shor, P. W. (1992). The rectilinear steiner arborescence problem. *Algorithmica*, 7:277–288.

Ren, H., Pan, D., Alpert, C., Villarrubia, P., and Nam, G.-J. (2007). Diffusion-based placement migration with application on legalization. *IEEE Trans. on Computer-Aided Design of Integrated Circuits and Systems*, 26(12):2158–2172.

Riess, B., Doll, K., and Johannes, F. (1994). Partitioning very large circuits using analytical placement techniques. *Proc. ACM Design Automation Conf.*, pages 646–651.

Robins, G. and Salowe, J. (1994). On the maximum degree of minimum spanning trees. In *Proc. of the Annual Symposium on Computational Geometry*, pages 250–258.

Sanchis, L. (1989). Multiple-way network partitioning. *IEEE Trans. on Computers*, 38(1):62–81.

Sapatnekar, S. (2000). A timing model incorporating the effect of crosstalk on delay and its application to optimal channel routing. *IEEE Trans. on Computer-Aided Design of Integrated Circuits and Systems*, 19(5):550–559.

Sassone, P. and Lim, S. K. (2006). Traffic: A novel geometric algorithm for fast wire-optimized floorplanning. *IEEE Trans. on Computer-Aided Design of Integrated Circuits and Systems*, 25(6):1075–1086.

Sechen, C. and Sangiovanni-Vincentelli, A. (1985). The TimberWolf placement and routing package. *IEEE Journal of Solid-State Circuits*, 20(2):510–522.

Shahookar, K. and Mazumder, P. (1991). VLSI cell placement techniques. *ACM Computing Survey*, 23(2):143–220.

Shahrokhi, F. and Matula, D. W. (1990). The maximum concurrent flow problem. *Journal of ACM*, 37(2):318–334.

Shi, W. (1995). An optimal algorithm for area minimization of slicing floorplans. *Proc. IEEE Int. Conf. on Computer-Aided Design*, pages 480–484.

Shi, W. and Su, C. (2000). The rectilinear Steiner arborescence problem is NP-complete. *Proc. ACM-SIAM Symposium on Discrete Algorithms*, pages 780–787.

Shragowitz, E. and Keel, S. (1987). A global router based on a multicommodity flow model. *Integration, the VLSI Journal*, 5(1):3–16.

Sigl, G., Doll, K., and Johannes, F. (1991a). Analytical placement: a linear or a quadratic objective function? In *Proc. ACM Design Automation Conf.*, pages 427–432.

Sigl, G., Doll, K., and Johannes, F. (1991b). Analytical placement: a linear or a quadratic objective function? *Proc. ACM Design Automation Conf.*, pages 427–432.

Stockmeyer, L. (1983). Optimal orientation of cells in slicing floorplan designs. *Information and Control*, 57(2–3):91–101.

Su, H., Hu, J., Sapatnekar, S., and Nassif, S. (2004). A methodology for the simultaneous design of supply and signal networks. *IEEE Trans. on Computer-Aided Design of Integrated Circuits and Systems*, 23(12):1614–1624.

Suaris, P. and Kedem, G. (1989). A quadrisection-based combined place and route scheme for standard cells. *IEEE Trans. on Computer-Aided Design of Integrated Circuits and Systems*, 8(3):234–244.

Sun, W.-J. and Sechen, C. (1995). Efficient and effective placement for very large circuits. *IEEE Trans. on Computer-Aided Design of Integrated Circuits and Systems*, 14(3):349–359.

Sun, W.-J. and Sechen, C. (1997). A parallel standard cell placement algorithm. *IEEE Trans. on Computer-Aided Design of Integrated Circuits and Systems*, 16(11):1342–1357.

Sutanthavibul, S., Shragowitz, E., and Rosen, J. (1991). An analytical approach to floorplan design and optimization. *IEEE Trans. on Computer-Aided Design of Integrated Circuits and Systems*, 10(6):761–769.

Swartz, W. and Sechen, C. (1990). New algorithms for the placement and routing of macro cells. In *Proc. IEEE Int. Conf. on Computer-Aided Design*, pages 336–339.

Tang, X., Tian, R., and Wong, D. F. (2001). Fast evaluation of sequence pair in block placement by longest common subsequence computation. *IEEE Trans. on Computer-Aided Design of Integrated Circuits and Systems*, 20(12):1406–1413.

Teslenko, M. and Dubrova, E. (2004). Hermes: LUT FPGA technology mapping algorithm for area minimization with optimum depth. *Proc. IEEE Int. Conf. on Computer-Aided Design*, pages 748–751.

Ting, B. and Tien, B. (1983). Routing techniques for gate array. *IEEE Trans. on Computer-Aided Design of Integrated Circuits and Systems*, 2(4):301–312.

Tsay, Y.-W. and Lin, Y.-L. (1995). A row-based cell placement method that utilizes circuit structural properties. *IEEE Trans. on Computer-Aided Design of Integrated Circuits and Systems*, 14(3):393–397.

Vaishnav, H. and Pedram, M. (1995). Delay optimal partitioning targeting low power VLSI circuits. *Proc. IEEE Int. Conf. on Computer-Aided Design*, pages 638–643.

Viswanathan, N. and Chu, C. (2005). Fastplace: efficient analytical placement using cell shifting, iterative local refinement, and a hybrid net model. *IEEE Trans. on Computer-Aided Design of Integrated Circuits and Systems*, 24(5):722–733.

Vygen, J. (1997). Algorithms for large-scale flat placement. In *Proc. ACM Design Automation Conf.*, pages 746–751.

W. Swartz, C. Sechen (1995). Timing driven placement for large standard cell circuits. *Proc. ACM Design Automation Conf.*, pages 211–215.

Wang, M., Yang, X., and Sarrafzadeh, M. (2000). Dragon2000: standard-cell placement tool for large industry circuits. *Proc. IEEE Int. Conf. on Computer-Aided Design*, pages 260–263.

Wei, Y.-C. and Cheng, C.-K. (1989). Towards efficient hierarchical designs by ratio cut partitioning. In *Proc. IEEE Int. Conf. on Computer-Aided Design*, pages 298–301.

Wong, D. F. and Liu, C. L. (1986). A new algorithm for floorplan design. In *Proc. ACM Design Automation Conf.*, pages 101–107.

Xiong, J. and He, L. (2005). Extended global routing with RLC crosstalk constraints. *IEEE Trans. on VLSI Systems*, 13(3):319–329.

Xiu, Z., Ma, J., Fowler, S., and Rutenbar, R. (2004). Large-scale placement by grid-warping. In *Proc. ACM Design Automation Conf.*, pages 351–356.

Yang, H. and Wong, D. F. (1994). Edge-map: optimal performance driven technology mapping for iterative LUT based Fpga designs. *Proc. IEEE Int. Conf. on Computer-Aided Design*, pages 150–155.

Yang, H. and Wong, D. F. (1995). New algorithms for min-cut replication in partitioned circuits. *Proc. IEEE Int. Conf. on Computer-Aided Design*, pages 216–222.

Yang, H. and Wong, D. F. (1996). Efficient network flow based min-cut balanced partitioning. *IEEE Trans. on Computer-Aided Design of Integrated Circuits and Systems*, 15(12):1533–1540.

Yang, H. and Wong, D. F. (1997). Circuit clustering for delay minimization under area and pinconstraints. *IEEE Trans. on Computer-Aided Design of Integrated Circuits and Systems*, 16(9):976–986.

Yang, H. and Wong, D. F. (1998). Optimal min-area min-cut replication in partitioned circuits. *IEEE Trans. on Computer-Aided Design of Integrated Circuits and Systems*, 17(11):1175–1183.

Yao, B., Chen, H., Cheng, C. K., and Graham, R. (2001). Revisiting floorplan representations. In *Proc. Int. Symp. on Physical Design*, pages 138–143.

Yildiz, M. and Madden, P. (2001). Improved cut sequences for partitioning based placement. *Proc. ACM Design Automation Conf.*, pages 776–779.

Yoshimura, T. and Kuh, E. (1982). Efficient algorithms for channel routing. *IEEE Trans. on Computer-Aided Design of Integrated Circuits and Systems*, 1(1):25–35.

Young, F.Y. and Wong, D.F. (1998). Slicing floorplans with pre-placed modules. *Proc. IEEE Int. Conf. on Computer-Aided Design*, pages 252–258.

Zarkesh-Ha, P., Davis, J., and Meindl, J. (2000). Prediction of net-length distribution for global interconnects in a heterogeneous system-on-a-chip. *IEEE Trans. on VLSI Systems*, 8(6):649–659.

Zhong, K. and Dutt, S. (2000). Effective partition-driven placement with simultaneous level processing and global net views. *Proc. IEEE Int. Conf. on Computer-Aided Design*, pages 254–259.

Zhou, H. (2004). Efficient steiner tree construction based on spanning graphs. *IEEE Trans. on Computer-Aided Design of Integrated Circuits and Systems*, 23(5):704–710.

Zhou, H., Shenoy, N., and Nicholls, W. (2002). Efficient spanning tree construction without delaney triangulation. *Information Processsing Letter*, 81(5):271–276.

About the Author

Dr. Sung Kyu Lim received the B.S., M.S., and Ph.D. degrees from the Computer Science Department, University of California, Los Angeles (UCLA), in 1994, 1997, and 2000, respectively. From 2000 to 2001, he was a Post-Doctoral Scholar at UCLA, and a Senior Engineer at Aplus Design Technologies, Inc. He joined the School of Electrical and Computer Engineering at the Georgia Institute of Technology in 2001, where he is currently an Associate Professor. His research focus is on the physical design automation for 3-D circuits, 3-D system-on-packages, microarchitectural physical planning, and field-programmable analog arrays.

Dr. Lim received the Design Automation Conference (DAC) Graduate Scholarship in 2003 and the National Science Foundation Faculty Early Career Development (NSF CAREER) Award in 2006. He was on the Advisory Board of the ACM Special Interest Group on Design Automation (SIGDA) during 2003–2008. He is an Associate Editor of the IEEE Transactions on Very Large Scale Integration Systems (TVLSI) and served as a Guest Editor for the ACM Transactions on Design Automation of Electronic Systems (TODAES). He has served the Technical Program Committee of several ACM and IEEE conferences on electronic design automation. He is a senior member of the IEEE.